Y0-EOO-857

BIOLOGICAL MACROMOLECULES AND ASSEMBLIES

Volume 1: Virus Structure

Contributors

PATRICK ARGOS, Purdue University, West Lafayette, Indiana

ROGER M. BURNETT, College of Physicians and Surgeons, Columbia University, New York, New York

KENNETH C. HOLMES, Max-Planck-Institut für Medizinische Forschung, Heidelberg, Federal Republic of Germany

JOHN E. JOHNSON, Purdue University, West Lafayette, Indiana

LARS LILJAS, Institute of Molecular Biology, University of Uppsala, Uppsala, Sweden

LEE MAKOWSKI, College of Physicians and Surgeons, Columbia University, New York, New York

IVAN RAYMENT, Rosenstiel Basic Medical Sciences Research Center, Brandeis University, Waltham, Massachusetts

MICHAEL G. ROSSMANN, Purdue University, West Lafayette, Indiana

J. J. SKEHEL, National Institute for Medical Research, London, England

BROR STRANDBERG, Institute of Molecular Biology, University of Uppsala, Uppsala, Sweden

GERALD STUBBS, Vanderbilt University, Nashville, Tennessee

D. C. WILEY, Harvard University, Cambridge, Massachusetts

I. A. WILSON, Research Institute of Scripps Clinic, La Jolla, California

BIOLOGICAL MACROMOLECULES AND ASSEMBLIES

Volume 1: Virus Structures

Editors

Frances A. Jurnak and Alexander McPherson

University of California
Riverside

RUTH SCARBOROUGH
LIBRARY
MAY 22 1985
SHEPHERD COLLEGE

A Wiley-Interscience Publication

JOHN WILEY & SONS

New York Chichester Brisbane Toronto Singapore

162815

Copyright © 1984 by John Wiley & Sons, Inc.

All rights reserved. Published simultaneously in Canada.

Reproduction or translation of any part of this work
beyond that permitted by Section 107 or 108 of the
1976 United States Copyright Act without the permission
of the copyright owner is unlawful. Requests for
permission or further information should be addressed to
the Permissions Department, John Wiley & Sons, Inc.

Library of Congress Cataloging in Publication Data:

Main entry under title:

Biological macromolecules and assemblies.

"A Wiley-Interscience publication."
Contents: v. 1. Virus structures.
Includes indexes.
1. Macromolecules—Collected works. I. Jurnak,
Frances A. II. McPherson, Alexander, 1944- . [DNLM:
1. Macromolecular systems. QD 381 B615]

QP801.P64B554 1984 574.19'24 83-21732
ISBN 0-471-87077-3 (v. 1)

Printed in the United States of America

10 9 8 7 6 5 4 3 2 1

?
01
P64
554
984
c.1

Preface

This volume presents what may well be the most impressive achievements of macromolecular analysis by X-ray diffraction methods, the structures of the viruses. Represented here are examples of icosahedral or spherical plant viruses, helical plant viruses, filamentous phages, and initial investigations of the more complex icosahedral animal viruses. In addition to studies of the intact viruses, two examples are included of high resolution analyses of the structures of proteins found on the surfaces of an adenovirus and the influenza virus. These lend significant new insights into the architectural principles involved in the design and assembly of viral particles and interactions with their hosts. The investigations described here establish a sound structural basis for interpretation of the vast accumulation of data involving viruses from other sources of biochemical, physical, and physiological research. They provide the most convincing confirmation of now widely accepted virus structure theory but at the same time make evident the likelihood that these principles will see substantial alteration and refinement as larger and more complex structures become available. A not insignificant feature of these achievements is that they serve as inspiring testament to the perseverance and ingenuity of diffractionists working at the leading edge of their science with success never assured, methods marginally adequate to the task, and specimens possessing the most complex and elusive properties.

Chapter 1 is an overview of the chemical and physical properties of the spherical plant viruses and their implications for general virus structure. As such, it establishes conditions and standards by which the structure analyses must be measured. Chapters 2 and 3 focus on two small icosahedral plant viruses whose structures have now been determined in near atomic detail, southern bean mosaic virus and satellite tobacco necrosis virus. These two structures, along with that of tomato bushy stunt virus, elegantly illustrate the principles of virus construction postulated by Caspar and Klug but at the same time suggest alternative bonding mechanisms by which the postulated quasi-equivalence might be achieved. Further, they beautifully illustrate the integration of protein structural principles into larger macromolecular assemblies, the economy of structure and function, and the consistency of design.

Tobacco mosaic virus, described in Chapters 4 and 5, represents a different class entirely, the viruses constructed according to precise helical symmetry. The structure of tobacco mosaic virus, the most exhaustively studied of all the viruses, illustrates the means by which small, identical protein subunits of relatively uncomplicated structure have encoded within their three-dimensional array of amino acids, the ability to recognize and bind their genomic nucleic acid, to order and protect it, and to organize themselves into a complete viral assembly with new and sophisticated properties.

In Chapter 6, the traditional crystallographic technique of isomorphous replacement is by necessity abandoned in the study of the filamentous bacteriophages. Using primarily physical constraints in conjunction with fiber diffraction data, visual images have been achieved that reveal the interaction between small α-helical subunits and the DNA genome at the core of the virus. Unlike the icosahedral and helical viruses, these bacteriophages are designed according to subtly different principles of symmetry and nonequivalence of interaction. They suggest that there may exist many variations or entirely new patterns of self-assembly in viruses.

Chapter 7 describes the low resolution analysis of an animal virus, the polyoma virus. Unexpectedly—perhaps because it so faithfully assumed the principles of Casper and Klug regarding icosahedral symmetry—the results of this study have suggested the most radical departure from those principles. Once again, the methodology was somewhat unorthodox in that it did not employ isomorphous replacement; nevertheless, it appears valid by current criteria. If the results are confirmed by other investigations or by higher resolution analyses of the polyoma virus itself, they will certainly require a reconsideration of our expectations for capsid structure and assembly in the $T = 7$ and larger virus particles.

Chapters 8 and 9 are elegant and impressive examples of traditional isomorphous replacement methods coupled with the clever application of molecular replacement and symmetry constraints. The structures emerging from these investigations, the capsid protein hexon from adenovirus and the influenza virus hemagglutinin, are striking in the manner by which recognized protein secondary structural patterns are extended and reorganized to produce unusual and unique protein molecules. These are molecules that not only contribute to our understanding of general protein structure but bring to light new principles and mechanisms for macromolecular interaction, assembly, and physiological activity.

The nine chapters presented here illustrate and describe in a reasonably comprehensive fashion what is currently known of virus structure at the atomic and near atomic level. The structures were derived principally by X-ray diffraction and the perspectives primarily are those of scientists versed in that technique. This should in no way be interpreted as ignoring the extensive achievements resulting from the many other methods that have been applied to virus structure analysis. Indeed, it should be noted that immediately following or closely entwined with the description of any structure is the correlation or rationalization of each feature in terms of benchmarks derived from biochemical and physical analyses. Perhaps the crystallographers may here repay their debt to the scientific community at large by suggesting new approaches and new experiments and at the same time provide a structural basis for their design and interpretation.

FRANCES A. JURNAK
ALEXANDER MCPHERSON

Riverside, California
March 1984

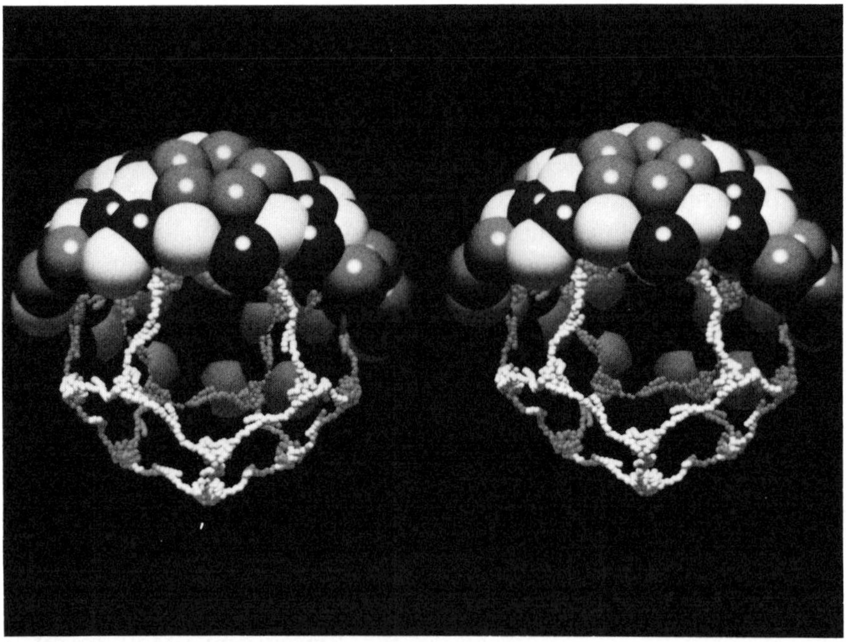

Sterographic illustration created by computer imaging of a portion of the capsid of tomato bushy stunt virus showing the disposition of the amino terminal peptides in the interior of the virus particle. Courtesy of A. Olsen, Research Institute of Scripps Clinic, La Jolla, California.

Contents

1

Chemical Stability in Simple Spherical Plant Viruses

PATRICK ARGOS
JOHN E. JOHNSON
Department of Biological Sciences
Purdue University
West Lafayette, Indiana

CONTENTS

1. OVERVIEW

1.1. Introduction

The relationship between molecular structure and function is a primary concern in molecular biology. The simple spherical plant viruses have been the subject of such study since the early 1950s. It was only in the 1970s, however, that an investigative level was reached that permits analysis of this relationship at atomic resolution. The tertiary structures of three spherical plant viruses and a helical plant virus have yielded substantial knowledge of subunit association and suggested the nature of protein–RNA interactions. The functional aspects of chemical stabilization and its relation to assembly processes have been exhaustively reviewed by Kaper (1975) through 1974. Since then, studies similar to those reported by Kaper have been extended and new methods of analysis have been brought to bear on the virus problem. These include neutron diffraction, nuclear magnetic resonance spectroscopy, circular dichroism, and primary structure analysis of viral proteins and nucleic acids.

The goal of this chapter is to describe plant virus structure and assembly processes established through biochemical and biophysical methods other than single crystal X-ray diffraction techniques. The viral systems chosen for discussion are taken from the simple spherical plant viruses, which have undergone extensive chemical study. The rod-shaped tobacco mosaic virus has been reviewed recently and exhaustively (Hirth and Richards, 1981). The first part of this chapter provides a comparative analysis of the virus systems; the second presents the detailed experimental evidence and related references.

1.2. Capsid Protein

Table 1 lists the gross structural parameters for the type members of various virus groups. In the native state all except alfalfa mosaic virus (AMV) are icosahedral with hexamer-pentamer clustering dominating the capsomeric distribution. With the exception of satellite tobacco necrosis virus (STNV), cowpea mosaic virus (CpMV), and AMV, all display a triangulation number of 3 and contain 180 identical subunits in the protein shell. The $T = 3$ viruses are approximately the same size ranging from 260 to 310 Å in diameter. CpMV particles fall in this range despite their

Table 1. Physical and Chemical Properties of Selected Plant Viruses

	Particle					
	Diameter (Å)	Triangulation Number	Sedimentation Coefficient (S)	State of Structure Determination	Multipartite	*In Vivo* Capsids
TYMV	280	3	54 (T)	Crystallized[a] 40 Å	No	Yes
			117			
CpMV	280	1	58 (T)	Crystallized	Yes	Yes
			95 (M)			
			115 (B)			
STNV	176	1	50	3.0 Å	No	No
SBMV	284	3	115	2.8 Å	No	No
TBSV	308	3	131	2.9 Å	No	No
BMV	260	3	87	None[b]	Yes	No
CMV	290	3	99	Crystallized	Yes	No
AMV	190[c]	1	94[d]	4.5 Å[c]	Yes	No
			82			
			73			
			66			

	Protein					
	Molecular Weight	Protein Sequence	Basic Amino Terminus	Tertiary Structure	State of Soluble Disassembled Protein	Polymorphism
TYMV	20,133	Known	No	?	?	No
CpMV	22,000	Known	?	?	?	No
	42,000	Known				
STNV	21,600	Known	Yes	β barrel	?	No
SBMV	28,200	Known	Yes	β barrel	dimer	Yes $T = 1, T = 3$
TBSV	40,000	Unknown	Yes	β barrel	dimer	Yes[e] $T = 1, T = 3$
BMV	20,300	Known	Yes	?	dimer	Yes $T = 1, T = 3$, tubes, shells
CMV	26,200	Known	Yes	?	?	No
AMV	24,240 (21,100)[e]	Known	Yes	β barrel	dimer	Yes $T = 1$

Table 1. *(Continued)*

	RNA					
	Subunit Molecular Weight (× 10^6)	3′ Terminus	5′ Terminus	Sequence	Secondary Structure	Subgenomic Messenger
TYMV	2.0	tRNA-like valine chargeable	7-Methyl-guanosine cap	Partial	~60%	Yes
CpMV	1.39	poly (A)	Protein bound	Partial	>60%	No
	2.39	poly (A)	Protein bound	Partial	>60%	
STNV	0.4	No special structure	No special structure	Known	?	Yes
SBMV	1.2	No special structure	Protein bound	Partial	~60%	Yes
TBSV	1.5	?	?	No	?	Yes
BMV	1.1	t-RNA like tyrosine chargeable	7-Methyl-guanosine cap	Partial	~60%	Yes
	1.0			Partial	~60%	
	0.7			Known	~60%	
	0.3			Known	~60%	
CMV	1.27	t-RNA like tyrosine chargeable	7-Methyl-guanosine cap	Partial	?	Yes
	1.13			Partial	?	
	0.82			Known	?	
	0.35			Known	?	
AMV	1.04	All 3′ are similar but not chargeable and no poly (A)	7-Methyl-guanosine cap	Partial	?	Yes
	0.73			Partial	?	
	0.62			Partial	?	
	0.28			Partial	?	

		Stability and Assembly					
		Primary Stabilizing Forces					
	Swelling	Hydrophobic	pH	Metal	Protein–RNA	Sensitivity to RNase	Reversible Disassociation
TYMV	?	Yes	No	No	No	No	No
CpMV	?	Yes	No	No	Yes	No	?
STNV	Yes	No	Yes	Yes	Yes	No	?
SBMV	Yes	No	Yes	Yes	Yes	No (swollen yes)	Yes
TBSV	Yes	No	Yes	Yes	Yes	No (swollen yes)	?
BMV	Yes	No	Yes	Yes	Yes	No (swollen yes)	Yes
CMV	No	No	No	No	Yes	Yes	Yes
AMV	No	No	No	No	Yes	Yes	Yes

[a] Belladonna mottle virus has been crystallized.
[b] Cowpea chlorotic mottle virus has been crystallized.
[c] A reassembled protein capsid.
[d] AMV contains four cylindrical components of different size and RNA content.
[e] TCV displays polymorphism.

$T = 1$ icosahedral symmetry. The thickness of the protein shell, approximately 40 Å, is also similar for the viruses. The structure, sequence data, and proteolytic cleavage patterns indicate that the N-terminal region of southern bean mosaic virus (SBMV), tomato bushy stunt virus (TBSV), STNV, AMV, cowpea chlorotic mottle virus (CCMV), and cucumber mosaic virus (CMV) is composed predominantly of basic residues associated with the nucleic acid. When the unassembled capsid protein is in solution, it generally exists as a dimer for all the viruses studied.

The capsid protein serves primarily a protective role. Table 1 lists the molecular weight and certain properties of the coat protein sequences. The molecular weights range from 19,600 for the bromoviruses to 44,000 for the large protein of CpMV. TBSV protein is the largest of the $T = 3$ viruses. The three-dimensional structures determined for TBSV, SBMV, and STNV reveal that a β-barrel formed by antiparallel strands is the predominant feature of their tertiary architectures. The molecular weight of the barrel is about 19,000 for all three viruses with any large variations accounted for by the length of the amino terminal arms (roughly 8000 for TBSV and SBMV and 3000 for STNV) and an additional barrel domain in TBSV (12,000). Circular dichroic studies of many of the other viral proteins suggest a large quantity of β-sheet; predictions of TYMV secondary structure indicate the β-fold. Many of the subunit molecular weights shown in Table 1 are near 20,000 daltons, suggesting a similar β-topology in their backbones. In the case of CpMV, which contains two polypeptide chains in the capsid, the large protein could consist of two domains similar to the basic β-fold and the small protein one such domain, resulting in a near $T = 3$ lattice of surface β-barrels.

1.3. Nucleic Acid

The RNA molecules for the spherical viruses fall into three size classes: 2.0×10^6 daltons or larger, $0.6–1.4 \times 10^6$ daltons, and less than 0.5×10^6 daltons. The smallest RNA molecules are likely monocistronic subgenomic species that are the functional messenger for the capsid protein. In AMV, CCMV, CMV, SBMV, and TYMV, the sequences for the coat protein are also present in a larger RNA; nonetheless, encapsidation of the subgenomic RNA occurs and yet is not essential for infection. The exception is CpMV, which does not generate a subgenomic messenger.

Many of the RNA molecules occupy only 20–30% of the volume available within the capsid, assuming a partial specific volume for RNA of

0.55 Å^3/dalton. The viruses containing a high percentage of RNA (CpMV bottom component and the TYMV group) have approximately half the capsid interior occupied.

The termini of the viral RNAs display a variety of chemical structures. The bromovirus and CMV molecules are capped with 7-methylguanosine at their 5′ termini, and their 3′ termini display substantial sequence homology and probably similar secondary structure. The 3′ termini can be charged with tyrosine in the presence of the appropriate tRNA synthetase. Tymovirus RNA can be charged with valine. SBMV and CpMV RNAs have proteins covalently linked to their 5′ termini. CpMV contains a sequence of polyadenylic acid at the 3′ terminus. SBMV contains 130 noncoding bases that do not fold into a particularly stable secondary structure. In fact, the bean and cowpea strains of SBMV show no sequence homology in this region. In contrast, STNV RNA does not display any of these various terminal structures. It seems probable that the diverse features of the RNA are closely related to the replication and translation strategy employed in their hosts.

The RNA molecules for most of the viruses have been the subject of *in vitro* translation studies. In general more than one protein is synthesized and the coat protein for all but the comoviruses is produced from the subgenomic messenger RNA. In all but STNV the RNA probably codes for a subunit of the RNA replicase molecule as well as the coat protein. CpMV RNA codes for a protease that cleaves a viral polyprotein into viable molecules.

The RNA is highly organized within the particle. Circular dichroism indicates at least 60% of the polynucleotide is ordered in secondary structures. Isolated SBMV RNA when stretched into a fiber gives an X-ray diffraction pattern consistent with A DNA, as expected for double helical RNA. While the RNA is highly organized, it is not necessarily in its most compact form. SBMV and TYMV RNA can undergo *in situ* rearrangements that make the nucleic acid more compact but less infectious. If the RNA is denatured, it regains full infectivity. Apparently the structure of encapsidated RNA provides a balance between minimum packing volume and rapid unfolding for processing.

A problem of RNA packaging at physiological pH is the cancellation of negative charge associated with phosphate groups. There are three sources of positive charge for neutralization: cations such as potassium or calcium, which are present in the cytoplasm, polyamines such as spermine and spermidine, and basic residues of the N-terminal arm of the coat protein.

1.4. Capsid Stabilization

Strong protein–protein interactions fall into three categories: metal ion mediated, pH dependent, and hydrophobic interactions. Those involving metal ions are also pH dependent. SBMV, STNV, TBSV, and the bromoviruses show stability from pH 3.0 to 10.0 in the native state; however, swelling of the particles generally occurs above neutrality with metal ion removal by EDTA. The tertiary structures of SBMV, STNV, and TBSV show the metal ions are associated with acidic residues at the subunit interfaces. Deprotonation of these residues in the absence of metal ions results in a local buildup of charge causing the subunits to separate with protein–RNA interactions primarily maintaining the structure. All viruses that contain removable metal ions display this structural transition.

The CpMV group shows pH dependence in capsid stability, but metal ions do not play a significant role. Above pH 8.0 an alteration takes place in the RNA-free particles, making them susceptible to disassembly by low concentrations of denaturants while the nucleoprotein particles remain stable. TYMV particles, also stabilized by very strong protein–protein interactions, will, under conditions of high salt and pH, release RNA from a momentarily expanded structure. A stable empty capsid nonetheless remains after RNA expulsion. However, another tymovirus, belladonna mottle virus (BDMV), will release its RNA at pH 7.0 under only mild buffering conditions. All evidence points to extended regions of hydrophobic interactions between the subunits as the stabilizing force in CPMV and TYMV.

An N-terminal region not involved in RNA interactions is important in establishing the architecture of some virions. The SBMV and TBSV X-ray structures show the formation of a β annulus structure about the icosahedral threefold axes which are quasi-sixfold axes. Using the nomenclature of Harrison, three of the C subunits contribute to the hydrogen bonded structure whereas the quasi-equivalent A and B subunits do not as they display a disordered structure in this region. Thus, considering only the C subunits and the β annuli, a $T = 1$ structure is formed. The functional role of this region is clearly demonstrated in SBMV and TBSV in which the protein will not assemble into a $T = 3$ structure if the arm is removed, whereas fully infectious particles can be formed from intact protein and nucleic acid.

The capsid proteins of SBMV, TBSV, turnip crinkle virus (TCV), AMV, and CCMV display a diverse polymorphism depending on the conditions of reassembly such as pH, cation concentration, and the presence or absence of the N-terminal arm and RNA. The subunits form $T = 1$ structures if the arm is removed and RNA is absent from the reassociation solution. Even the AMV protein forms such capsids in spite of the minimum protein–protein interactions that stabilize the cylindrically shaped nucleoprotein components *in vivo*. Recent SBMV assembly studies suggest that the charge configuration of certain carboxyl clusters also regulate subunit interactions and modes of assembly ($T = 1$ vs. $T = 3$).

1.5. Protein–RNA Interactions

Protein–RNA interactions are evident in all the nucleoprotein particles. In AMV and CMV these play the major role in maintaining particle integrity, while they display a supporting role in the other viruses. Two models have emerged for protein–RNA interactions in virus structures. The first invokes that proposed for histone–DNA interaction in which a helix with positively charged residues on one side can interact with nucleic acid phosphate groups while the other side forms a surface favorable for interaction with double helical DNA through side group hydrogen bonding with the bases. Diffraction studies show that protamines adopt helical configurations when interacting with a double stranded region of tRNA (Warrant and Kim, 1978). The structure of *cro* repressor protein and its modeled interaction with DNA result in a nearly perfect fit between protein helical side chains and the DNA major groove. Helical wheels of the N-terminal regions of STNV, brome mosaic virus (BMV), CCMV, and AMV can be formed with basic residues predominantly distributed on one side. A second model first proposed by Kaper involves protein–RNA interactions via hydrogen bonds between acidic side chain carboxyl groups and cytidine residues. NMR studies of BDMV at various pH values strongly support this model for tymoviruses.

The more general protein–RNA interaction is likely to involve the basic residues. Metal ion stabilized viruses in the swollen form as well as AMV and CMV behave very much like polyelectrolyte complexes. All of these particles disassemble in high salt with pH values somewhat above neutrality. It is possible to virtually titrate the RNA of SBMV with capsid protein as a function of ionic strength. In many respects CMV and AMV can be regarded as swollen forms of the metal ion stabilized viruses.

1.6. Particle Assembly

All of the successful reassembly experiments have been performed with the viruses that do not form empty capsids *in vivo*. This suggests there may be a different mechanism of assembly for the capsids primarily stabilized by strong protein–protein interactions, namely, CpMV and TYMV. In the other viruses, the initial step in assembly is probably an ionic interaction between dimers of coat protein and the nucleic acid. *In vitro* this association is not highly specific as the capsid protein of one virus can often encapsidate the RNA of another or even DNA and, in some cases, sodium dextran sulfate molecules. Despite the lack of specificity, some studies have shown that there is preferential interaction between the coat protein and corresponding viral RNA during competitive reassembly. Particles containing foreign nucleic acid are less stable than native virions and the nucleic acid is more susceptible to nuclease digestion.

The problems associated with reassembly of viruses exhibiting strong protein–protein interactions may arise for two reasons. The most obvious is that harsh conditions are generally required to disassemble the virus, irreversibly altering the tertiary structure of the subunits. Alternatively the viruses may require some type of membrane template to organize and solubilize the individual subunits or they may initially assemble in a precursor stage with subsequent removal of a portion of the protein critical for assembly. It is certainly possible, as well, that the proper conditions for disassembly and assembly have just not yet been found.

2. SOUTHERN BEAN MOSAIC VIRUS GROUP

2.1. Introduction

SBMV was first reported in 1942 in bean plants of the southern United States (Zaumeyer and Harter, 1942) and crystallized in 1943 (Price, 1943; Price, 1946). Because of its stability SBMV was the subject of a number of early biophysical studies (Miller and Price, 1946; Laufer et al., 1952). Four strains of SBMV have been reported: the type or bean strain originally isolated (Zaumeyer and Harter, 1942), the cowpea strain (Shepherd and Fulton, 1962) serologically related to the bean strain but lacking a common host, the Mexican or severe bean strain (Grogan and Kimball, 1964) serologically related to the latter two but with different host range and symptomatic expression, and the cowpea strain from Ghana (Lamptey and Hamilton, 1974) serologically related to the other strains and capable of infecting both bean and cowpea hosts. Almost all studies have utilized either the bean or cowpea strain, which share similar RNA base and coat protein amino acid compositions. The nucleotide sequences coding for the capsid protein are also similar although there is virtually no homology in the noncoding region (Mang et al., 1982).

SBMV is the type member of the southern bean mosaic virus group (Hull, 1977b) and is by far the most intensely studied (Ghabrial, Shepherd, and Grogan, 1967). The properties that characterize this group are an isometric particle of diameter 280 Å containing 180 protein subunits each of molecular weight around 30,000; a total particle molecular weight of 6.0 to 6.5 $\times$ 10^6 with a sedimentation coefficient of 110–120 S; a protein capsid stabilized by pH dependent protein–protein interactions, divalent cations, and protein–RNA interactions; a single RNA molecule of weight near 1.4 $\times$ 10^6; virions with multiple bands in cesium sulfate gradients; and virion crystalline inclusion bodies located in the nucleus and cytoplasm of infected plants. Additional viruses in the group include cocksfoot mottle virus (CfMV), rice yellow mosaic virus (RYMV), sowbane mosaic virus (SoMV), and turnip rosette virus (TrosV).

2.2. Virion Stability

SBMV particles are stable and infectious between pH 2.5 and 9.5 in moderate ionic strength buffers (Sehgal et al., 1979), between pH 3.5 and 9.0 in 0.1% SDS, and between pH 4.0 and 6.0 in 2.0% SDS (Sehgal, 1980). In 4 *M* urea the particles are stable between pH 3.5 and 6.5 and in 8 *M* urea the stability range is pH 4.0–6.0. This level of structural integrity suggests strong protein–protein interactions in the capsid (Kaper, 1973). At pH 8.25, in the presence of 10 m*M* EDTA, the particle undergoes a structural transition ("swelling") characterized by a decrease in sedimentation coefficient from 115 to 100 S (Wells and Sisler, 1969) and an increase in diameter of 10% (Krüse et al., 1982). The swelling above pH 7.0 in 10 m*M* EDTA is reversible by adjusting the pH back to 6.0 (Sehgal, 1980; Hull, 1977a). The swollen particles are susceptible to degradation by low concentrations of urea or SDS (Sehgal et al., 1979; Sehgal, 1980) and disassemble into free protein and nucleic acid in 1 *M* sodium chloride (Tremaine and Ronald, 1977). Disassembly produces a series of specific nucleoprotein complexes ranging from the full complement of protein (180 subunits) and RNA at low ionic strength to only 10 subunits and RNA in 0.5 *M* NaCl (Tremaine et al., 1982). One of these subviral entities sedimenting at 50 S and composed of 60 protein subunits and the full size RNA could be a structurally significant intermediate in the assembly process (Sehgal and Sinha, 1974).

The native SBMV particle contains between 110 and 130 Mg^{2+} ions and 75 to 100 Ca^{2+} ions per particle as determined in studies by Hsu et al. (1976). Hull (1978) however reports there are 300–400 Mg^{2+} ions and 200–300 Ca^{2+} ions. Exhaustive dialysis against 10 m*M* EDTA removes all the metal ions. Binding studies and competition experiments at various pHs on a number of divalent and trivalent metal ions show that Ca^{2+} is by far the most effective stabilizing agent, even though Mg^{2+} is present in the native state (Hull, 1978). Circular dichroic studies of the swollen virus showed that Ca^{2+} returned the spectra to that of the native state at far lower concentrations than either Mg^{2+} or Sr^{2+} (Odumosu et al., 1981). Hull (1978) suggests that Ca^{2+} mediates the protein–protein interactions and not the protein–RNA interactions and that the likely Ca^{2+} ligands are carboxylate groups (Abdel-Meguid et al., 1981) with additional binding sites becoming available at elevated pH (up to three per subunit). Thus, the SBMV capsid is primarily stabilized by metal ion mediated protein–protein interactions, whereas in the metal-free state at high pH, protein–RNA interactions dominate (Kaper, 1973). Swollen SBMV is sensitive to proteolytic and nucleolytic digestion (Hsu et al., 1976; Tremaine and Ronald, 1978; Tremaine et al., 1981). In electron micrographs the particles appear penetrated by negative stain. Low levels of trypsin cause the coat protein in the swollen particle to be degraded in a stepwise manner with products of molecular weights ranging from that of the intact subunit to 22,100. The latter product results from cleavage of 68 N-terminal amino acids of the capsid protein (Tremaine and Ronald, 1978). Structural studies of native SBMV (Abad-Zapatero et al., 1980) suggest that the N-terminal region of the protein is most likely to interact with the RNA (see Rossmann, Chapter 2). Nevertheless this region is available for proteolytic attack in the swollen virus. Subsequent studies (Tremaine et al., 1981) have shown that the N-terminal polypeptide binds RNA, DNA, and sodium dextran sulfate; however, antibodies made against swollen SBMV did not cross-react with the N-terminal residues. Low angle neutron diffraction studies of the swollen form indicate that the protein and RNA do not reorganize. The density of the protein shell is greatly reduced, probably due to clefts formed between the subunits (Krüse et al., 1982). NMR studies support the neutron experiments since there was no variation in the mobility of the amino acids associated with the N-terminal arm in passing from the native to swollen to partially dissociated state (McCain et al., 1982).

2.3. Behavior in Cesium Salt Gradients

Cesium chloride density gradient centrifugation produces a single band of particles with a buoyant density of 1.36 g/ml for all the SBMV strains, whereas in a cesium sulfate gradient, at least two bands of buoyant densities 1.29 g/ml and 1.32 g/ml result (Hull, 1977a). The bean strain displays equal quantities of material in both bands whereas the cowpea strain is confined mainly to the light one. Analysis of the sulfate bands revealed no differences in RNA, infectivity, or electrophoretic mobility. When the metal ions were removed for both strains at pH 7.0, the particles sedimented entirely in the heavy band in sulfate gradients. If the pH was raised to 8.25, the particles disassembled. The sulfate anion is apparently responsible for the capsid heterogeneity, possibly creating a more chaotropic environment than the chloride anion. The results are analogous to those observed for TYMV where it is suggested that the cesium ions replace the normal cations neutralizing the RNA (Noort et al., 1982).

2.4. Assembly of Southern Bean Mosaic Virus

Assembly of SBMV has been reported from disassociated RNA and protein (Tremaine and Ronald, 1977; Savithri and Erickson, 1982) and from the 50 S subviral entity and free viral protein (Hsu et al., 1977). RNA and protein were dialysed from a high to low ionic strength buffer containing Ca^{2+} and Mg^{2+} producing fully infectious particles that were resistant to SDS and nucleases (Tremaine and Ronald, 1977). SBMV protein was found to encapsidate SoMV RNA, producing infectious particles giving typical SoMV symptoms on susceptible hosts; similarly, SoMV protein would encapsidate SBMV RNA. In fact both proteins encapsidated sodium dextran sulfate; however, these particles did not display the stability of the virions when exposed to SDS (Tremaine and Ronald, 1977).

SBMV protein and RNA form a series of nucleoprotein complexes during assembly if the ionic strength is carefully controlled. The complexes are analogous to those observed during disassembly (Tremaine et al., 1982). SBMV protein is capable of forming either $T = 1$ or $T = 3$ particles depending on the state of capsid protein cleavage, pH, and the presence of divalent ions (Savithri and Erickson, 1983; Erickson and Rossmann, 1982). The various chemical requirements for the SBMV assembly modes suggest carboxyl group clusters in a putative dimer initiation complex which regulate subunit interactions and ($T = 1$ vs. $T = 3$) association. These charge clusters are different from those involved in swelling of the virus. Thus far it has not been possible to reassemble SBMV protein into a $T = 3$ capsid without RNA.

2.5. Southern Bean Mosaic Virus Nucleic Acid

SBMV RNA has a molecular weight of 1.4×10^6 and contains nearly equal molar quantities of the four nucleotides for all strains (Tremaine, 1966). The RNA is highly ordered in particles of SBMV (Odumosu et al., 1981) and TrosV (Denloye et al., 1978) with 70% of the bases paired. The free RNA displays more secondary structure than in the particle. In addition to the genomic RNA within the capsid, there are a number of lower molecular weight RNA species (0.3 to 0.4×10^6) that are closely associated with the genomic RNA and function as the messenger for the coat protein (Rutgers et al., 1980). The quantity of these subgenomic RNA molecules, which are derived from the genomic RNA (Ghosh et al., 1981), increase with time of infection (Weber and Sehgal, 1982). When SBMV particles are heated to 65°C for 10 min, the RNA forms a highly stable condensed structure that is not infectious; upon denaturation the RNA regains infectivity (Veerisetty et al., 1981).

A 10,000 molecular weight protein essential for infectivity (Veerisetty and Sehgal, 1980) is covalently linked to the 5′ terminus of the SBMV genomic RNA which contains no poly(A) region at the 3′ end (Ghosh et al., 1979). *In vitro* translation of the genomic RNA produces three proteins of molecular weight 105,000, 75,000, and 14,000. The largest probably results from a leaky stop codon for the 75,000 molecular weight protein. The function of the proteins is unknown.

The first 400 3′-proximal nucleotides have been determined for the RNA of the bean and cowpea strains including a substantial 3′-terminal region of the coat protein gene. The RNA of both strains contain approximately 130 noncoding bases that show no homology, while the genes coding for the coat protein display a high degree of homology (Mang et al., 1982). Recently the complete amino acid sequence for the coat protein of the cowpea strain has been reported (Hermodson et al., 1982) and successfully placed in the electron density map derived from crystallographic studies (Abad-Zapatero et al., 1980). The relationship between the teritiary SBMV structure and many of the properties reported here are discussed in detail in Chapter 2 by Rossmann.

3. BROMOVIRUSES

3.1. Introduction

The bromoviruses (Lane, 1974, 1979; Harrison et al., 1971) consist of three members that have been studied extensively: brome (grass) mosaic virus (BMV), cowpea chlorotic mottle virus (CCMV), and broad bean mottle virus (BBMV). The most recent review of bromovirus research is that of Lane (1980).

These divided-genome viruses use three identical protein shells to encapsidate four RNA species with respective molecular weights of 1.20, 1.07, 0.81, and 0.25 $\times$ 10^6 (Bancroft and Flack, 1972; Reijenders et al., 1974). The two largest RNA molecules are enveloped separately in a heavy (RNA-1) and a light (RNA-2) nucleoprotein particle while the medium-dense virion contains species RNA-3 and RNA-4 (Bancroft and Flack, 1972). RNA-1, -2, and -3 comprise the viral genome necessary for infection while RNA-4 is the subgenomic coat protein messenger (Shih et al., 1972; Bastin et al., 1976). The protein capsid consists of 180 identical protomers (T = 3) each with molecular weight near 20,000 and arranged on an icosahedrally symmetric surface lattice (Caspar and Klug, 1962). Electron microscopic image reconstruction of the CCMV virion shows 12 pentamers at vertices and 20 hexamers on the faces of an icosahedron

(Horne et al., 1975; Steven et al., 1978). The tripartite genome of CCMV codes for the coat protein (RNA-4 and the 3′ end of RNA-3), a 35,000 dalton protein likely to be a subunit of a viral replicase (5′ end of RNA-3), and a near 100,000 dalton protein of unknown function (RNA-1 or RNA-2) (Sakai et al., 1977). Protein(s) coded by the remaining genome ($\sim 1 \times 10^6$ daltons) have not been isolated.

3.2. *In Vitro* Disassociation and Association

Bromoviruses are stable around pH 5.0 and display a homogeneous Svedberg coefficient of 88 S. At pH 7.5 and high salt concentration (ionic strength greater than 0.5), the virions disassociate into RNA–protein complexes and protein subunits (Hiebert, 1969; Bancroft, 1967). The isolated coat protein can reassociate under various conditions of ionic strength, pH, and temperature into a variety of quaternary structures, which include sheets, tubes, empty protein shells, double shells, and rosettes (cf. Bancroft et al., 1969a, 1969b; Adolph and Butler, 1974; Pfeiffer and Hirth, 1974). At pH 7.5 and 1 *M* NaCl virion disassociation results in RNA and protein dimers (Bancroft and Hiebert, 1967; Verduin, 1974; Adolph and Butler, 1977). Dialysis of the dimers to pH 5.0 with 1 *M* NaCl allows the formation of *in vitro* empty protein shells (pseudo-top) (Verduin, 1974) that display an architecture similar to the *in vivo* nucleoprotein coat (Finch and Bancroft, 1968) and yet possess a different surface charge (Bancroft et al., 1968). Cleavage of the 25 N-terminal residues in CCMV does not prevent formation of the pseudo-top component (Bancroft et al., 1967; Childlow and Tremaine, 1971). In the absence of RNA, BMV protein will form $T = 1$ particles at pH 7.6 after degradation with trypsin which cleaves the N-terminal 63 residues (Cuillel et al., 1981). BBMV differs from CCMV and BMV in that an empty protein shell will not form from disassociation products that are a heterogeneous population of protein subunit oligomers dominated by a monomer–trimer equilibrium (Bancroft, 1970).

CCMV protein dimers show a different conformation ("loose") than the "compact" folding of the subunit in the $T = 3$ empty shells. The transition from compact to loose is governed by deprotonation, strongly time dependent, and reversible (Verduin, 1974; Johnson et al., 1973; Jacrot, 1975). Hydrophobic interactions are largely responsible for the dimeric formation as evidenced by their stability under high salt conditions, temperature dependence, and calorimetry studies (Verduin, 1978). Circular dichroic ellipticities suggest little helical content and mostly β-secondary structures for the coat protein in the native state (Verduin, 1978).

3.3. Swelling of Bromoviruses

At low ionic strength and with an increase in pH to 7.5, the sedimentation coefficient of the bromoviruses is altered from 88 to 78 S (cf. Bancroft et al., 1968; Incardona and Kaesberg, 1964) along with an increased sensitivity to nucleolytic and proteolytic enzymes (Adolph, 1975; Pfeiffer, 1980). Low angle neutron scattering (Chauvin et al., 1978b) and intensity fluctuation spectroscopy (Zulauf, 1977) for BMV show a distinct increase in the hydrodynamic radius of the virion resulting from a radial extension of the protein and RNA ("swelling"). The BMV radius increases from 134 Å at pH 5.0 to 155 Å at pH 7.5 in 0.2 *M* KCl and 10 m*M* EDTA (Chauvin et al., 1978b). Reversibility of the swelling process cannot be accomplished upon back-titration to pH 5.0 as evidenced by a resulting 84 S particle with radius at 137 Å. In the presence of 10 m*M* $MgCl_2$, the process appears reversible; however, the swelling does not achieve its fullest extent at pH 7.5 (Chauvin et al., 1978b).

A distinct polymorphism in the extent of RNA and protein radial swelling and conformation is observed for various chemical environments of pH, ionic strength, temperature, and the presence of divalent ions and spermine (Incardona and Kaesberg, 1964; Zulauf, 1977; Incardona et al., 1973). The neutron investigations (Chauvin et al., 1978b) on BMV are particularly revealing in this regard. At pH 5.5 and 1.5 *M* KCl, the RNA compacts without a change in capsid size. The addition of divalent ions reduces the extent of viral swelling with the greatest reduction for the RNA. Spermine condenses the RNA even more efficiently. Swelling does not significantly alter the thickness of the protein shell. In the case of the BMV pseudo-top component, the neutron studies show a radius of 147 Å and a scattering curve very different from the normal nucleoprotein particle. Fluorescence studies of CCMV modified with pyridoxal 5′-phosphate also point to the pseudo-top structural alterations (Krüse et al., 1980). Pfeiffer (1980) has proposed a tentative model to explain the bromovirus polymorphism. The model is based on some exposure of the RNA upon swelling, allowing nucleolytic cleavage, which is negated upon cation, or polyamine addition, which prevents RNA expansion and protrusion from the protein surface.

Coat protein Ca^{2+} binding is also possible as observed in the tertiary structures of SBMV (Abad-Zapatero et al., 1980), TBSV (Harrison et al., 1978), and STNV (Liljas et al., 1982). The presence of Ca^{2+} in the SBMV virion at alkaline pH prevents swelling (Hsu et al., 1976; Durham, 1978). It has been speculated that the reduced Ca^{2+} concentration in the host cell cytoplasm provides the free energy for virion penetration of the cell membrane and disassembly (Durham, 1978; Durham et al., 1977). Verduin (1978) suggests a protein subunit carboxyl cage of glutamates and aspartates that bind Ca^{2+}; upon cation release, the carboxyl replusion acts as a spring inducing protein conformational changes, swelling, and RNA release. Similar proposals have been made for tobacco mosaic virus (Stubbs et al., 1977). Titration of CCMV and BMV cause the release of about seven protons per subunit; back-titration also results in the absorption of the same number of protons though the pH return shows hysteresis properties (Jacrot, 1975; Pfeiffer and Durham, 1977).

3.4. Protein–RNA Interactions

The mixing of bromovirus coat protein and nucleic acid followed by dialysis at pH 7.4, low ionic strength, and the presence of Mg^{2+} results in nucleoproteins predominantly sedimenting at around 80 S with other fractions showing coefficients greater than 110 S (Bancroft and Hiebert, 1967; Adolph and Butler, 1977; Verduin, 1978). SDS polyacrylamide gel electrophoresis revealed that RNA-3 was primarily encapsidated whereas the >110 S particles mainly contained RNA-3 and -4 (Verduin, 1978; Herzog and Hirth, 1978). The protein itself will not associate under these conditions, suggesting the significance of the RNA in achieving proper protein interaction for viral assembly under neutral conditions. The BMV nucleoproteins will also assemble in the presence of yeast tRNA, AMV 12 S RNA, and oligo(U) (Adolph and Butler, 1977; Cuillel et al., 1981; Bancroft et al., 1969a); however, competition experiments show that the favored nucleic acid for envelopment is that of BMV (Cuillel et al., 1981), suggesting preferred base sequences in recognition.

Verduin (1978) has observed through rate zonal centrifugation and fractionation of disassociation products that, at pH 7.5 and 1 *M* NaCl, the four intact CCMV RNAs each had four to eight bound coat protein subunits; he also found protein dimers in solution. Verduin thus proposes an assembly model of protein dimer association with the nucleic acid which directs the protein conformational change from loose to compact, followed by further coat protein association into an icosahedral capsid. Verduin also observes through circular dichroism that the CCMV *in situ* RNA has less base stacking than the isolated nucleic acid; apparently the protein–RNA interaction results in some melting of the nucleic acid structure. Different conformational states of the CCMV RNA components has also been noted in the presence of polyanions (Dickerson and Trim, 1978).

Recent proton NMR investigations of CCMV virions and pseudo-top components confirm the importance of the 25 N-terminal residues for RNA interaction as the observed mobility of the N-terminal span was considerably lessened in the virion. Furthermore, side chain assignments in the empty capsid signal correlated well with the CCMV N-terminal amino acid composition which consists of nine basic residues (Vriend et al., 1981, 1982). Argos (1981) has proposed through helical wheels and secondary structure prediction algorithms an α-helical structure for this region with a hydrophobic side for possible interaction in the grooves of double-stranded RNA and a basic hydrophilic side for phosphate group recognition. In STNV a helical configuration has been observed in the N-terminal arm structure (Liljas et al., 1982).

Crystals of CCMV have been recently obtained (Heuss et al., 1981b). The tertiary structure determination of the virus by X-ray diffraction techniques will hopefully elucidate the nature of the coat protein conformation and its interaction with nucleic acid.

3.5. Nucleotide Sequencing of Bromovirus RNA

The work of Kaesberg and colleagues has resulted in the complete nucleotide sequences of CCMV and BMV RNA-4; BMV RNA-3; and the 3′-terminal sequences of RNA-1, -2, -3, and -4 from CCMV, BMV, BBMV, and CMV. The BMV RNA-4 is 876 bases in length with a 300-nucleotide 3′ noncoding region and a 9-nucleotide 5′ noncoding span; CCMV RNA-4 at 824 bases shows a 244-nucleotide 3′ noncoding area and only 10 nucleotides for the 5′ noncoding span (Dasgupta et al., 1980; Dasgupta and Kaesberg, 1982). The 188 and 189 amino acids of the respective BMV and CCMV coat proteins align with a near 70% homology.

The nucleotide sequence match is similarly impressive with a 75% homology in first codon positions, 83% in the second, and only 37% for the third as would be expected for divergent evolution. The random third-base correlation is consistent with the observed 60% lack of similar codon usage for the homologous amino acids. Comparison of the bromovirus mRNAs with those from other plant viruses, TYMV (Guilley and Briand, 1978), TMV (Guilley et al., 1979), and AMV (Brederode et al., 1980), showed no significant sequence relationship.

The BMV RNA-3 sequence displays a 91-base 5′ noncoding region, a 909-base cistron for a 32,480 molecular weight protein, a 250-base intercistronic noncoding region, and 867 bases for the BMV coat protein (Ahlquist et al., 1981a). Furthermore an internal poly (A) region (16–22 bases) occurs 20 bases 5′ to the beginning of the RNA-4 sequence. This latter region may be important for the recognition of the RNA-4 juncture in the RNA-3 genome.

The predicted RNA secondary structures bear striking resemblances to the 3′-terminal sequences of RNA-1, -2, -3, and -4 from BMV, CCMV, BBMV, and CMV despite the widely different serological, host range, and symptomatic properties of the viruses (Ahlquist et al., 1981b). Kaesberg and co-workers suggest that these conserved, tRNA-like structures are essential for synthetase recognition and therefore for initiation of replication. Several chemical investigations support their contention (Ahlquist et al., 1979).

Ahlquist et al. (1979) have observed that two ribosomes can bind to RNA-3 in cell-free extracts of wheat embryo. The disome formation is not related to the dicistronic organization of RNA-3 but is more likely a result of a two-step process where the ribosome just binds the mRNA at an entry site 5′ to the initiation codon and then moves, perhaps through an intervening holding site, to the initiation site of translation.

4. TYMOVIRUSES

4.1. Introduction

The presently known members of the plant tymovirus class (Matthews, 1980) include TYMV, eggplant mosaic virus (EGMV), BDMV, and erysium latent virus (ESLV). They are characterized by an icosahedrally symmetric protein shell consisting of 180 copies of identical subunits each with molecular weight approximating 20,000. The capsid that envelops about 2.0×10^6 daltons of RNA comprising 37% of the virion weight shows great physical stability, has a diameter near 300 Å, and occurs *in vivo* in large quantities without RNA. The empty protein shell has been named the "top component" in contrast to the faster sedimenting particle called the "bottom component," which is the actual infectious virion due to its RNA. The nucleic acid is high in cytosine content (approximately 38%).

The most extensively studied member of the group is TYMV, first discovered in 1949 by Markham and Smith (1949). The stability, easy isolation in gram quantities from Chinese cabbage hosts, and *in vivo* top–bottom production made the virus a prime target for extensive scientific research. However, its major asset, stability, has prevented its use in reassembly studies. The virion could be degraded with relative ease through many pathways but the end products, a heterogenous mix of protein aggregates and generally fragmented RNA, defied complete virion reconstruction. Nonetheless, the chemical disassembly analyses aided recently by a known primary structure (Guilley and Briand, 1978) have provided significant information on protein–protein, protein–RNA, and RNA–RNA interactions.

The present scenario of at least viral disassembly can be summarized in the following manner. The TYMV bottom component can be disassociated into empty capsids, RNA (which may be degraded), and smaller protein units. The disassembly process requires high ionic strength (1 *M* KCl) as well as one of several treatments: (1) pH 11.5–12.0; (2) 8 *M* urea, pH 7.0; (3) 12 *M* formamide, pH 7.0; or (4) moderate heating around 45°C. At around pH 7.0 protein–RNA interactions are believed to be disrupted; namely, hydrogen bonds are broken between the carboxyl groups of acidic amino acid side chains (glutamates and aspartates) and the N-3 atoms and C-2 keto oxygens of cytosine. At increasingly alkaline pH the nucleic acid expands due to interchain repulsions. A drastic RNA structural rearrangement now results along with possible increased cleavage of the phosphodiester backbone. By pH 11.5, the radius of the particle increases from 146 to 152 Å within 30 sec, providing pores in the capsid for the RNA to escape along with its structurally stabilizing spermidine within 3–10 min. It is also possible that five to eight coat protein subunits are released along with the RNA and may even be associated with it. It must be emphasized that the scenario of tymovirus disassembly has been derived from rather harsh chemical studies. Some other members of the group do not demand such "clubbing" techniques to induce RNA release. The following paragraphs will provide details that have led to this model over years of investigation. An excellent review of the TYMV studies covering research to about 1974 can be found in Kaper's (1975) text on virus structure.

4.2. Protein–Protein Interactions

Image reconstruction from electron micrographs showed the TYMV protein capsid to be of triangulation number three, indicated by hexamer and pentamer subunit protrusions from an essentially spherical shell (Klug et al., 1966; Mellema and Amos, 1972). The diameter of the TYMV particles is near 300 Å. Packing arrangements in the top and bottom component crystals of BDMV (Heuss et al., 1981c) and ESLV (Coleman et al., 1978) also suggest this dimension for the tymoviruses generally.

The *in vivo* capsids are composed of the same protein subunits as those of the bottom component as supported by electron microscopy (Cosentino et al., 1956), isoelectric point, diffusion constants, hydration diameters (Markham, 1951), amino acid compositions (Fraser and Cosentino, 1957) and isomorphous space groups for top and bottom crystals yielding similar X-ray diffraction patterns at high resolution (J. K. MohanaRoa, K. L. Heuss, and P. Argos, unpublished). Even the *in vitro* capsids (produced by release of RNA from complete virions) bear striking resemblances to the *in vivo* shells as they display similar diffraction patterns (Finch and Klug, 1960; Longley and Leberman, 1966) and the same number of sulfhydryl groups in the unoxidized state (Kaper and Houwing, 1962; Jonard, 1972). For BDMV, similar proton NMR and analytic ultracentrifugation profiles can be obtained for both *in vivo* and *in vitro* capsids (Virudachalam et al., 1982a,b).

Several techniques have been used to degrade TYMV or induce RNA release with preserved capsids: alkaline pH and controlled ionic strength (Kaper, 1964, 1968), addition of high molarity urea or formamide (Jonard et al., 1972), heating (Lyttleton and Matthews, 1958; Kaper, 1971), freezing followed by thawing (Kaper and Alting Siberg, 1969b), use of mercurials (Kaper and Jenifer, 1967), ultraviolet light irradiation (Kaper, 1971), addition of quaternary ammonium salts (Stols and Veldstra, 1965), and RNA extraction with ethanol and phenol (Markham and Smith, 1949;

Diener and Schneider, 1968). The results from such studies point to one dominant fact: protein–RNA linkages are broken in a pH range of 6.0–7.0. At high ionic strength (0.75 *M* KCl) and pH 11.5, the virion can be disassociated into capsids and RNA while at lower ionic strength (0.15 *M* KCl), protein subunit aggregates and RNA predominate (Kaper, 1964, 1968). However, the resultant capsids remain stable at low or high ionic strengths, suggesting that the small protein aggregates are a result of virion and not capsid degradation. The urea studies are reminiscent (Jonard et al., 1972).

Several investigations involving the interactions of the protein sulfhydryls with various mercurials have been attempted. Knowledge of the TYMV amino acid sequence (Guilley and Briand, 1978) has shown that disulfide bridges are not a likely part of the protein fold. It appears that one of four of the cysteine residues is exposed for mercurial binding while the others are apparently buried as high concentrations of the metals will invariably unfold the protein (Kaper, 1975; Kaper and Jenifer, 1968; Dorne et al., 1971; Goldschalk and Veldstra, 1966).

Hydrophobic binding appears to be the major stabilizing force in capsid stability (Goldschalk and Veldstra, 1965). In 8 *M* urea the capsid stability is increased at high ionic strength and at ambient temperatures. Because urea favors chain–solvent interactions, cold and low salt cannot aid hydrophobic stability. Sensitivity of TYMV capsid structural integrity to freezing, phenol, ethanol, and quaternary ammonium salts also support the significance of the hydrophobic interactions (Kaper, 1975).

The coat protein topology for three plant viruses (Abad-Zapatero et al., 1980; Harrison et al., 1978; Liljas et al., 1982) whose tertiary structures have been determined by X-ray diffraction techniques is that of a "jelly roll" β-barrel. Secondary structural prediction algorithms suggest a similar fold for TYMV (Argos, 1981). Circular dichroic analyses also support the predominance of β-structure in the TYMV fold (Tamburro et al., 1978).

4.3. Protein–RNA Interactions

Experiments discussed previously as well as titration studies on the TYMV-*p*-mercuribenzoate complex at various ionic strength conditions (Kaper and Alting Siberg, 1969a) suggested protein–RNA linkages disassociated between pH 6 and 7. Direct evidence has been provided by recent NMR studies on BDMV where the RNA is released merely by increasing the pH in a 0.05 *M* phosphate buffer (Virudachalam et al., 1982a) with the pH midpoint at 6.8. ^{31}P NMR investigations display a dramatic increase in the RNA mobility around the same midpoint pH. Proton NMR spectra for the BDMV bottom component show few mobile side chains at pH 5.5, whereas many nonrigid groups are observed for the *in vivo* and *in vitro* top components whose profiles bear a striking resemblance (Vriend et al., 1982; Virudachalam et al., 1983a,b). Tentative assignments of side chain groups gaining mobility include residues contained in the first 15 N-terminal amino acids of TYMV, suggesting their importance in linking the nucleic acid to the capsid (Virudachalam et al., 1982b). This N-terminal phenomenon is observed in other plant viruses with basic groups being the prime linkers (Abad-Zapatero et al., 1980; Harrison et al., 1978; Liljas et al., 1982; Vriend et al., 1981). Recent ^{13}C NMR results (Virudachalam et al., 1983b) on BDMV point to the loss of rigidity for glutamic and aspartic acids and yet maintenance of nonmobility for aromatic groups. The proton NMR spectrum of *in vivo* TYMV top component is strikingly similar to that of the *in vitro* BDMV top particles. A general model for protein–RNA interactions in tymoviruses as first proposed by Kaper (1972, 1975) and Jonard (1972) would thus involve hydrogen bonding of the carboxyl side chains with the N-3 atom of cytosine and the C-2 keto oxygen of the base. It is noteworthy that four acidic residues are contained within the 11 N-terminal amino acids of TYMV (Gulley and Briand, 1978). *In situ* cross-linking studies on TYMV have also pointed to the importance of the N-terminal span for nucleic acid interaction as well as two other primary structural regions centering near residues 40 and 140 (Ehresmann et al., 1980). Recent low-angle neutron scattering investigations of TYMV (Jacrot, 1981) suggest that the RNA does not penetrate the capsid as previously thought from X-ray diffraction results (Klug et al., 1966; Finch and Klug, 1966).

Cesium chloride density gradient centrifugation shows that there are at least three other TYMV components in addition to the top and bottom particles (Higgins et al., 1978). They contain varying amounts of genomic and/or subgenomic RNA which is less than that of the native virion. These nucleoprotein fractions do not swell upon RNA release (Keeling et al., 1979). Apparently the electrostatic repulsion of the fewer-than-normal phosphate groups is insufficient to induce capsid expansion.

Photon correlation spectroscopy provided the first direct evidence for

TYMV swelling under alkaline conditions in 1 *M* KCl (Keeling et al., 1979). The particle radius increases from 146 to 152 Å within 30 sec and RNA is released within 3–10 min. Keeling and Matthews (1982) show that a specific pentamer or hexamer protein cluster is loosened allowing RNA release. The protein subunits may well be associated with the free RNA through specific recognition of a base sequence.

4.4. RNA–RNA Interactions

Since TYMV RNA is 38% cytodylic acid in content (Matthews, 1980), any double helical arrangement is likely to involve cytosine base hydrogen bonding (Mitra and Kaesberg, 1965). Titration studies on poly (C) have indicated the feasibility of hydrogen bonds between N-3 atoms of a cytosine pair (Hartman and Rich, 1965).

Pleij et al. (1977) have shown that the *in situ* TYMV RNA undergoes a drastic structural rearrangement under alkaline conditions and high ionic strength. The normal 28 S sedimentation state becomes 38 S, suggesting a compaction of the RNA. Fragmentation of the RNA noted by previous researchers was not observed. However, CD studies (Tamburro et al., 1978), which confirm considerable base pairing for the RNA, indicate little structural change for *in situ* or free RNA.

Gas chromatography and mass spectroscopy demonstrate clearly that spermidine is the major polyamine binding the TYMV RNA *in situ* (Cohen and Greenberg, 1981). In fact 200–700 molecules are observed per virion. Under neutral conditions the virus is impermeable to exogenous spermidine and does not exchange or leak internal polyamines, implying spermidine packaging during formation of the virions in the plant hosts. Recent proton NMR results confirm the presence of spermidine in TYMV (K. Sitaraman, K. L. Heuss, R. Virudachalam, J. L. Markley, and P. Argos, unpublished). However, similar studies on BDMV do not yield the spermidine spectra and yet addition of 5 m*M* spermine or spermidine prevents release of BDMV RNA under conditions (0.05 *M* Tris buffer, pH 8.5) that normally empty all the virions in a few days. The BDMV protein shell may be porous to polyamines (perhaps during momentary swelling) and the BDMV RNA is not packaged with the polyamine. Furthermore, stabilization of the BDMV RNA by the positively charged amide groups and subsequent prevention of RNA release emphasizes the significance of phosphate group repulsion in RNA escape. In fact 5 m*M* Ca^{2+}, 5 m*M*

Mg^{2+}, or 1 *M* KCl in a Tris buffer at pH 8.5 prevent BDMV RNA release (K. Sitaraman, K. L. Heuss, R. Virudachalam, J. L. Markley, and P. Argos, unpublished work).

Attempts to reassemble TYMV have been unsuccessful. Jonard et al. (1972) disassociated TYMV at pH 7 in 8 *M* urea and 1 *M* NaCl; subsequent dialysis to pH 4.2 at low ionic strength with small amounts of Mg^{2+} or spermidine result in nucleoprotein particles with sedimentation rates of 70 and 105 S which contrast with the top and bottom values of 54 and 115 S (Matthews, 1980). The particles were ribonuclease sensitive probably due to the exterior location of the RNA on the protein shell. Furthermore the experiments were not reproducible using *in vivo* empty capsids and TYMV RNA, indicating a subtle alteration in the *in vitro* produced protein shells which is perhaps related to the protein associated with the released RNA.

TYMV particles contain two main RNA species (Pleij et al., 1976; Klein et al., 1976): the genomic RNA necessary for infection and the coat protein messenger RNA which corresponds to the 3′ region of the genomic RNA. RNA can be enzymatically aminoacylated with valine *in vitro* (Pinck et al., 1970; Yot et al., 1970; Joshi et al., 1982) and most probably *in vivo* by valyl-tRNA synthetase (Joshi et al., 1978; Loesch-Fries and Hall, 1982). The genomic RNA appears to be divided into two cistrons separated by only 14 nucleotides, allowing the possibility of *in vivo* "read-through" events (Morch et al., 1982).

5. OTHER VIRUSES

5.1. Introduction

Many plant viruses have been the subject of stability studies in addition to those previously discussed. These viruses are also reviewed as their structure has been determined or sufficient data become available to provide some picture of the stabilizing forces. They include viruses primarily stabilized by protein–RNA interactions (AMV and CMV), by metal ion mediated protein–protein interactions and protein–RNA interactions (TBSV), and by pH dependent protein–protein interactions as well as protein–RNA interactions (the comoviruses).

5.2. Cucumber Mosaic Virus Group

The cucumber mosaic virus group contains two members in addition to CMV: tomato aspermy virus (TAV) and peanut stunt virus (PSV). Each of these viruses has numerous strains that are serologically related to the CMV type strain. The CMV group has recently been reviewed by Kaper and Waterworth (1981) and Kaper (1981).

The cucumber viruses display the same $T = 3$ icosahedral architecture observed for other plant viruses. The particle radius is approximately 140 Å with protein subunits of molecular weight 24,500 clustered in groups of hexamers and pentamers (Van Regenmortel et al., 1972; Habili and Francki, 1974a,b; Finch et al., 1967). Members of the CMV group sediment at approximately 100 S (Francki et al., 1966).

CMV is a multipartite virus organized like the bromoviruses. The heaviest two of the four RNA molecules (1.27, 1.13, 0.82, and 0.35 $\times 10^6$ daltons) are individually encapsidated while a third virion contains the two smaller RNA molecules (Pinder et al., 1974). It is presently impossible to distinguish the particle types on the basis of centrifugation or electrophoresis (Van Regenmortel, 1967).

Members of the CMV group are clearly stabilized by protein–RNA interactions (Kaper, 1973; Habili and Francki, 1974a,b) as evidenced by the lack of empty capsid formation *in vivo* or *in vitro*. The dissociation of CMV above pH 10.0, the likely lysine pK_a, strongly suggests that interactions between positively charged basic residues of the protein and negatively charged phosphate groups of the RNA provide the primary means of stabilizing the particle (Kaper and Greelen, 1971). At very low SDS concentrations ($<0.01\%$) and high molar salt (e.g., 1.5 *M* LiCl) the particle also disassembles (Kaper, 1976; Boatman and Kaper, 1976). The kinetics of disassociation with the chemical modifying agent trinitrobenzesulfonic acid show that three lysine groups in each subunit react readily and are essential for stability, while nine other lysines react more slowly and apparently do not play as significant a role. Virus particles disassembled at elevated pH or high salt concentrations yield intact RNA and capsid protein which can be reassembled as infectious particles by dialysis against solutions at neutral pH and containing low salt (Kaper and Geelen, 1971). Protein produced by disassembly with lysine chemical modifiers will not reassemble.

Metal ions do not play a significant role in stabilizing the CMV capsid which maintains structural integrity in the presence of EDTA (Habili and Francki, 1974a). When exposed to magnesium chloride, CMV precipitates while TAV remains in solution (Habili and Francki, 1974a). CMV does not undergo swelling at elevated pH and yet sensitivity to nucleases and proteases is similar to that of the bromo and SBMV viruses in the swollen form, suggesting that the mature CMV virion may exist in the swollen state. Further properties shared by CMV and the bromoviruses include the presence of four major RNA species, similar subunit molecular weight and particle size, sequence homology at the 3′-OH noncoding region of the RNA molecules (Ahlquist et al., 1981b), and a high percentage of basic residues in the capsid protein. As in BMV, the CMV basic residues are concentrated near the amino terminus (Gould and Symons, 1982). Contrast masking neutron diffraction analysis indicates that both CMV and swollen BMV display significant penetration of the RNA between the protein subunits (Jacrot et al., 1977).

5.3. Alfalfa Mosaic Virus

The structure of AMV particles is significantly different from the simple spherical viruses discussed as the coat protein and nucleic acid can form four nucleoprotein particles of bacilliform shape, each containing an RNA molecule of given size (Jaspars, 1974). In solution, the disassembled 24,000 dalton coat proteins exist almost exclusively as dimers (Driedonks et al., 1977), which can be assembled into nucleoprotein particles with a hexameric lattice presumably sealed at its ends with pentameric caps (Driedonks et al., 1977; Mellema and van den Berg, 1974; Mellema, 1975). Particle stabilization depends greatly on protein–nucleic acid interactions, which can be disrupted by high salt (Bol and Kruseman, 1969). The nucleic acid even establishes the length of the particles in most strains of the virus; however, the VRU strain polymerizes without nucleic acid and frequently encapsidates more than one RNA molecule (Driedonks et al., 1978b). The capsid is very open, permitting nucleic acid degradation at pH 7.0 (Bol and Veldstra, 1969) and protein digestion with trypsin (Bol et al., 1974). The coat protein of AMV is also functional in the infection process as it binds specifically and with high affinity at a site near the RNA 3′-OH end (Bol et al., 1971; Staker et al., 1980), a likely initiation region for particle assembly and elongation (Driedonks et al., 1978a).

The capsid subunits can be cleaved *in situ* with trypsin, removing 27 amino acids from the protein N-terminus (Bol et al., 1974). This cleavage prevents interaction between protein and nucleic acid and inhibits the activating potential of the protein for replication. The specifically cleaved protein assembles into $T = 1$ spherical particles that have been studied crystallographically at 4.5 Å resolution (Fukuyama et al., 1982). The subunits are tightly associated at the dimer and trimer axes. Large holes, present in the capsid at the fivefold axes, are consistent with electron micrographs of elongated structures composed of hexameric clusters with large central pores. The electron density distribution is consistent with the β-barrel fold observed for other viral protein structures though the low resolution does not allow tracing of the peptide backbone.

Like CMV, AMV does not undergo swelling at elevated pH (Oostergetel et al., 1981), does not depend on cations for structural integrity, and disassembles into protein–RNA complexes above pH 9.0 (Verhagen and Bol, 1972). These properties prompted the suggestion that AMV exists in a swollen state at neutral pH. Recent proton NMR studies of the virions and isolated protein suggest the capsid subunit is composed of a stable shell domain beginning at residue 37 and extending to the carboxy terminus as well as a 36-residue amino-terminal arm which interacts extensively with RNA and shows a large degree of mobility in disassembled protein (Kan et al., 1982).

AMV protein has been shown to effectively encapsidate DNA in addition to RNA (Driedonks et al., 1978a). The polymorphic nature of AMV protein and its ability to conform to the size of nucleic acid molecules may be of practical value if viral DNA is used as a vehicle for introducing specific genes into plants. The DNA of cauliflower mosaic virus can be inserted into plant chromosomes (Gronenborn et al., 1981).

5.4. Tomato Bushy Stunt Virus

TBSV is the type number of the tombusvirus group, which also includes turnip crinkle virus (TCV) and possibly other members (Martelli, 1981). TBSV has been the subject of intense structural investigation (Harrison et al., 1978) whereas TCV has been more exhaustively studied chemically. A recent review of the tombusviruses covers their pathology (Martelli, 1981).

TBSV displays a $T = 3$ capsid with obvious dimer clustering of protein subunits (Finch et al., 1970). The subunit molecular weight is 40,000 (Michelin-Lausarot et al., 1970; Weber et al., 1970) and that of the RNA is 1.5×10^6. The sedimentation coefficient for the virus is 135 S; the spherically averaged particle diameter is 310 Å (Tremaine, 1970; Harrison, 1969; Chauvin et al., 1978a). TBSV and TCV appear to be very similar to the SBMV group in many of their biochemical properties. TBSV and TCV are stabilized by metal cations. If these are removed by dialysis against EDTA and the pH elevated above 7.5, the particles swell approximately 10%. In this state TCV is susceptible to proteolysis (Golden and Harrison, 1982). The structure of TBSV in the expanded state reveals separation of the quasi-threefold released subunits and yet maintenance of the interactions at the hexamers and pentamers. The protruding domains of the subunits undergo a rotation about the twofold and quasi-twofold axes (Robinson and Harrison, 1982).

TCV protein is capable of polymorphism. When the virus is degraded at pH 10.5 in 0.2 *M* NaCl, the particles disassemble and then reassemble into $T = 1$ particles (Leberman and Finch, 1970). Electron microscopy shows that the dimer clustering of protruding domains is maintained (Finch et al., 1970). Further studies demonstrate a reduction in the molecular weight of the $T = 1$ protein subunits, suggesting cleavage (Munowitz et al., 1980) of the N-terminal arm for their formation, the $T = 1$ particles having been crystallized (Leberman and Longley, 1970; Butler, 1970).

The protein subunit of TBSV is composed of two major domains that fold into β-barrel structures (Harrison et al., 1978), suggesting gene duplication in their evolution (Argos et al., 1980). A smaller domain, composed of about 100 N-terminal residues, interacts with the nucleic acid and forms a portion of the intersubunit interaction between 60 of the proteins. The TCV subunit can be cleaved using trypsin or chymotrypsin at two sites, presumably domain interfaces observed in the TBSV structure. TCV under mild conditions shows that the dimer interactions are maintained in solution. Like SBMV, the N-terminal arm of the TCV protein can be cleaved when the virus is in the swollen state (Golden and Harrison, 1982). ^{31}P NMR studies of TBSV show that the RNA is relatively immobile within the native virion. However, during swelling of the particle, dynamic disorder of the nucleic acid is detectable (Munowitz et al., 1980).

5.5. Cowpea Mosaic Virus

Recent reviews of the CpMV group include reports by Bruening (1977), Van Kammen and Mellema (1977), and Stace-Smith (1981). Though there are 11 members of the group, most of the structural biochemistry has been performed on CpMV and red clover mosaic virus (RCMV).

CPMV displays three bands when subjected to sucrose gradient analysis. The diameters of the particles extracted from each band are approximately 280 Å (Schmidt et al., 1983). These particles display identical electrophoretic mobility below pH 7.5 (Van Kammen and Mellema, 1977). The RNA content of the particles by weight is 35% for the bottom component (115 S), 25% for middle component (98 S), and 0% for top component (58 S). The two ribonucleoprotein components are required for infection (Bruening and Agrawal, 1967; Van Kammen, 1968). The single-stranded RNA molecules have molecular weights of 2.39×10^6 and 1.39×10^6 for bottom and middle components respectively (Murant et al., 1981). Both RNA molecules are polyadenylated at their 3′ termini (El Manna and Bruening, 1973) and have a 10,000 dalton protein covalently linked to the 5′ end (Daubert et al., 1978; Stanley et al., 1978). The coat proteins of molecular weight 42,000 and 22,000 are produced by post-translational cleavage of a polyprotein synthesized from the middle component RNA (Pelham, 1979). The protease is coded by the bottom RNA as are the molecules essential for RNA replication (Franssen et al., 1982; Goldbach et al., 1980).

The structure of CpMV has been investigated by three-dimensional image reconstruction (Crowther et al., 1974) and shown to have icosahedral symmetry with protrusions at the five- and threefold axes. The capsid structure of each component appears identical containing sixty copies of both the small and large proteins arranged with $T = 1$ symmetry. Analysis of the virus using circular dichroism indicates that the proteins contain very little α helix and that 60% of the RNA in both components is double helical (Verduin, unpublished). The virus is currently the subject of study by single crystal X-ray diffraction (White and Johnson, 1980; Johnson and Hollingshead, 1981).

The stability of CpMV and RCMV is derived from both protein–protein and protein–RNA interactions. Empty capsids are found *in vivo*. The sedimentation rate of RCMV decreases slowly from pH 3.0 to 6.0 where it passes through a minimum and then increases slowly reaching its maximum value at pH 8.0. If hydrodynamic behavior reflects particle size, the virion reaches its maximum size at pH 6.0 (Abdelmoeti and Oxelfelt, 1982). This behavior is in marked contrast to the other spherical viruses where swelling occurs above pH 7.0. At pH 9.0 the empty capsids of RCMV and CpMV disassemble, whereas the nucleoprotein components are stable to pH 10.0 (J. White, J. E. Johnson, V. MacKenzie, G. Bruening, and L. Russel, unpublished; Abdelmoeti and Oxelfelt, 1982). Treatment of the CpMV empty capsid or nucleoprotein component with lysine specific chemical modifying agents causes particle disassembly at pH 7.0 (L. Russel, V. MacKenzie, G. Bruening, and J. E. Johnson, unpublished). At pH 9.0 the top component disassembles yielding a discrete product sedimenting at 13 S. The oligomer contains both the large and small protein and is likely to be a pentamer of the two protein types. At pH 7.0 the empty capsid is stable in 4.0 *M* formamide. The nucleoprotein components are stable at pH 10.0 in 2.0 *M* formamide.

Proton magnetic resonance spectra of the middle and top components are virtually identical (Vriend et al., 1982); thus, the region of the protein interacting with RNA does not increase in mobility when the RNA is removed. This is in contrast to the behavior observed for TYMV and BDMV where substantial motion is found for a portion of the protein in the empty capsid that is apparently fixed in the nucleoprotein component.

Metal ions do not play a significant role in stabilizing the capsid protein structure in RCMV or CpMV (Abdelmoeti and Oxelfelt, 1982). Dialysis against 0.1 *M* EDTA at various pH values had no effect on the sedimentation behavior of RCMV and addition of 0.1 *M* $MgCl_2$ or $CaCl_2$ caused limited aggregation of the virus.

6. CONCLUSIONS

Many of the properties for the small spherical plant viruses can be explained by only two assumptions: (1) the capsid proteins of all the viruses are derived from an ancestral gene coding for the basic β-barrel fold which allows formation of icosahedral capsids with discrete volumes and (2) the amount of RNA encapsidated dictates the triangulation number and the mode of protein–protein and protein–RNA interactions. In STNV, the quantity of RNA is small; a $T = 1$ capsid is therefore used with sufficient volume remaining to allow for a basic, interior N-terminal arm for protein–RNA interactions and charge neutralization of many of the nucleic acid phosphate groups. This tethering of the protein shell to the RNA requires modulation of the disassembly process through calcium mediation and swelling. Viruses such as SBMV and TBSV must package more RNA and therefore have undergone evolutionary modification to maintain a $T = 3$ shell which provides a larger but discrete capsid volume. Sufficient interior space remains to utilize an arm and display consequent swelling properties. To encapsidate even more RNA in a $T = 3$ shell, TYMV and CpMV must eliminate the basic arm to provide space for the packaged RNA, modify the β-barrel side groups for strong protein–protein interactions to contain the RNA, use cations or polyamines to cancel the nucleotide charge, and possibly skew the RNA base composition to promote secondary structure for more efficient packaging (38% cytosine for TYMV and 32% uridine for CpMV). A by-product of the strong protein–protein interactions are the formation of *in vivo* top capsids. The protein–RNA interactions are likely to be more specific such as the use of glutamates in tymoviruses as well as less flexible than those provided by an extensive N-terminal arm. Calcium utilization and subsequent swelling are unnecessary. Bromoviruses package more RNA through the use of the multipartite genome encapsidated in three $T = 3$ shells; adequate space remains for maintenance of the arm, swelling, and calcium disassembly modulation. Apparently construction of a particle with a higher triangulation number requires excessive evolutionary effort. It must be emphasized that this hypothesis is speculative and awaits testing by more detailed information for other plant viruses as well as those discussed presently.

Crystals that diffract to atomic resolution for small spherical plant viruses were once considered a rare phenomenon. The list now includes TBSV (Harrison and Jack, 1975), STNV (Fridborg et al., 1965), SBMV (Akimoto et al., 1975), CpMV (White and Johnson, 1980; Johnson and Hollingshead, 1981), ESLV (Coleman et al., 1978), BDMV (Heuss et al., 1981b), CCMV (Heuss et al., 1981c), the $T = 1$ capsid of AMV (Fukuyama et al., 1981), the $T = 1$ capsid of SBMV (Erickson and Rossmann, 1982), and tobacco ringspot virus (Heuss et al., 1981a). The plant virus groups represented are bromoviruses, comoviruses, nepoviruses, southern bean mosaic viruses, tombusviruses, and tymoviruses. The atomic resolution structures for SBMV (Abad-Zapatero et al., 1980), TBSV (Harrison et al., 1978), and STNV (Liljas et al., 1982) are known; a low resolution view of $T = 1$ AMV has also recently been determined (Fukuyama et al., 1982). The structural determination of the crystallized viruses, the sophisticated biophysical probes used for viral analysis, the extensive knowledge of viral nucleotide sequences, and site-specific mutations in the nucleic acid should all contribute to a better and more general understanding of the structure–function relationship in plant viruses in the reasonably near future.

REFERENCES

Abad-Zapatero, C., Abdel-Meguid, S. S., Johnson, J. E., Leslie, A. G. W., Rayment, I., Rossmann, M. G., Suck, D., and Tsukihara, T. (1980). *Nature (London)* **286,** 33–39.

Abdel-Meguid, S. S., Yamane, T., Fukuyama, K., and Rossmann, M. G. (1981). *Virology* **114,** 81–85.

Abdelmoeti, M., and Oxelfelt, P. (1982). *Phytopathol. Z.* **103,** 55–65.

Adolph, K. W. (1975). *Eur. J. Biochem.* **53,** 449–455.

Adolph, K. W., and Butler, P. J. G. (1974). *J. Mol. Biol.* **88,** 327–341.

Adolph, K. W., and Butler, P. J. G. (1977). *J. Mol. Biol.* **109,** 345–357.

Ahlquist, P., Dasgupta, R., Shih, D. S., Zimmern, D., and Kaesberg, P. (1979). *Nature (London)* **281,** 277–282.

Ahlquist, P., Luckow, V., and Kaesberg, P. (1981a). *J. Mol. Biol.* **153,** 23–38.

Ahlquist, P., Dasgupta, R., and Kaesberg, P. (1981b). *Cell* **23,** 183–189.

Akimoto, T. Wagner, M. A., Johnson, J. E., and Rossmann, M. G. (1975). *J. Ultrastruct. Res.* **53,** 306–318.

Argos, P. (1981). *Virology* **110,** 55–62.

Argos, P., Tsukihara, T., and Rossmann, M. G. (1980). *J. Mol. Evol.* **15,** 169–179.

Bastin, M., Dasgupta, R., Hall, T. C., and Kaesberg, P. (1976). *J. Mol. Biol.* **103,** 737–745.

Bancroft, J. B. (1970). *Adv. Virus Res.* **16,** 99–134.

Bancroft, J. B., and Flack, I. H. (1972). *J. Gen. Virol.* **15,** 247–251.

Bancroft, J. B., and Hiebert, E. (1967). *Virology* **32,** 354–356.

Bancroft, J. B., Hills, G. J., and Markham, R. (1967). *Virology* **31,** 354–379.

Bancroft, J. B., Hiebert, E., Rees, M. W., and Markham, R. (1968). *Virology* **34,** 224–239.

Bancroft, J. B., Hiebert, E., and Bracker, C. E. (1969a). *Virology* **39,** 924–930.

Bancroft, J. B., Bracker, C. E., and Wagner, G. W. (1969b). *Virology* **38,** 324–335.

Boatman, S., and Kaper, J. M. (1976). *Virology* **70,** 1–16.

Bol, J. F., and Kruseman (1969). *Virology* **37,** 485–488.

Bol, J. F., and Veldstra, H. (1969). *Virology* **37,** 74–85.

Bol, J. F., Van Vloten-Doting, L., and Jaspars, E. M. J. (1971). *Virology* **46,** 73–85.

Bol, J. F., Kraal, B., and Brederode, F. Th. (1974). *Virology* **58,** 101–110.

Brederode, F. Th., Kaper-Zwarthoff, E. C., and Bol, J. F. (1980). *Nucl. Acids Res.* **8,** 2213–2223.

Bruening, G. (1977). In *Comprehensive Virology*, H. Fraenkel-Conrad and R. R. Wagner, Eds., Vol. 11. Plenum, New York, pp. 55–141.

Bruening, G., and Agrawal, H. (1967). *Virology* **32,** 306–320.

Butler, P. (1970). *J. Mol. Biol.* **52,** 589–593.

Caspar, D. L. D., and Klug, A. (1962). *Cold Spring Harbor Symp. Quant. Biol.* **27,** 1–24.

Chauvin, C., Witz, J., and Jacrot, B. (1978a). *J. Mol. Biol.* **124,** 641–651.

Chauvin, C., Pfeiffer, P., Witz, J., and Jacrot, B. (1978b). *Virology* **88,** 138–148.

Childlow, J., and Tremaine, J. H. (1971). *Virology* **43,** 267–278.

Cohen, S. R., and Greenberg, M. L. (1981). *Proc. Natl. Acad. Sci. U.S.A.* **78,** 5474–5479.

Coleman, P. M., Tulloch, P. A., Shukla, D. D., and Gough, K. H. (1978). *J. Mol. Biol.* **142,** 263–268.

Cosentino, V., Paigen, K., and Steere, R. L. (1956). *Virology* **2,** 139–148.

Crowther, R. A., Geelen, J. L. M., and Mellema, J. E. (1974). *Virology* **57,** 20–27.

Cuillel, M., Jacrot, B., and Zulauf, M. (1981). *Virology* **110,** 63–72.

Dasgupta, R., and Kaesberg, P. (1982). *Nuc. Acids Res.* **10,** 703–713.

Dasgupta, R., Ahlquist, P., and Kaesberg, P. (1980). *Virology* **104,** 339–346.

Daubert, S., Bruening, G., and Najarian, R. (1978). *Eur. J. Biochem.* **92,** 45–51.

Denloye, A. O., Hamer, R. B., and Hull, R. (1978). *J. Gen. Virol.* **41,** 77–85.

Dickerson, P. E., and Trim, A. R. (1978). *Nucl. Acids Res.* **5,** 987–998.

Diener, T. O., and Schneider, I. R. (1968). *Arch. Biochem. Biophys.* **124,** 401–412.

Dorne, B., Jonard, G., Witz, J., and Hirth, L. (1971). *Virology* **43,** 279–290.

Driedonks, R. A., Krijgsman, P. C. J., and Mellema, J. E. (1977). *J. Mol. Biol.* **113,** 123–140.

Driedonks, R. A., Krijgsman, P. C. J., and Mellema, J. E. (1978a). *Eur. J. Biochem.* **82,** 405–417.

Driedonks, R. A., Krijgsman, P. C. J., and Mellema, J. E. (1978b). *J. Mol. Biol.* **124,** 713–719.

Durham, A. C. H. (1978). *Biomedicine* **28,** 307–314.

Durham, A. C. H., Hendry, D. A., and Von Wechmar, M. B. (1977). *Virology* **77,** 524–533.

Ehresmann, B., Briand, J. P., Reinbolt, J., and Witz, J. (1980). *Eur. J. Biochem.* **108,** 123–129.

El Manna, M. M. and Bruening, G. (1973). *Virology* **56,** 198–206.

Erickson, J., and Rossmann, M. G. (1982). *Virology* **116,** 128–136.

Finch, J. T., and Klug, A. (1960). *J. Mol. Biol.* **2,** 434–435.

Finch, J. T., and Klug, A. (1966). *J. Mol. Biol.* **15,** 344–364.

Finch, J., Klug, A., and Van Regenmortel, M. (1967). *J. Mol. Biol.* **24,** 303–305.

Finch, J. T., and Bancroft, J. B. (1968). *Nature (London)* **220,** 815–816.

Finch, J., Klug, A., and Leberman, R. (1970). *J. Mol. Biol.* **50,** 215–222.

Francki, R., Randles, J., Chambers, T., and Wilson, S. B. (1966). *Virology* **28,** 729–741.

Franssen, H., Goldbach, R., Broekhuysen, M., Moerman, M., and Van Kammen, A. (1982). *J. Virol.* **41,** 8–17.

Fraser, D., and Cosentino, V. (1957). *Virology* **4,** 126–129.

Fridborg, K., Hjerten, S., Hoglund, S., Liljas, A., Lundberg, B. K. S., Oxelfelt, P., Phibpson, L., and Strandberg, B. (1965). *Proc. Natl. Acad. Sci. U.S.A.* **54,** 513–521.

Fukuyama, K., Abdel-Meguid, S. S., and Rossmann, M. G. (1981). *J. Mol. Biol.* **150,** 33–41.

Fukuyama, K., Abdel-Meguid, S. S., Johnson, J. E., and Rossmann, M. G. (1982). *J. Mol. Biol.* **167,** 873–894.

Ghabrial, S. A., Shepherd, R. J., and Grogan, R. G. (1967). *Virology* **33,** 17–25.

Ghosh, A., Dasgupta, R., Salerno-Rife, T., Rutgers, T., and Kaesberg, P. (1979). *Nuc. Acids Res.* **7,** 2137–2146.

Ghosh, A., Rutgers, T., Ke-Qiang, M., and Kaesberg, P. (1981). *J. Virol.* **39,** 87–92.

Goldbach, R., Rezelman, G., and Van Kammen, A. (1980). *Nature (London)* **286,** 297–300.

Golden, J., and Harrison, S. (1982). *Biochemistry* **21,** 3862–3866.

Goldschalk, W., and Veldstra, H. (1965). *Arch. Biochem. Biophys.* **111,** 161–173.

Goldschalk, W., and Veldstra, H. (1966). *Arch. Biochem. Biophys.* **113,** 347–355.

Grogan, R. G., and Kimball, K. A. (1964). *Phytopathology* **54,** 75–78.

Gronenborn, B., Gardner, R., Schaefer, S., and Shepherd, R. (1981). *Nature (London)* **294,** 773–776.

Gould, A. R., and Symons, R. H. (1982). *Eur. J. Biochem.* **126,** 217–226.

Guilley, H., and Briand, J. P. (1978). *Cell* **15,** 113–122.

Guilley, H., Jonard, G., Kukla, B., and Richards, K. E. (1979). *Nucl. Acids Res.* **6,** 1287–1308.

Habili, N., and Francki, R. I. B. (1974a). *Virology* **60,** 20–36.

Habili, N., and Francki, R. I. B. (1974b). *Virology* **57,** 392–401.

Harrison, B. D., Finch, J. J., Gibbs, A. J., Hollings, M., Shepherd, R. J., Valenta, V., and Wetter, C. (1971). *Virology* **45,** 356–363.

Harrison, S. C. (1969). *J. Mol. Biol.* **42,** 457–483.

Harrison, S. C., and Jack, A. (1975). *J. Mol. Biol.* **97,** 173–191.

Harrison, S. C., Olson, A. J., Schutt, C. E., Winkler, F. K., and Bricogne, C. (1978). *Nature* (*London*) **276,** 368–373.

Hartman, K. A., and Rich, A. (1965). *J. Am. Chem. Soc.* **87,** 2033–2039.

Hermodson, M., Abad-Zapatero, C., Abdel-Meguid, S. S., Pundak, S., Rossmann, M. G., and Tremaine, J. (1982). *Virology* **119,** 133–149.

Herzog, M., and Hirth, L. (1978). *Virology* **86,** 48–56.

Heuss, K. L., Murthy, M. R. N., and Argos, P. (1981a). *J. Mol. Biol.* **153,** 1161–1168.

Heuss, K. L., MohanaRao, J. K., and Argos, P. (1981b). *J. Mol. Biol.* **146,** 635–640.

Heuss, K. L., MohanaRao, J. K., and Argos, P. (1981c). *J. Mol. Biol.* **146,** 629–634.

Hiebert, E. (1969). Ph.D. thesis, Purdue University, West Lafayette, Indiana.

Higgins, T. J. V., Whitfeld, P. R., and Matthews, R. E. F. (1978). *Virology* **84,** 153–161.

Hirth, L., and Richards, K. (1981). *Adv. Virus Res.* **26,** 145–199.

Horne, R. W., Hobart, J. M., and Pasquali-Ronchetti, I. (1975). *J. Ultrastruct. Res.* **53,** 319–330.

Hsu, C. H., Sehgal, O. P., and Pickett, E. E. (1976). *Virology.* **69,** 587–595.

Hsu, C. H., White, J. A., and Sehgal, O. P. (1977). *Virology* **81,** 471–475.

Hull, R. (1977a). *Virology* **79,** 50–57.

Hull, R. (1977b). *J. Gen. Virol.* **36,** 289–295.

Hull, R. (1978). *Virology* **89,** 418–422.

Incardona, N. L., and Kaesberg, P. (1964). *Biophys. J.* **4,** 11–21.

Incardona, N. L., McKee, S., and Flanegan, J. B. (1973). *Virology* **53,** 204–214.

Jacrot, B. (1975). *J. Mol. Biol.* **95,** 433–446.

Jacrot, B., Chawin, C., and Witz, B. (1977). *Nature* (*London*) **266,** 417–421.

Jacrot, B. (1981). *Compr. Virol.* **17,** 129–181.

Jaspars, E. M. J. (1974). *Adv. Virus Res.* **19,** 37–140.

Johnson, J. E., and Hollingshead, C. (1981). *J. Ultrastruct. Res.* **74,** 223–231.

Johnson, M. W., Wagner, G. W., and Bancroft, J. B. (1973). *J. Gen. Virology.* **19,** 263–273.

Jonard, G. (1972). Ph.D. thesis, University of Strasbourg.

Jonard, G., Witz, J., and Hirth, L. (1972). *J. Mol. Biol.* **66,** 165–169.

Joshi, S., Haenni, A. L., Hubert, E., Huez, G., and Marbaix, G. (1978). *Nature (London)* **275,** 339–341.

Joshi, S., Chapeville, F., and Haenni, A. L. (1982). *Nucl. Acids Res.* **10,** 1947–1961.

Kan, J. H., Andree, P. J., Kouijzer, L. C., and Mellema, J. E. (1982). *Eur. J. Biochem.* **126,** 29–33.

Kaper, J. M. (1964). *Biochemistry* **3,** 486–493.

Kaper, J. M. (1968). In *Molecular Basis of Virology* H. Fraenkel-Conrat, Ed., Reinhold, New York, pp. 1–100.

Kaper, J. M. (1971). *J. Mol. Biol.* **56,** 259–276.

Kaper, J. M. (1972). In *RNA Viruses: Replication and Structure*, E. M. J. Jaspars and A. Van Kammen, Eds., FEBS Symp. Vol. 27. American Elsevier, New York, pp. 19–41.

Kaper, J. M. (1973). *Virology* **55,** 299–304.

Kaper, J. M. (1975). In *The Chemical Bases of Virus Structure, Disassociation and Reassembly*, American Elsevier, New York, pp. 273–320.

Kaper, J. M. (1976). *Virology* **71,** 185–198.

Kaper, J. (1981). In *Handbook of Plant Virus Infections and Comparative Diagnosis* E. Kurstak, Ed. Elsevier, Amsterdam, pp. 296–332.

Kaper, J. M., and Alting Siberg, R. (1969a). *Cryobiology* **5,** 366–371.

Kaper, J. M., and Alting Siberg, R. (1969b). *Virology* **38,** 407–415.

Kaper, J. M., and Geelen (1971). *J. Mol. Biol.* **56,** 277–294.

Kaper, J. M., and Houwing, C. (1962). *Arch. Biochem. Biophys.* **96,** 125–138.

Kaper, J. M., and Jenifer, F. G. (1967). *Biochemistry* **6,** 440–446.

Kaper, J. M., and Jenifer, F. G. (1968). *Virology* **35,** 71–81.

Kaper, J., and Waterworth, H. (1981). In *Handbook of Plant Virus Infections and Comparative Diagnosis*, E. Kurstak, Ed. Elsevier, Amsterdam, pp. 257–296.

Keeling, J., and Matthews, R. E. F. (1982). *Virology* **119,** 214–218.

Keeling, J., Collins, E. R., and Matthews, R. E. F. (1979). *Virology* **97,** 100–111.

Klein, C., Fritsch, C., Briand, J. P., Richards, K. E., Jonard, C., and Hirth, L. (1976). *Nucl. Acids Res.* **3,** 3043–3061.

Klug, A., Longley, W., and Leberman, R. (1966). *J. Mol. Biol.* **15,** 315–343.

Krüse, J., Verduin, B. J. M., and Visser, A. J. W. G. (1980). *Eur. J. Biochem.* **105,** 395–401.

Krüse, J., Timmins, P. A., and Witz, J. (1982). *Virology* **119,** 42–50.

Lamptey, P. N. L., and Hamilton, R. I. (1974). *Phytopathology* **64,** 1100–1104.

Lane, L. C. (1974). *Adv. Virus Res.* **19,** 151–220.

Lane, L. C. (1979). In *Descriptions of Plant Viruses*, No. 215. Commonwealth Mycological Institute and Association of Applied Biologists, Surrey.

Lane, L. C. (1980). In *Plant Virus Infections* E. Krustak, Ed. North Holland, Amsterdam, pp. 333–376.

Langenberg, W. G., and Schroeder, H. F. (1975). *J. Ultrastruct. Res.* **51,** 166–175.

Laufer, M., Taylor, N., and Wunder, C. (1952). *Arch. Biochem.* **40,** 453–471.

Leberman, R., and Finch, J. T. (1970). *J. Mol. Biol.* **50,** 209–212.

Leberman, R., and Longley, W. (1970). *J. Mol. Biol.* **50,** 213.

Liljas, L., Unge, T., Jones, T. A., Fridborg, K., Lovgren, S., Skoglund, V., and Strandberg, B. (1982). *J. Mol. Biol.* **159,** 93–108.

Loesch-Fries, L. S., and Hall, T. C. (1982). *Nature* (*London*) **298,** 771–772.

Longley, W., and Leberman, R. (1966). *J. Mol. Biol.* **19,** 223–225.

Lyttleton, J. W., and Matthews, R. E. F. (1958). *Virology* **6,** 460–471.

McCain, D. C., Virudachalam, R., Markley, J., Abdel-Meguid, S. S., and Rossmann, M. G. (1982). *Virology* **117,** 501–503.

Mang, R., Ghosh, A., and Kaesberg, P. (1982). *Virology* **116,** 264–274.

Markham, R. (1951). *Discuss. Faraday Soc.* **11,** 221–227.

Markham, R., and Smith, K. M. (1949). *Parasitology* **39,** 330–342.

Martelli, G. P. (1981). In *Handbook of Plant Virus Infections and Comparative Diagnosis* E. Kurstak, Ed., Elsevier, Amsterdam, pp. 61–90.

Matthews, R. E. F. (1980). In *Descriptions of Plant Viruses*, No. 230. Commonwealth Mycological Institute and Association of Applied Biologists, Surrey.

Mellema, J. E. (1975). *J. Mol. Biol.* **94,** 643–648.

Mellema, J. E., and Amos, L. A. (1972). *J. Mol. Biol.* **72,** 819–822.

Mellema, J. E., and van den Berg, H. J. N. (1974). *J. Supramol. Struct.* **2,** 17–31.

Michelin-Lausarot, P., Ambrosince, C., Steere, R. L., and Reichman, M. E. (1970). *Virology* **41,** 160–165.

Miller, G. L., and Price, W. C. (1946). *Arch. Biochem.* **10,** 467–477.

Mitra, S., and Kaesberg, P. (1965). *J. Mol. Biol.* **14,** 558–569.

Morch, M. D., Drugeon, G., and Benicourut, C. (1982). *Virology* **119,** 193–198.

Munowitz, M. G., Dobson, C. M., Griffin, R. G., and Harrison, S. C. (1980). *J. Mol. Biol.* **141,** 327–333.

Murant, A., Taylor, M., Duncan, G., and Roschke, J. H. (1981). *J. Gen. Virology.* **53,** 321–332.

Noort, A., van den Dries, C., Pleij, C., Jaspars, E., and Bosch, L. (1982). *Virology* **120,** 412–421.

Odumosu, A. O., Homer, R. B., and Hull, R. (1981). *J. Gen. Virology* **53,** 193–196.

Oostergetel, G. T., Krijgsman, P. C. J., Mellema, J. E., Cusack, S., and Miller, A. (1981). *Virology* **109,** 206–210.

Pelham, H. R. B. (1979). *Virology* **96,** 463–477.

Pfeiffer, P. (1980). *Virology* **102,** 54–61.

Pfeiffer, P., and Durham, A. C. H. (1977). *Virology* **81,** 419–432.

Pfeiffer, P., and Hirth, L. (1974). *Virology* **61,** 160–167.

Pfeiffer, P., and Hirth, L. (1975). *FEBS Lett.* **56,** 144–148.

Pinck, M., Yot, P., Chapeville, F., and Duranton, H. M. (1970). *Nature* (*London*) **226,** 954–956.

Pinder, J., Staynov, D., and Gratyor, W. (1974). *Biochemistry* **13,** 5373–5378.

Pleij, C. W. A., Neeleman, A., Van Vloten-Doting, L., and Bosch, L. (1976). *Proc. Natl. Acad. Sci. U.S.A.* **73,** 4437–4441.

Pleij, C. W. A., Eecen, H. G., Bosch, L., and Mandel, M. (1977). *Virology* **76,** 781–786.

Price, W. C. (1943). *Science* **101,** 515–517.

Price, W. C. (1946). *Am. J. Bot.* **33,** 45–54.

Reijenders, L., Aalbers, A. M. J., Van Kammen, A., and Thuring, R. W. J. (1974). *Virology* **60,** 515–521.

Robinson, J., and Harrison, S. C. (1982). *Nature* (*London*) **267,** 563–568.

Rutgers, T., Salerno-Rife, T., and Kaesberg, P. (1980). *Virology* **104,** 506–509.

Sakai, F., Watts, J. W., Dawson, J. R. O., and Bancroft, J. B. (1977). *J. Gen. Virol.* **34,** 285–293.

Savithri, H. S., and Erickson, J. W. (1983). *Virology* **126,** 328–335.

Schmidt, J., Johnson, J. E., and Phillips, W. (1983). *Virology* **127,** 65–73.

Sehgal, O. P. (1980). *Phytopathology* **70,** 342–348.

Sehgal, O. P., and Sinha, R. C. (1974). *Virology* **59,** 499–508.

Sehgal, O. P., Van, M., and White, J. (1979). *Virology* **94,** 479–483.

Shepherd, R. J., and Fulton, R. W. (1962). *Phytopathology* **52,** 489–493.

Shih, D. S., Lane, L. C., and Kaesberg, P. (1972). *J. Mol. Biol.* **64,** 353–362.

Stace-Smith, R. (1981). In *Handbook of Plant Virus Infections and Comparative Diagnosis* E. Kurstak, Ed. Elsevier, Amsterdam, pp. 171–195.

Staker, K., Kaper-Zwarthoff, E. C., Bol, J. F., and Jaspars, E. M. J. (1980). *FEBS Lett.* **121,** 123–126.

Stanely, S., Rottier, P., Davies, J., Zabel, P., and Van Kammen, A. (1978). *Nucl. Acids Res.* **5,** 4505–4522.

Steven, A. C., Smith, P. R., and Horne, R. W. (1978). *J. Ultrastruct. Res.* **64,** 63–73.

Stols, A. L. H., and Veldstra, H. (1965). *Virology* **25,** 508–515.

Stubbs, G., Warren, S., and Holmes, K. (1977). *Nature* (*London*) **267,** 216–221.

Tamburro, A. M., Guantieri, V., Piazzolla, P., and Gallitelli, D. (1978). *J. Gen. Virol.* **40,** 337–344.

Tremaine, J. H. (1966). *Virology* **30,** 348–354.

Tremaine, J. (1970). *Phytopathology* **60,** 454–460.

Tremaine, J. H., and Ronald, W. P. (1977). *Can. J. Bot.* **55,** 2274–2277.

Tremaine, J. H., and Ronald, W. P. (1978). *Virology* **91,** 164–172.

Tremaine, J. H., Ronald, W. P., and Kelley, E. M. (1981). *Can. J. Microbiol.* **27,** 654–663.

Tremaine, J. H., Ronald, W. P., and Kelley, E. M. (1982). *Virology* **118,** 35–44.

Van Kammen, A. (1968). *Virology* **34,** 312–318.

Van Kammen, A., and Mellema, J. E. (1977). In *The Atlas of Insect and Plant Viruses* K. Maramiorosch, Ed. Academic Press, New York, pp. 167–179.

Van Regenmortel, M. (1967). *Virology* **31,** 391–396.

Van Regenmortel, M., Hendry, D., and Balty, T. (1972). *Virology* **49,** 647–653.

Veerisetty, V., and Sehgal, O. P. (1980). *Phytopathology* **70,** 282–284.

Verrisetty, V., El Hassan, S., and Sehgal, O. P. (1981). *Virology* **108,** 268–296.

Verduin, B. J. M. (1974). *FEBS Lett.* **45,** 50–54.

Verduin, B. J. M. (1978). PhD. thesis, Laboratory for Virology, Wageninger.

Verhagen, W., and Bol, J. F. (1972). *Virology* **50,** 431–439.

Virudachalam, R., Heuss, K. L., Argos, P., and Markley, J. L. (1983a). *Virology* **130,** 360–371.

Virudachalam, R., Heuss, K. L., Markley, J. L., and Argos, P. (1983b). *Virology* **130,** 351–359.

Vriend, G., Hemminga, M. A., Verduin, B. J. M., deWit, J. L. T., and Schaafsma, T. J. (1981). *FEBS Lett.* **134,** 167–171.

Vriend, G., Verduin, B. J. M., Hemminga, M. A., and Schaafsma, T. J. (1982). *FEBS Lett.* **145,** 49–52.

Warrant, R. W., and Kim, S. H. (1978). *Nature* (*London*) **271,** 130–135.

Weber, K. A., and Sehgal, O. P. (1982). *Phytopathology* **72,** 909–913.

Weber, K., Rosenbusch, J., and Harrison, S. C. (1970). *Virology* **41,** 763–765.

Wells, J. M., and Sisler, H. D. (1969). *Virology* **37,** 227–236.

White, J., and Johnson, J. E. (1980). *Virology* **101,** 319–324.

Yot, P., Pinck, M., Haenni, A. L., Duranton, H. M., and Chapeville, F. (1970). *Proc. Natl. Acad. Sci. U.S.A.* **67,** 1345–1352.

Zaumeyer, W. J., and Harter, L. L. (1942). *Phytopathology* **32,** 438–439.

Zulauf, M. (1977). *J. Mol. Biol.* **114,** 259–266.

2

The Structure of Southern Bean Mosaic Virus

MICHAEL G. ROSSMANN
Department of Biological Sciences
Purdue University
West Lafayette, Indiana

CONTENTS

1. INTRODUCTION

Southern bean mosaic virus (SBMV) derives its name from being the pathogen of a disease first discovered in the southern states of the United States (Miller and Price, 1946; Zaumeyer and Harter, 1942) where it is a horticultural problem. Unlike many other plant viruses, it has a rather narrow host specificity, being confined primarily to legumes.

Turnip rosette, cocksfoot mottle, rice yellow mottle, and sowbane mosaic virus (Hull, 1977a), as well as perhaps velvet tobacco mottle virus (Randles et al., 1981), have similar morphological and chemical properties to SBMV. The roughly spherical viral particles have a diameter of about 300 Å (Leonard et al., 1953; Johnson et al., 1976) and appear with a smooth surface in electron micrographs (Akimoto et al., 1975) (Fig. 1). The molecular weight of the virus is 6.6×10^6 (Miller and Price, 1946; Yphantis,

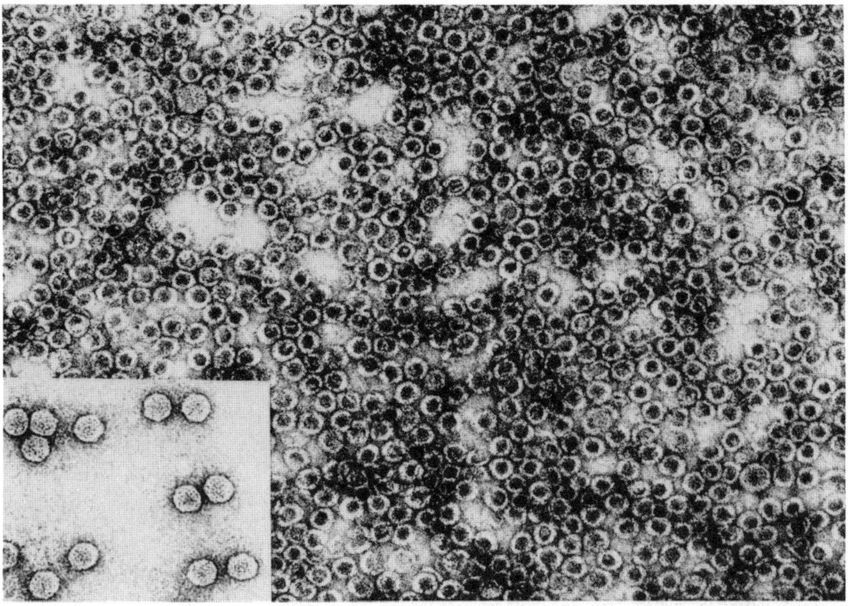

Figure 1. Electron micrograph of reassembled $T = 1$ particles (approximate diameter 190 Å) of SBMV protein compared (inset) with native virions (diameter 300 Å).

1964; Rayment et al., 1978) and it contains 21% of RNA by weight. The viral coat consists of 180 copies of a single protein of molecular weight 28,200. The sequence of the 260 amino acids in the coat protein has recently been determined (Hermodson et al., 1982; Tremaine et al., 1981; Dietrich et al., 1981) by combining the techniques of protein sequencing, RNA sequencing (Mang et al., 1982), and X-ray crystallography. The viral coat has icosahedral symmetry with a $T = 3$ surface lattice (Caspar and Klug, 1962). However, the environments of the three quasi-equivalent subunits do not conform to the assumptions made by Caspar and Klug in their analyses of possible subunit organizations.

The three-dimensional structure determination of the virus (Abad-Zapatero et al., 1980) demonstrated a remarkable similarity to tomato bushy stunt virus (TBSV; Harrison et al., 1978) and satellite tobacco necrosis virus (STNV; Liljas et al., 1982). Apart from turnip crinkle virus (J. M. Hogle, personal communication), a "tombus" virus whose structural similarity to TBSV is not surprising, these are the only icosahedral plant viruses whose structure is known in atomic detail at this time. However, of these viruses, SBMV has been far more extensively investigated primarily due to the work of Om Sehgal in Columbia, Missouri, Roger Hull in Norwich, England, Jack Tremaine in Vancouver, Canada, and Paul Kaesberg in Madison, Wisconsin.

The assembly properties of SBMV are reported in detail by Johnson and Argos (see Chapter 1) and, hence, only a brief summary is given here. The bean strain of SBMV swells on removal of cations (in particular Ca^{2+} and Mg^{2+}) at alkaline pH (Hull, 1977b; Hsu et al., 1976) and can then be readily dissociated (Sehgal et al., 1979a) or attacked enzymatically (Sehgal et al., 1979b). Similar results have been obtained for the cowpea strain (Rayment et al., 1979). Hull (1977c) and Sehgal et al. (1979a) concluded that the SBMV virion is stabilized by pH-dependent bonds. Sehgal and Hsu (1977) isolated covalently associated dimers of SBMV coat protein, whereas Verduin and Erickson (unpublished, 1982) have shown that the protein normally aggregates as dimers. Hsu et al. (1977) and Hull (personal communication) have formed swollen ribonucleoprotein complexes containing about one-third of the protein of the native virions. Tremaine et al. (1982) have studied formation of intermediates in the assembly of the virus from its component protein and RNA (Tremaine and Ronald, 1977) under varying concentrations of NaCl. Savithri and Erickson (1983) reassembled the coat protein and RNA to form $T = 1$ or $T = 3$ particles depending on pH and divalent cation concentration. It is possible to convert swollen SBMV particles into $T = 1$ particles by digestion with trypsin, which removes the basic amino-terminal segment of the coat protein (Sehgal et al., 1979b; Tremaine and Ronald, 1978). Erickson and Rossmann (1982) have shown that RNA is not required for the formation of $T = 1$ particles (Fig. 1).

2. THE CRYSTAL STRUCTURE DETERMINATION

Beatrice Magdoff Fairchild was the first to investigate crystals of SBMV (Magdoff, 1960). Johnson et al. (1974) reported a rhombohedral form of SBMV but these crystals (type I, Table 1) could never be reproduced. However, the type II crystals (Akimoto et al., 1975), grown at very high virus concentration, gave a beautiful diffraction pattern. The type III crystals, also reported by Akimoto et al. (1975), gave a good, high resolution

Table 1. Known Crystal Forms of SBMV

Type	Space Group	Cell Lengths (Å) a	b	c	Angle	Setting	Z	V_M ($Å^3$/dalton)	Fraction of Virus in Asymmetric Unit	Morphology	Reference
Magdoff's	$C222_1$	295	508	474	—	Orthorhombic	4	2.7	$\frac{1}{2}$		Magdoff (1960)
Type I	$R32$	542 923	542 923	542 299	$\alpha = 116°42'$ $\gamma = 120°$	Rhombohedral Hexagonal	3 9	3.6	$\frac{1}{2}$	Rhombs with threefold axis through obtuse corner	Johnson et al. (1974)
Type II	$R32$	318 334.3	318 334.3	318 757.5	$\alpha = 64°0'$ $\gamma = 120°$	Rhombohedral Hexagonal	1 3	3.6	$\frac{1}{6}$	Rhombs with threefold axis through acute corner	Akimoto et al. (1975)
Type III	$C222_1$	522	330	530	—	Orthorhombic	4	3.5	$\frac{1}{2}$	Diamond shaped plates with the (001) face parallel to the plate	Akimoto et al. (1975)

pattern but showed signs of disorder. Hence, the type II R32 crystals were selected for intensive studies. These crystals were grown from SBMV (cowpea strain) at a concentration of 75–110 mg/ml placed in a small Lucite dialysis cell. The virus solution was dialyzed against 0.015 *M* $(NH_4)_2SO_4$. The orthorhombic type III crystals usually appeared after a few days, as rather thin rectangular parallelepipeds. After a week or more, a few type II crystals could be recognized by their rhombohedral shapes (Fig. 2). Seeding with small type II crystals reduced the formation of unwanted type III crystals. The unique axis of the type II crystals was parallel to the body diagonal through the acute vertex.

The rhombohedral unit cell of the type II crystals contained one particle positioned on the intersection of the crystallographic threefold and two-fold axes (Fig. 3). Thus, the asymmetric unit of the crystals contained one-sixth of the virus or 10 icosahedral units. This redundancy of information was essential to all stages of the structure determination. Two

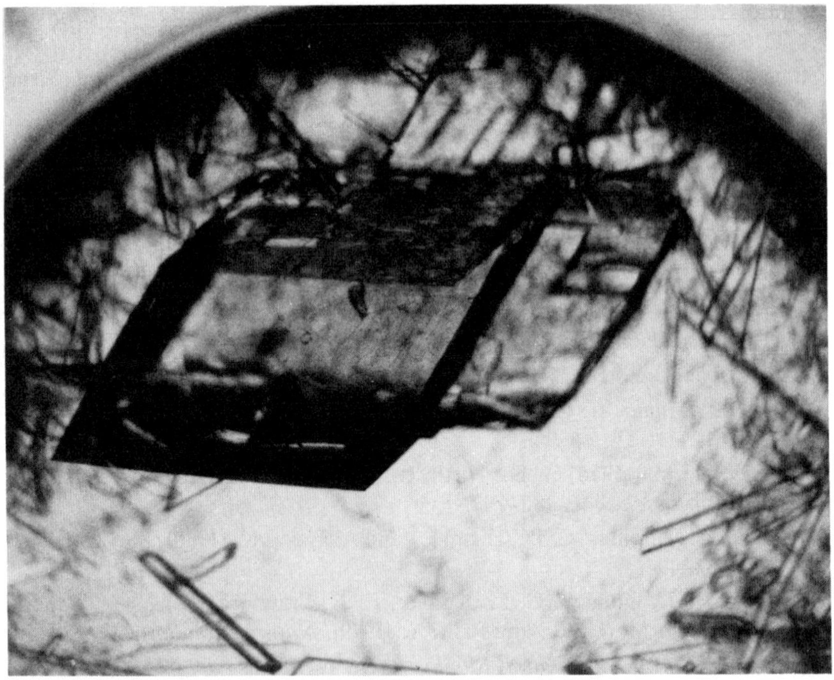

Figure 2. View of a type II SBMV crystal grown in a dialysis cell. Largest body diagonal is approximately 1 mm.

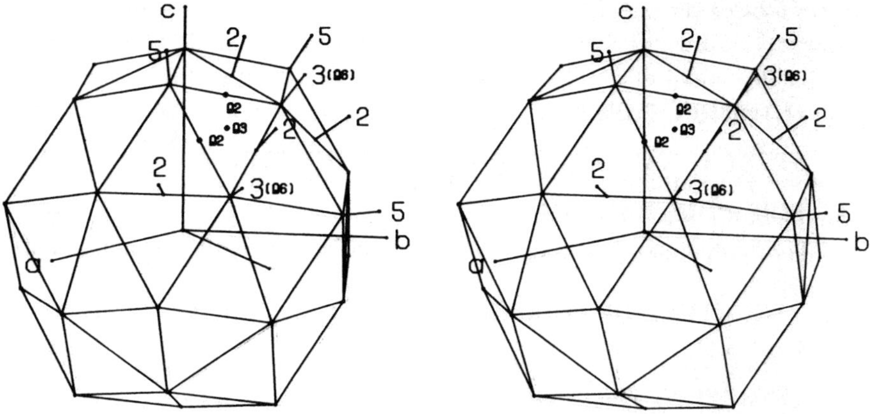

Figure 3. Stereo view of the icosahedral symmetry elements of one viral particle with respect to the crystal cell. Quasi-symmetry axes of the $T = 3$ surface lattice are also shown in one icosahedral asymmetric unit.

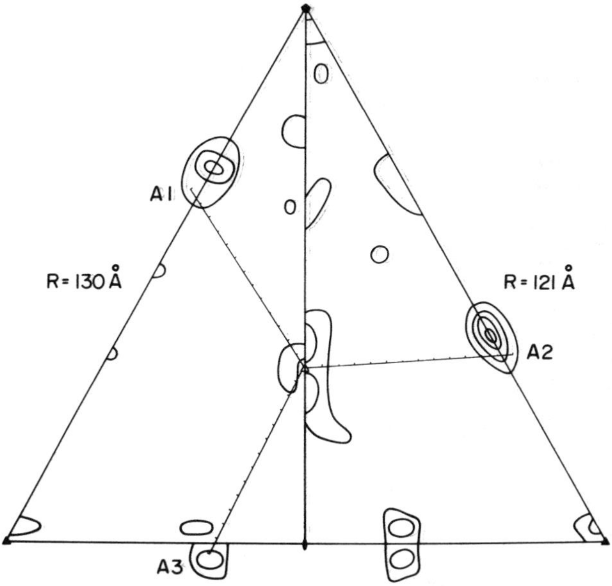

Figure 4. Patterson search function results for radii of 130 Å (left) and 121 Å (right). The three major peaks are related by an approximate threefold axis positioned near the center of the icosahedral asymmetric unit.

possible orientations of the particle, consistent with the crystallographic symmetry, were differentiated by a rotation function study at 11 Å resolution (Rayment et al., 1978). Data for the first heavy atom derivative (K_2HgI_4) were collected on a precession camera and yielded an interpretable heavy atom difference Patterson (Rayment et al., 1978). Unfortunately, the Patterson search function (Fig. 4) showed the three quasi-equivalent sites to be in rather special positions along the edges of the icosahedral symmetry. Hence, the calculated structure phases contained

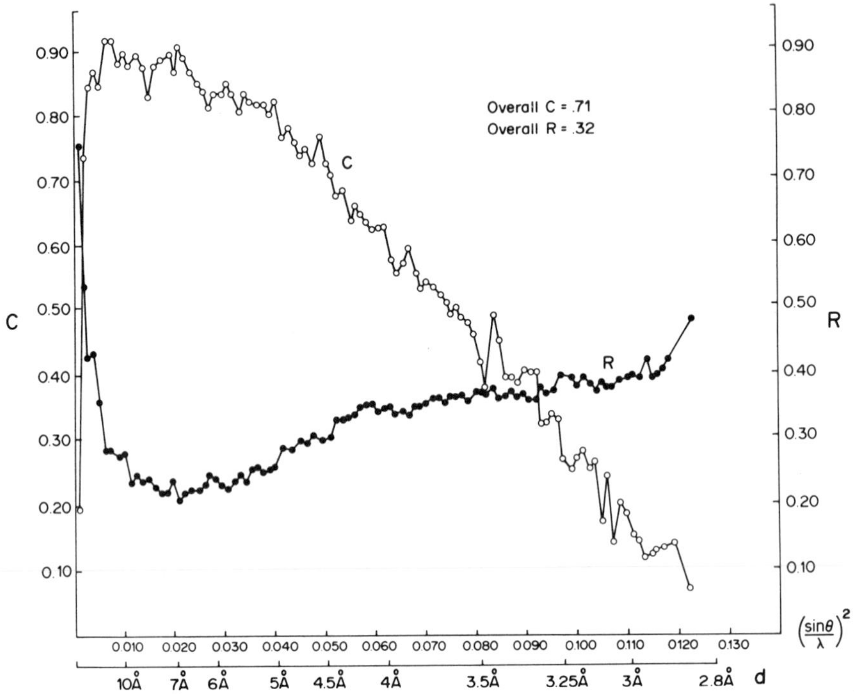

Figure 5. Plot of the *R* factor and *C* correlation factor as a function of resolution for the third molecular replacement refinement cycle where

$$R = \frac{\sum |(|F_o| - |F_c|)|}{\sum |F_o|} \times 100$$

and

$$C = \frac{\sum (F_o - \bar{F}_o)(F_c - \bar{F}_c)}{[\sum (F_o - \bar{F}_o)^2 (F_c - \bar{F}_c)^2]^{1/2}}$$

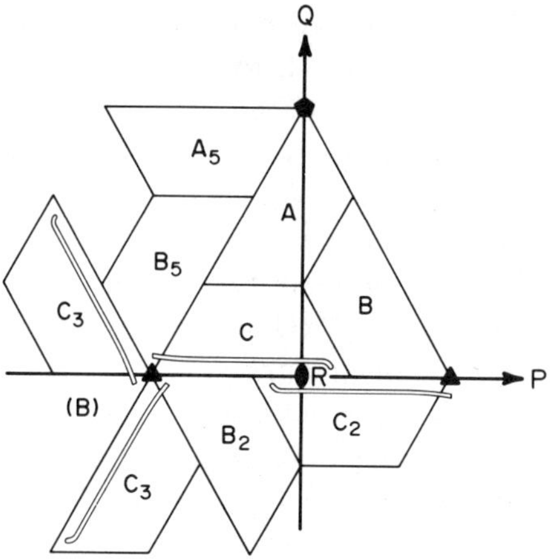

Figure 6. Arrangement of the quasi-equivalent A, B, and C subunits within one icosahedral asymmetric unit and the right-handed coordinate system *P*, *Q*, *R* to which all atomic coordinates have been referred. The origin is at the virus center. Diagram also shows the rough outline of subunits and the nomenclature used here to reference them. The view is from outside the virus looking toward the center. Hence, *R* is coming up, out of the page. The reference subunits A, B, and C are unsubscripted. The subscript 5 implies an anticlockwise rotation of 72° about the shown fivefold axis when viewed from the origin. The subscript 2 implies a 180° rotation about the twofold axis (*R*) and the subscript 3 refers to an anticlockwise rotation about the threefold axis on the left half of the diagram.

a center of symmetry at the cell origin giving rise to images of the structure and its enantiomorph. This problem continued with the discovery of other heavy atom derivatives and was only solved by the sum of relatively small deviations from the special positions. The K_2HgI_4 derivative was found to cause crystal damage and could not be used for high resolution work. However, it provided the initial phasing necessary for solving the more complex heavy atom arrangements in other compounds.

Further high resolution data collection required the development of oscillation photography (Arndt and Wonacott, 1977; Rossmann, 1979; Rossmann et al., 1979; Leslie and Tsukihara, 1980). Data were collected to 2.8 Å resolution from the native crystals and two heavy atom derivatives. Two further derivatives were collected only to 3.5 Å resolution (Abad-Zapatero et al., 1981). Eventually, a multiple isomorphous map was calculated. The resultant electron density map was averaged among the 10 noncrystallographically equivalent icosahedral asymmetric units

(Johnson, 1978) using a procedure similar to that of Bricogne (1976). The improved averaged map was used for the calculation of new structure phases. This iterative process was repeated for three cycles (Fig. 5). The electron density map was interpreted after the first cycle (Abad-Zapatero et al., 1980) in terms of the polypeptide fold. It was then used to help in the overlapping of sequenced peptides (Hermodson et al., 1982).

The final electron density map was viewed in the MMS-X computer graphics system using Jones's FRODO program (Jones, 1978). The structures of the three quasi-equivalent polypeptides (A, B, and C) have been accurately built in this manner and are the basis of the structural analysis given here. Figure 6 shows the organization of these three subunits within the asymmetric unit of an icosahedron. At this time, the atomic positions are being used to compute structure factors for further molecular replacement averaging. This procedure has yielded excellent results for STNV (T. A. Jones, L. Liljas, and B. Strandberg, personal communication) and is a powerful method for maintaining exact noncrystallographic symmetry in very large structures.

3. THE POLYPEPTIDE FOLD AND AMINO ACID DISTRIBUTION

There are two domains in each subunit: the R (random) and S (shell or surface) domain. The S domain bears a close resemblance to the S domain in TBSV and is moderately similar to the S domain in STNV (see below). The position and orientation of the S domain within the framework of the icosahedral symmetry elements are identical in SBMV and TBSV. The latter has an additional third carboxy-terminal P (protruding) domain whose fold is similar to the S domain (Argos et al., 1980). The nomenclature for naming the different subunits is given in Figure 6.

The first 38 residues of the C subunit, the first 62 residues of subunit A, and the first 64 residues in subunit B are disordered and associated with the RNA. The disordered portion is rich in arginines, lysines, prolines, and glutamines but lacks acidic and aromatic groups. The S domain, starting at residue 65, is an eight-stranded β-barrel (Fig. 7). The residues 38 to 52 form a "β-annulus" about the icosahedral threefold axes by hydrogen bonding between symmetry-related C subunits. Residues 53 to 64 form an extended arm in the C subunit which is ordered in the cavity

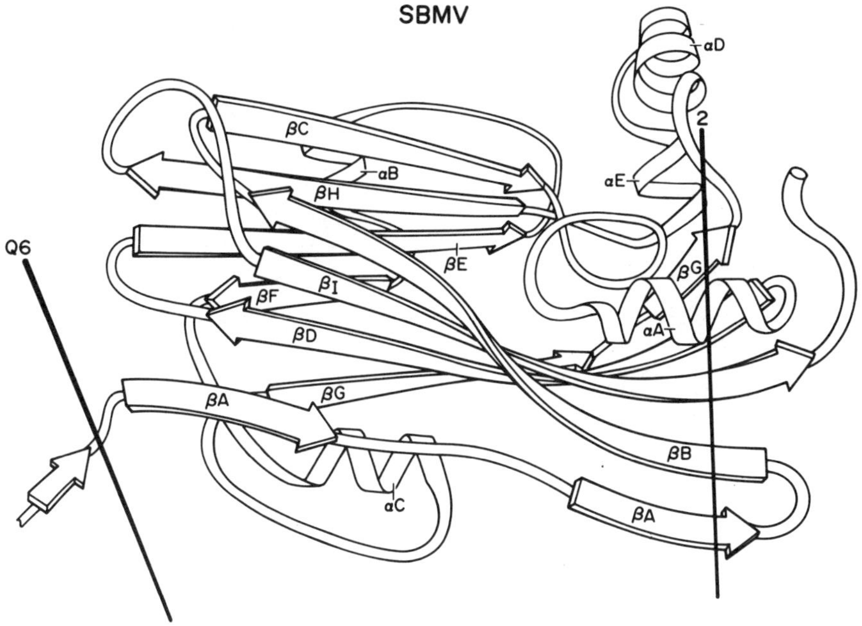

Figure 7. Diagrammatic representation of one subunit of SBMV coat protein. Secondary structural elements are shown. The diagram is based on an original drawing by Jane Richardson.

between the B and C_2 subunits (Fig. 8). This cavity is created because the quasi-threefold axis is parallel to its neighboring icosahedral twofold axes. In contrast, the quasi-twofold axes converge with the quasi-threefold axes leaving no corresponding space between the A and B_5 subunits. An identical situation occurs in TBSV where Harrison et al. (1978) recognized the substantial difference in the C subunit interactions, compared to the A and B subunits, due to the additional ordered arm interposed between the C and B subunits. In this respect, the assembly of SBMV and TBSV deviate from the requirements of quasi-symmetry (Caspar and Klug, 1962).

The S domain has the topology of a "jelly roll" where a pair of consecutive antiparallel strands are wound into a right-handed helix. The domain is narrower at one end, giving it the shape of a wedge, thus allowing the close approach of the subunits to the fivefold and quasi-sixfold axes. The main chain hydrogen bonding and amino acid sequence are

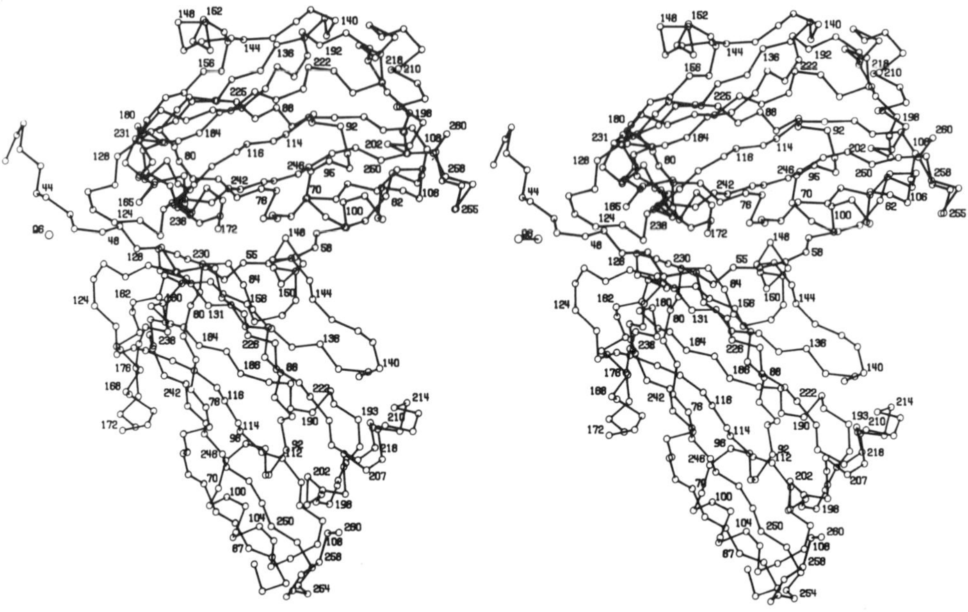

Figure 8. C_α backbone stereo diagram of subunits C and B_2 showing the quasi-sixfold contacts where the additionally ordered βA arm of subunit C is interspersed.

shown in Figure 9. The β-strands have been designated as βA, βB, . . . , βI and the α-helices as αA, αB, αD, and αE along the length of the polypeptide (Figs. 7 and 9). The A and B subunits have 33% β structure while the C subunit, with its additional ordered βA arm and β-annulus, has 38% β structure. All subunits have 16% α structure. This compares well with circular dichroism data for SBMV-like turnip rosette virus ($\alpha \simeq 10\%$, $\beta \simeq 20\%$ for subunits with Ca^{2+} or native virions) or for SBMV ($\alpha = 14\%$, $\beta \simeq 26$ or 36%, respectively) (Denloye et al., 1978; Odumosu et al., 1981).

The lack of uniformity of amino acid distribution (Table 2) has already been referred to above in relation to the properties of the random domain. In addition, basic groups are concentrated at the protein–RNA surface (Fig. 10). Serines, threonines, and, to some extent, valines are much in evidence at the sharp end of the wedge in the βB-βC, βD-βE, and βH-βI bends, consistent with the observations of Lifson and Sander (1979) of other antiparallel sheet structures. The occurrence of a disulfide bond between residues Cys 169 and Cys 176 is unexpected. This S–S bridge lies in a protein region, well buried within the RNA and occurs at an insertion of SBMV relative to TBSV.

Hsu et al. (1976) reported that there are 80 calcium and 120 magnesium ions firmly bound to each virion, although Hull (1978) found 280 calcium and 380 magnesium ions per virion. Abdel-Meguid et al. (1981) obtain intermediate results. Upon removal of these ions at basic pH, the virus swells by about 7% in diameter (Rayment et al., 1979, 1982). Abdel-Meguid et al. (1981) have shown that one of the calcium sites is on the quasi-threefold axis (Fig. 11) coordinated by the three carboxylate groups of Glu 194 in the A, B, and C subunits. This site accounts for 60 Ca^{2+} ions per virion. A second calcium ion site is between quasi-threefold-related subunits. It is generated by Asp 138 and Asp 141 of the C, A, and B units

Table 2. Spatial Distribution of Amino Acids in SBMV Coat Protein

Residue	Amino-Terminal Domain	Facing RNA	External	Internal	Subunit Contacts[a]	Total
Ala	7	3	6	6	6	28
Cys		2		1	1	4
Asp		2	1		3	6
Glu		1	2		5	8
Phe			½	3½		4
Gly	1		3	7	4	15
His					2	2
Ile	1	1	½	5½	5	13
Lys	5	5	2			12
Leu	3		2½	10½	6	22
Met		2	1	3	2	8
Asn	2	2	4	1	4	13
Pro	5	1	1	4	6	17
Gln	5	1	1		3	10
Arg	7	6	3			16
Ser		4	3	2	8	17
Thr	4	5	11½	5½	3	29
Val	2	2	2½	11½	5	23
Trp				1	3	4
Tyr				7	2	9
Total	42	37	44½	68½	68	260

[a] These do not include those residues that are also associated with RNA, such as some of the residues involved in the charge clusters shown in Table 8.

SBMV HYDROGEN BONDING IN SUBUNIT C

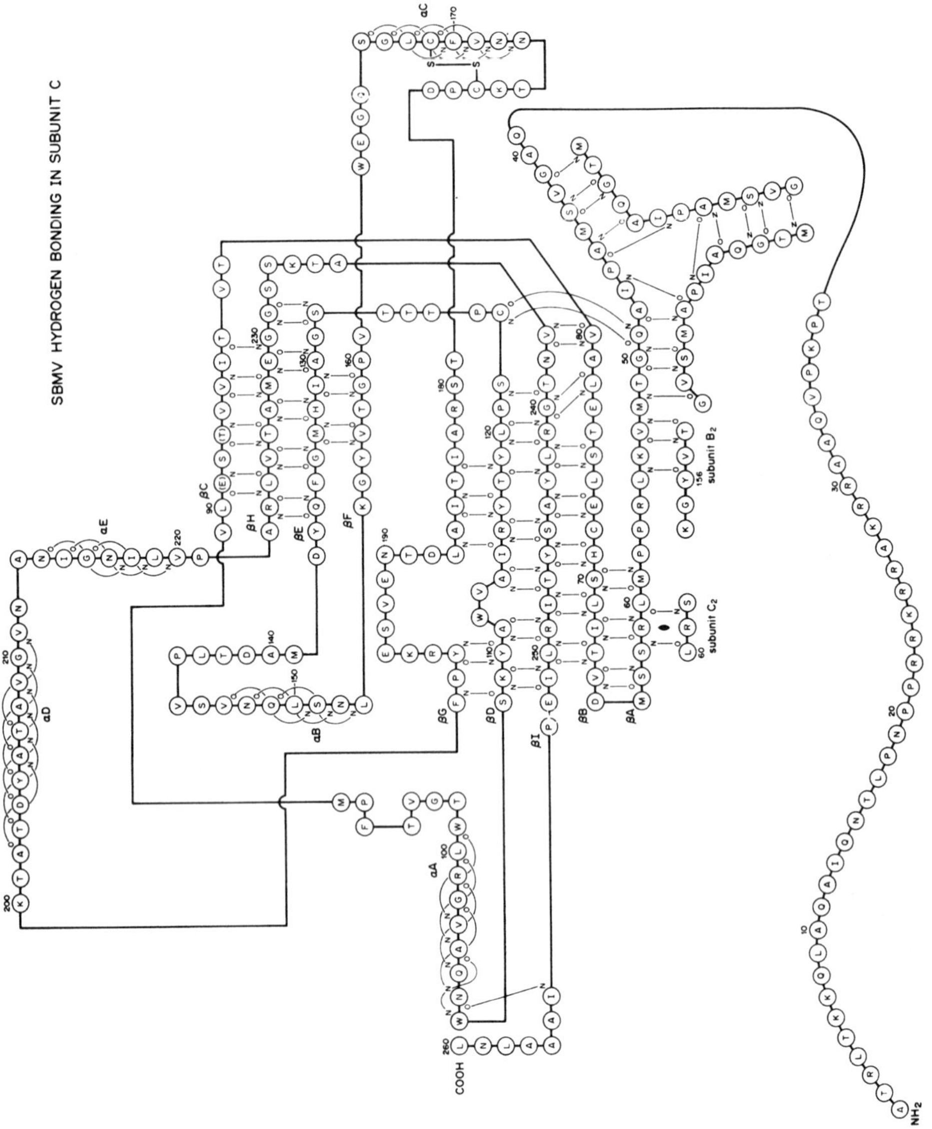

together with the carbonyl of residue 199 and 259 of the A, B, and C units, respectively. The second site accounts for another 180 Ca^{2+} ions per capsid and corresponds to an equivalent Ca^{2+} binding site in TBSV (Harrison, 1980). Both sites together account for 240 bound Ca^{2+} atoms in one virion. Only the Ca^{2+} at the first site, on the quasi-threefold axis, could be removed in crystals grown at low pH by dialysis against EDTA. The identity of the second site as Ca^{2+} rather than Mg^{2+} is dependent on the size of the electron peak and the nature of the protein ligands. Erickson (personal communication) has pointed out that the charge cluster (see below) His 132, Glu 229, Glu 77 in the BC_2 interface might be an ideal place for the binding of Mg^{2+} (Williams, 1977).

The binding of Ca^{2+} to STNV (T. A. Jones, L. Liljas, and B. Strandberg, personal communication) occurs in nonequivalent positions but all sites are at subunit interfaces in SBMV, TBSV, and STNV. In TBSV as in SBMV, removal of metal ions causes swelling (Robinson and Harrison, 1982) at alkaline pH. In TBSV, the swelling is associated with rotation of the P domains relative to the S domains. Hence, the binding of the vestigial hinge in SBMV (and presumably also in TBSV) to Ca^{2+} suggests a mechanism for P domain rotation on removal of Ca^{2+}. Other viruses have also been observed to swell at alkaline pH (Bancroft et al., 1967). Durham and co-workers (Durham, 1977; Durham et al., 1977) have suggested that the 10^3-fold lower calcium ion concentration in cytoplasm as compared to that in the extracellular fluid is a contributing factor in the disassembly of virions *in vivo*.

4. THE RNA–PROTEIN INTERFACE

Extracted RNA from SBMV consists of a spectrum of molecules, including the full-size infectious genome (Rutgers et al., 1980), the smaller molecules being partial copies of the complete genome. The coat protein is not translated from the complete genome but from a subgenomic copy (Weber and Sehgal, 1982). The RNA within the crystalline virus is mostly

Figure 9. (*Opposite*) Amino acid sequence of the SBMV coat protein subunit. The diagram shows the main chain H bonding organization and the nomenclature of the secondary structural elements. Residues 1–38 are disordered and buried in the RNA in all subunits. Residues 39–62 are ordered only in the C subunit.

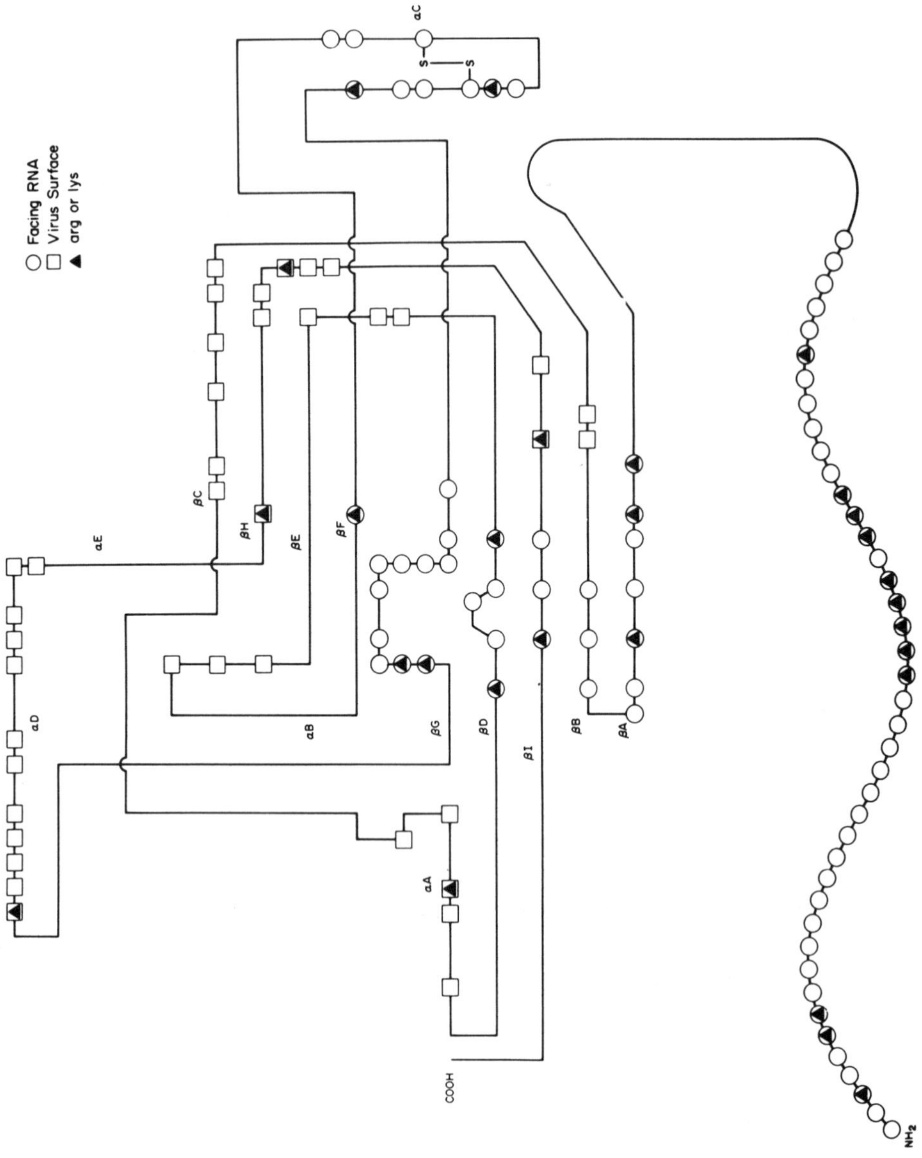

Facing RNA
Virus Surface
arg or lys
αC
S—S
αE
βC
βH
βE
βF
αD
αB
βG
βD
βI
βB
βA
αA
COOH
NH_2

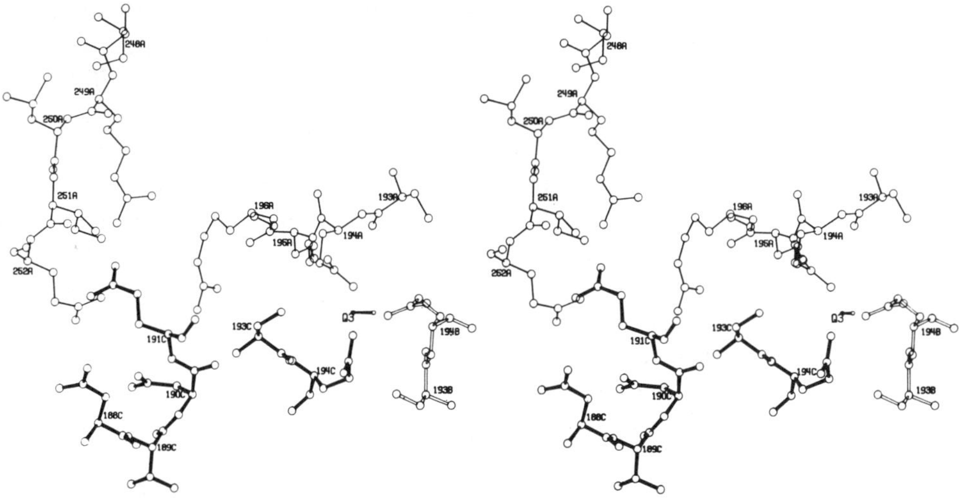

Figure 11. Stereo view of calcium binding site on the quasi-threefold axis.

randomly oriented relative to the coat protein and cannot be seen in the electron density. Nevertheless, it is likely to have a fairly discrete structure (McCain et al., 1982a,b) as is also the case for TBSV (Munowitz et al., 1980). Swelling of the virus by removal of the Ca^{2+} and Mg^{2+} or the removal of RNA causes the polypeptide in association with the RNA to become flexible as observed by ^{13}C NMR. [McCain et al. (1982a) suggest that the amino acids on the viral surface are responsible for the decreased line width. However, these conclusions were made before the complete amino acid sequence was available. The strong Arg and Lys resonance speaks in favor of greater flexibility of the random domain in the swollen virus.] Similar observations have been made for cowpea chlorotic mottle virus (Vriend et al., 1981), alfalfa mosaic virus (AMV; Kan et al., 1982), and TBSV (Munowitz et al., 1980). Conversely, the binding of RNA to the random domains has the effect of ordering both nucleic acid and protein, albeit not into icosahedral symmetry.

The amino-terminal region of many spherical and bacillus-shaped plant virus coat proteins is often strongly basic and predicts as an α-helix where

Figure 10. (*Opposite*) Distribution of arginine and lysine residues in a subunit of SBMV coat protein. The majority of these basic residues are in contact with RNA either at the base of the β-barrel or in the disordered amino-terminal portion.

the amino acid sequence is known (Argos, 1981). Indeed, in STNV some of the arm is ordered although buried in the RNA where it can be seen to form an α-helix with the basic groups at the RNA interface. In addition, the S domains of SBMV, STNV (Liljas et al., 1982), and TBSV (S. C. Harrison, personal communication) have a very basic surface facing the RNA (Fig. 10). For SBMV, there are a total of 23 positive charges per subunit, or 4140 per virion, directly in contact with the RNA, which nearly balances the roughly 4800 negative charges of the RNA. It appears that the basic amino-terminal ends have a histonelike function of folding the RNA during viral assembly. This might be compared with the basic amino-terminal ends of histones H2A, H2B, H3, and H4 and the ordering of these arms in H3 and H4 on binding to the linker DNA (Lilley et al., 1976; Moss et al., 1976; Thomas, 1978). Similarly, α-helices are formed when protamines associate with tRNA or probably also DNA (Warrant and Kim, 1978) and when tobacco mosaic virus protein is associated with its RNA (Butler and Klug, 1978; Stubbs et al., 1979; Holmes, 1980).

Odumosu et al. (1981) have used circular dichroism to show that about 70% of SBMV RNA is double helical but that this percentage diminishes when the virus swells. Rossmann et al. (1983a) came to the same conclusion by using X-ray diffraction of SBMV RNA fibers. These patterns are not unlike those from fragments of yeast ribosomal RNA (Fuller et al., 1967), which, like high molecular weight double-stranded viral RNAs (Langridge et al., 1964; Tomita and Rich, 1964), have structures similar to A-DNA. Krüse et al. (1982) use low angle neutron diffraction to show that the amino-terminal arm is closely associated with the RNA at all radii less than that of the ordered protein.

The basic residues on the SBMV protein surface facing the RNA are on a β-sheet (Arg 61, Arg 116, Lys 155, Lys 195, Arg 196, and Arg 249), on the nonhydrogen bonded portion of the random arm βA (Arg 56), and on the helix αC (Lys 175 and Arg 181). The residues on the β-sheet are very regularly stacked forming two roughly parallel lines (Lys 195, Arg 196, Arg 249, Arg 61 forming one line and Lys 154, Arg 116 forming the other). These lines are about 17 Å apart. The PO_4 groups in the A-RNA conformation (Arnott et al., 1976) are separated by the same distance both across the major and minor grooves (Fig. 12). Furthermore, the distances between the ε-amino or guanidinium groups correspond well to the separation of successive phosphates on the nucleic acid backbone. Hence, an attempt was made (Rossmann et al., 1983a) to "dock" the A-

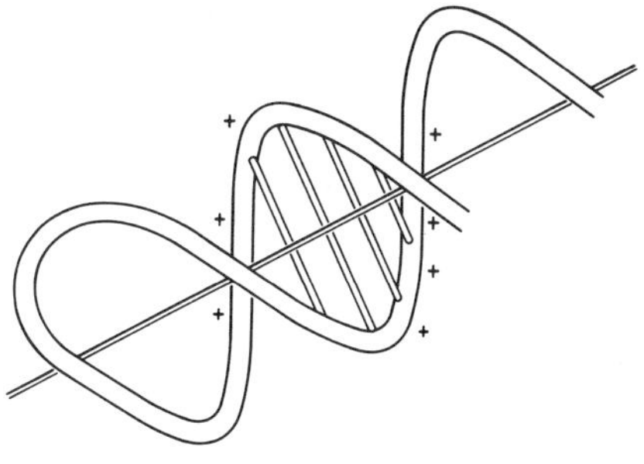

Figure 12. Diagrammatic representation of the docking between positively charged basic groups on the SBMV protein with double-helical A-RNA.

RNA double helix against SBMV protein. Two possibilities existed, either by docking the large major or small minor groove against the protein. However, the major groove generated an unlikely hole between protein and RNA. In contrast, the minor groove made a good fit, even allowing rather accurate interactions between bases and polar groups on the protein (Fig. 13; Table 3). Furthermore, it was reinforced by the remarkable conservation of the same pattern of basic residues in the similar STNV structure (Liljas et al., 1982; Rossmann et al., 1983b) and by the extension of the pattern into the neighboring, quasi-threefold-related subunit of SBMV. The docked RNA structure corresponds to nine base pairs, or three-fourths of a complete turn. Perhaps this is one of a small number of ways in which SBMV RNA can interact with its coat protein.

In AMV, there is a specific binding site on the RNA for AMV coat protein (Koper-Zwarthoff and Bol, 1980; Houwing and Jaspars, 1980). The amino-terminal arm must be present for specific binding. A similar and well characterized situation occurs in tobacco mosaic virus, which has a specific RNA sequence for binding the initial protein subunits from which the assembly of the virion then progresses (Zimmern, 1977; Jonard et al., 1977). Whether SBMV has a specific RNA-coat protein binding site is unknown, although there must be nonspecific interactions.

Four neighboring bases of the docked RNA were found to be well positioned to make hydrogen bonding interactions with polar groups (Ser 63, Arg 249, Lys 195 in subunit A and Glu 191, Thr 189, Asn 190 in subunit C) at the AC subunit boundary. The polar polypeptide 188 to 196, which includes the Ca^{2+} binding ligand Glu 194, is a large bulge in strand βG and protrudes into the RNA. It provides some of these specific interactions with the RNA. Hence, if there is any specific binding of RNA to SBMV protein, it might be dependent on this "recognition" sequence of the SBMV protein structure, although in AMV the flexible amino-terminal arm is of crucial importance. Nevertheless, "recognition" in the minor groove cannot differentiate between AU and UA nor between GC and CG as the base pairs have local twofold symmetry on this side with respect to their available hydrogen donor and acceptor positions (Rosenberg and Greene, 1982).

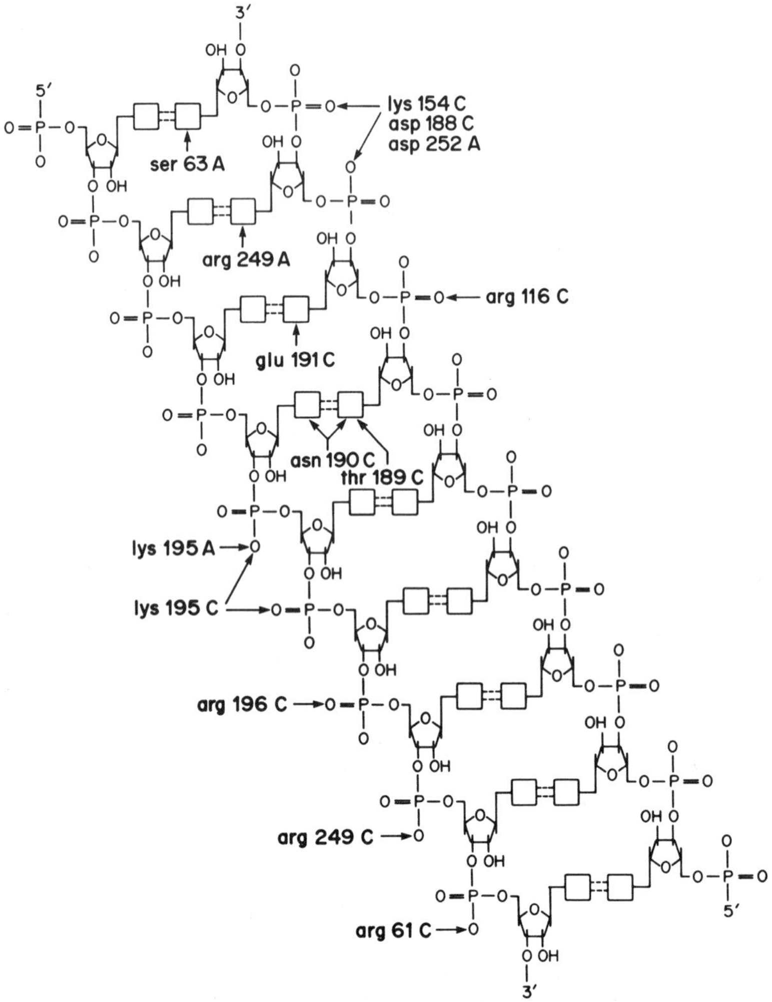

Figure 13. Chemical representation of docked residues of SBMV protein and nine base pairs of A-RNA. The docked amino acids are referred to by sequence number and a letter designating their subunit (C or A).

Table 3. SBMV–RNA Contacts in Proposed Docking[a]

		RNA Residue	Atom	Protein Residue	Atom	Distance (Å)
up	3′ ↑	1	N3 purine	63A	Ser OG	2.0
		1	O2P	154C	Lys NZ	4.1
		2	O2P	252A	Glu OE1	5.4
		2	O2P	154C	Lys NZ	4.0
		2	O5	65A	Asp OD2	Overlapping
		2	N3 purine	249A	Arg NEH2	3.7
		2	O2P	188C	Asp OD2	4.6
		2	O6	191C	Glu OE1	3.0
		3	N3 purine	191C	Glu OE1	4.4
		3	N3 purine	191C	Glu OE2	4.4
		3	O6	190C	Asn ND2	3.1
		3	O2P	187C	Leu O	4.6
		4	O2P	116C	Arg NEH2	5.3
		4	N3 purine	190C	Asn OD1	3.0
		4	N3 purine	189C	Thr O	3.3
		4	O6	190C	Asn OD1	5.3
	5′	4	O6	187C	Leu O	4.0
down	5′	4	O2 pyrimidine	190C	Asn ND2	4.6
		5	O2P	195A	Lys NZ	3.2
		5	O2P	195C	Lys NZ	2.8
		5	O6	190C	Asn O	4.3
		6	O1P	195C	Lys NZ	3.4
		6	O2P	196C	Arg NEH2	3.1
		6	O6	113C	Val CG2	3.0
		7	O1P	249C	Arg NEH1	3.0
		7	O2P	61C	Arg NEH1	3.3
	↓	7	O2P	68C	Ile CD1	3.4
	3′	8	O1P	61C	Arg NEH2	3.6

[a] From Rossmann et al. (1983a).

5. SUBUNIT–SUBUNIT INTERACTIONS

The C_α coordinates of quasi-equivalent subunits were compared (Rossmann et al., 1983c) using the procedure developed in this laboratory (Rao and Rossmann, 1973; Rossmann and Argos, 1975, 1976). The superimposed 190 equivalenced residues gave root-mean-square deviations between corresponding C_α positions ranging from 0.73 Å (A on B) to 1.17

Å (B on C). The rotation matrices are given in Table 4. By far the largest differences occur in the final six residues of the C subunit relative to the A and B conformations. This is the "hinge" region in TBSV connecting the S and P domains. The angle between the P and S domains differs in the C subunit as compared to the A and B subunit in TBSV. Thus, SBMV has a vestigial "hinge" which accommodates itself to the available space. Smaller differences occur near the wedge-shaped end of the β-barrel, in the helix αC within the RNA and the external helix αD. It will be shown that these regions are all involved in accommodating the quasi-symmetry.

Inspection of Figure 6 shows that the following types of subunit contacts exist.

1. Subunits related by the quasi-threefold axes: AB, BC, CA. These three contact regions were found to be very similar, although BC has some small differences compared to AB and CA.
2. Subunits related by the icosahedral fivefold axes: AA_5. These contacts were found to be very similar to the B_5C interactions.

Table 4. Comparison of the Backbone Conformation for the Quasi-Equivalent A, B, and C Subunits[a]

1. Subunit A onto Subunit B:

$$B = \begin{pmatrix} -0.4753 & 0.8789 & 0.0408 \\ -0.8797 & -0.4743 & -0.0335 \\ -0.0100 & -0.0518 & 0.9986 \end{pmatrix} A + \begin{pmatrix} -29.31 \\ 43.08 \\ 1.62 \end{pmatrix}$$

rms deviation between C_α atoms = 0.73 Å

2. Subunit B onto Subunit C:

$$C = \begin{pmatrix} -0.5454 & 0.8382 & -0.0006 \\ -0.8381 & -0.5453 & 0.0146 \\ 0.0119 & 0.0085 & 0.9999 \end{pmatrix} B + \begin{pmatrix} -22.59 \\ 38.88 \\ -0.35 \end{pmatrix}$$

rms deviation between C_α atoms = 1.17 Å

3. Subunit C onto Subunit A:

$$A = \begin{pmatrix} -0.4749 & 0.8797 & -0.0237 \\ -0.8783 & -0.4754 & -0.0508 \\ -0.0559 & -0.0033 & 0.9984 \end{pmatrix} C + \begin{pmatrix} -20.84 \\ 45.64 \\ 0.23 \end{pmatrix}$$

rms deviation between C_α atoms = 1.11 Å

[a] The position vectors *A*, *B*, and *C* imply coordinates of atoms in the respective subunits measured in Å along the *P*, *Q*, and *R* axes (Fig. 6).

3. Subunits related by the quasi-sixfold axes: B_5C, CB_2. These bear little resemblance to each other. In the CB_2 interactions, the additionally ordered βA arm of the C subunit is inserted between the subunits. In the B_5C contacts, this arm is absent.
4. Subunits related by the icosahedral threefold axes: CC_3. The CC_3 interface is generated by the β-annulus structure. There is no contact between B_5 and B_2.
5. Subunits related by the icosahedral twofold axes: CC_2. These contacts are sparse, unlike in TBSV where the P domains provide a strong interaction.
6. Subunits related by the quasi-twofold axes: AB_5. Both the AB_5 and CC_2 interactions are mediated by a hydrophobic pocket in one subunit into which is inserted Trp 99 of the other subunit. Nevertheless, there are large differences between the AB_5 and CC_2 contacts.

The rotation and translation elements of the different subunit interactions can be calculated (Table 5) by combining the quasi-equivalent threefold rotations given by the matrices in Table 4 and the icosahedral symmetry elements. Table 5 shows the polar coordinates (κ, ψ, ϕ) as described by Rossmann and Blow (1962) for the rotation axes relating any two subunits. All the subunit contacts (Table 5) are pure rotations within experimental error, the largest computed translation being 0.6 Å for the quasi-twofold contacts. The quasi-threefold rotation axes relating the A, B, and C subunits are all close to 120° rotation about an axis parallel to the icosahedral twofold R-axis passing through the point at $P = 0$ and $Q = 26.6$ Å. However, the rotation between subunits A and B and between C and A are 2° less than 120° whereas the rotation between the subunits B and C is 3° greater than 120°. These latter two subunits are involved in forming the interface which contains the additionally ordered arm of the C subunit. The quasi-sixfold contacts, B_5C and CB_2, require rotations of 68.4° and 57.0°, respectively, about axes whose directions differ substantially in orientation but not position from the icosahedral threefold axis. The contacts B_5C have a rotation close to the 72° of the AA_5 pentamer association and, indeed, the B_5C and AA_5 interactions are almost identical. In contrast, the rotation relating subunits C and B_2 is 3° less than 60°. Finally, the quasi-twofold association is a 180° rotation within experimental error, but the actual contact region is different than in the icosahedral twofold contacts CC_2.

Table 5. Rotation, Translation, Orientation, and Position of Symmetry Axes between Subunits in Contact with Each Other[a,b]

Type of contact S_1S_2[c]	Axial Type[d]	κ(°)	ψ(°)	φ(°)	u[e]	v[e]	w[e]	Δ(Å)[c]	P[f]	Q[f]
AB	Q3	−118.4	88.4	−89.4	0.0104	0.0289	0.9995	−0.07	−0.5	26.7
BC	Q3	−123.1	90.1	−89.8	0.0036	−0.0075	1.0000	0.14	−0.3	26.5
CA	Q3	−118.4	89.0	−91.6	−0.0270	0.0183	0.9995	0.06	−0.2	26.7
AA_5	I5	−72.0	58.3	−90.0	0.0000	0.5258	0.8506	0.00	0.0	77.3
B_5C	Q6	−68.4	75.5	−119.9	−0.4822	0.2499	0.8397	−0.52	−46.7	0.7
CB_2	Q6	−57.0	90.4	−89.2	0.0138	−0.0067	0.9999	0.22	−48.9	0.5
CC_3	I3	−120.0	90.0	−110.9	−0.3567	0.0000	0.9342	0.00	−47.8	0.0
CC_2	I2	−180.0	90.0	−90.0	0.0000	0.0000	1.0000	0.00	0.0	0.0
AB_5	Q2	−179.3	82.5	−105.8	−0.2700	0.1305	0.9539	−0.61	−23.1	40.4
Idealized[g]	Q3	120.0	78.5	−90.0	0.0000	0.1999	0.9798	0.00	0.0	25.5
Idealized[g]	Q2	180.0	73.0	−100.9	−0.1803	0.2930	0.9390	0.00	−24.0	39.0

[a] The rotation matrices relating the A, B, and C subunits are given in Table 4. The relation between subunits S_1 and S_2 is defined as $S_2 = (H)S_1 + d$ where (H) is a rotation matrix and d a translation vector with respect to the chosen coordinate system.

[b] The definition of the subunits and coordinate system relative to the icosahedron is shown in Figure 6.

[c] The type of interactions S_1S_2 signifies that S_1 is rotated by κ and shifted by Δ away from the origin to coincide with S_2.

[d] The axial type is identified by Q (quasi) or I (icosahedral) and a number designating a two-, three-, five-, or sixfold axis.

[e] u, v, w are the direction cosines of the rotation axes ψ, φ.

[f] The position of the instantaneous axis is given by (P, Q) where it intersects the plane $R = 125.0$.

[g] The "idealized" direction of the Q3 axis is taken from the centroid of the triangle (Fig. 6) in a plane R = constant. The "idealized" direction of the Q2 axis is taken as the locus of the points joining the midpoint of the sides of the triangle (Fig. 6) as R increases.

Let S_1S_2 represent the interaction between subunits S_1 and S_2. Similarly, S_3S_4 represents the interaction between subunits S_3 and S_4. Let us suppose that the interaction S_1S_2 is quasi-equivalent to S_3S_4. Let the relationship between S_1 and S_3 be given by

$$S_1 = (H)S_3$$

where (H) represents a 3×3 rotation matrix and a 1×3 translation vector. The relationship (H) between any two subunits can always be determined from the icosahedral symmetry plus the quasi-symmetry operations given in Table 4. By now applying the operation (H) to S_4, it is possible to determine how well it superimposes onto S_2. If it superimposes perfectly, then the interactions S_1S_2 and S_3S_4 are essentially identical. A shorthand for this test will be written as

$$\begin{array}{ccc} & S_1 & - \; S_2 \\ (H) & \uparrow & \\ & S_3 & - \; S_4. \end{array}$$

The results on applying this test are shown in Table 6. The quality of fit is given by comparing the rms differences between equivalent atoms in subunits S_1 and S_3 (which depends on the magnitude of the conformational

Table 6. Superposition of Pairs of Subunits and rms Deviation of Superimposed Atoms[a]

$$\begin{array}{ccc} S_1 & - & S_2 \\ \Uparrow & & \\ S_3 & - & S_4 \end{array}$$

	S_1	S_2	S_3	S_4	rms (S_1–S_3)	rms (S_2–S_4)	Contact
(*a*)	A	B	C	A	1.40	1.34	Q3, Q3
(*b*)	C	A	B	C	1.37	2.86	Q3, Q3
(*c*)	B	C	A	B	0.92	2.41	Q3, Q3
(*d*)	A_5	A	C	B_5	1.40	1.49	I5, Q6
(*e*)	A	A_5	C	B_2	1.40	9.21	I5, Q6
(*f*)	A	B_5	C	C_2	1.40	7.86	Q2, I2

[a] The interactions of subunits S_1 and S_2 are compared with those between S_3 and S_4. Subunit S_3 is superimposed onto subunit S_1 in order to determine how well S_4 then superimposes onto S_2. Computations are over main chain and side chain atoms omitting the random domain.

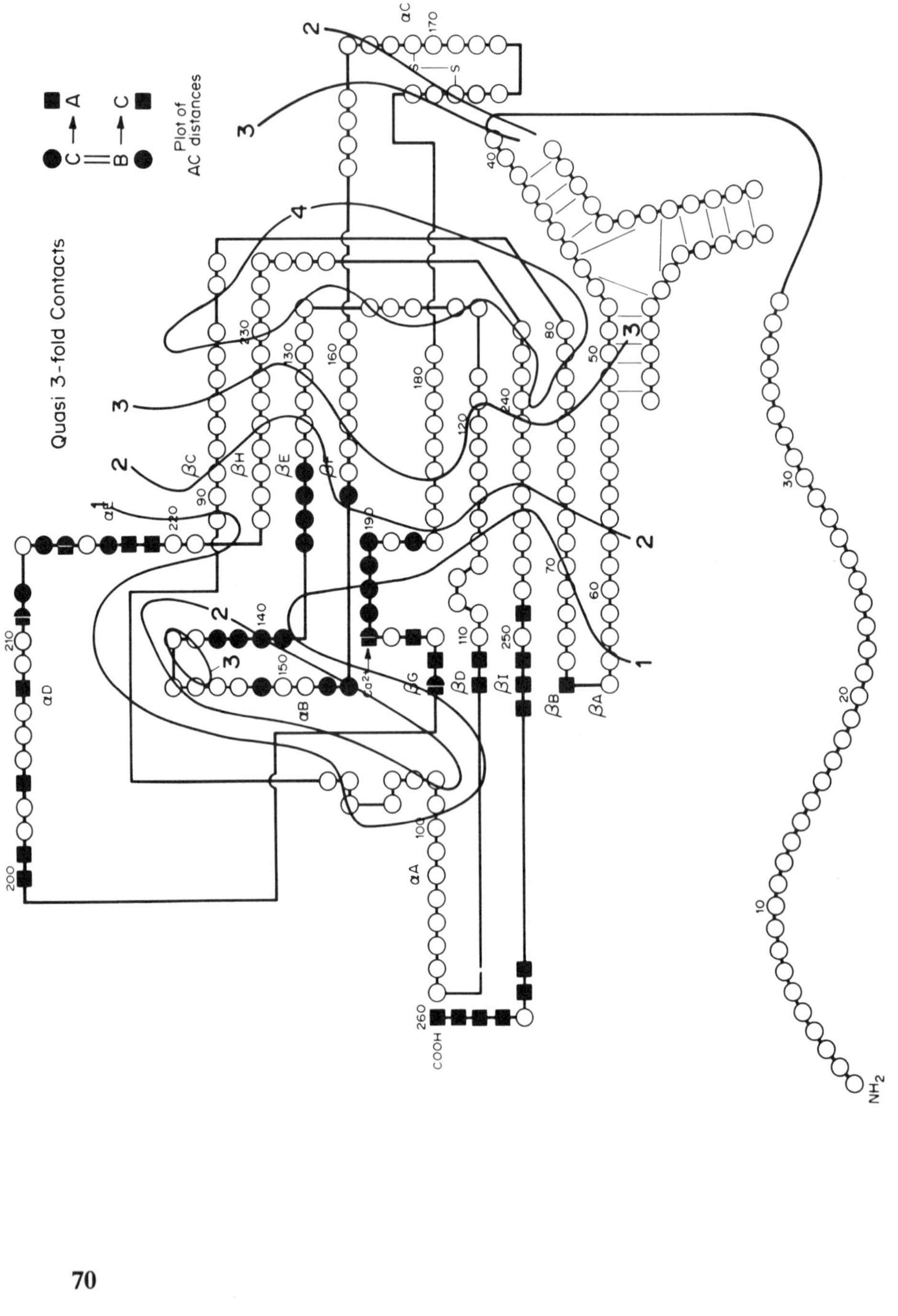

Quasi 3-fold Contacts
C → A
B → C
Plot of AC distances
αA
αB
αC
αD
βA
βB
βC
βD
βE
βF
βG
βH
βI
Ca2+
COOH
NH2

(b)

differences between these subunits) and the rms differences between equivalent atoms in subunits S_2 and S_4. These rms values are taken over all 1501 atoms in the subunit starting at residue number 63.

The first three comparisons in Table 6 relate to the quasi-threefold axis. The AB contact and CA contact are identical to within experimental error; however, BC is a little different. When the BC contact is compared to CA or AB, the rms values between equivalent atoms goes up to 2.4–2.9 Å, somewhat larger than the 1.4 Å rms value of exactly superimposed subunits. Comparison of the icosahedral fivefold subunit interactions A_5A with the quasi-sixfold interactions CB_5 (where no βA arm is interposed) shows no significant difference between these contacts.

The situation is, however, totally different on comparing the fivefold interactions AA_5 with the quasi-sixfold interaction CB_2. The CB_2 contacts have the hydrophobic βA arm interposed between the subunits. Clearly, these contacts are very different with an rms of 9.2 Å between S_2 and S_4 equivalent atoms. A similar situation pertains to the comparison of the quasi-twofold interactions AB_5 with the icosahedral contacts CC_2.

The distances between equivalent C_α atoms of subunits S_2 and S_4 are plotted in Figure 14 on the hydrogen bonding diagram. The amino acids involved in the subunit contacts are also shown. The smallest S_2S_4 distances for the quasi-threefold generated interactions occur near the contact regions, while the largest are found far away at the wedge-shaped end of the subunits (Figs. 14*b* and 15). Therefore, this is an excellent example of small deformations as suggested by Caspar and Klug (1962). Here the differences are amplified by the lever effect along the length of the subunit. Comparison of the icosahedral fivefold and closely similar quasi-sixfold CB_5 contacts does not clearly show any significant trend (Table 6; Fig. 14*d*).

The magnitude of the S_2–S_4 differences is far greater in the comparisons shown in Figure 14*e* and *f* which involve contacts with the interposed βA arm. However, there are local areas where the differences are less than 2 Å. In Figure 14*e*, which compares the icosahedral fivefold, AA_5, and the quasi-sixfold, CB_2, contacts, there are two areas, around C_α 128 and

Figure 14. (*Opposite and following pages*) The contours show the distances in Å between equivalent C_α atoms of subunits S_2 and S_4 when subunit S_3 is superimposed on subunit S_1. The diagrams are labeled (*b*), (*d*), (*e*), and (*f*) corresponding to the entries in Table 6. The amino acids that participate in the interactions S_1S_2 and S_3S_4 are also shown.

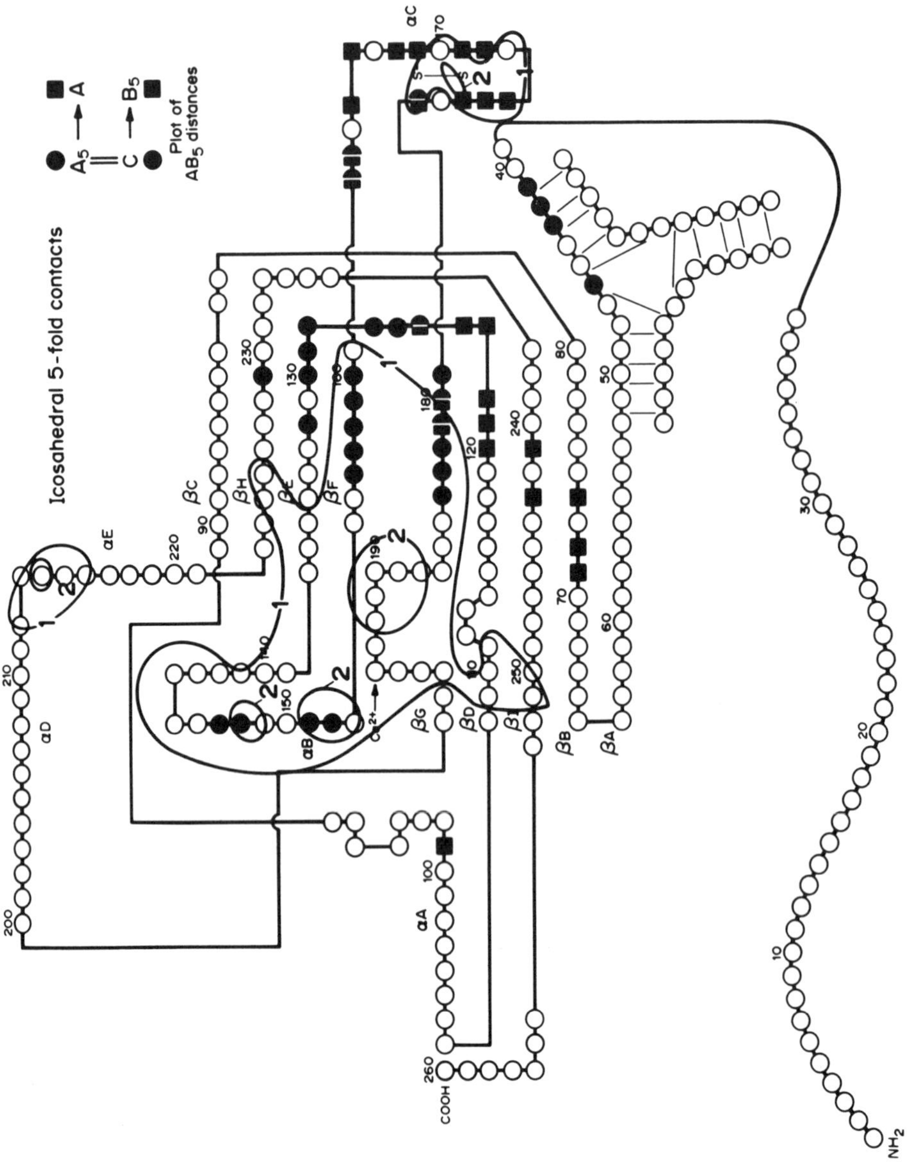
Icosahedral 5-fold contacts
A5 → A
C
Plot of AB5 distances
→ B5
αA
αB
αC
αD
αE
βA
βB
βC
βD
βE
βF
βG
βH
βI
COOH
NH2

(d)

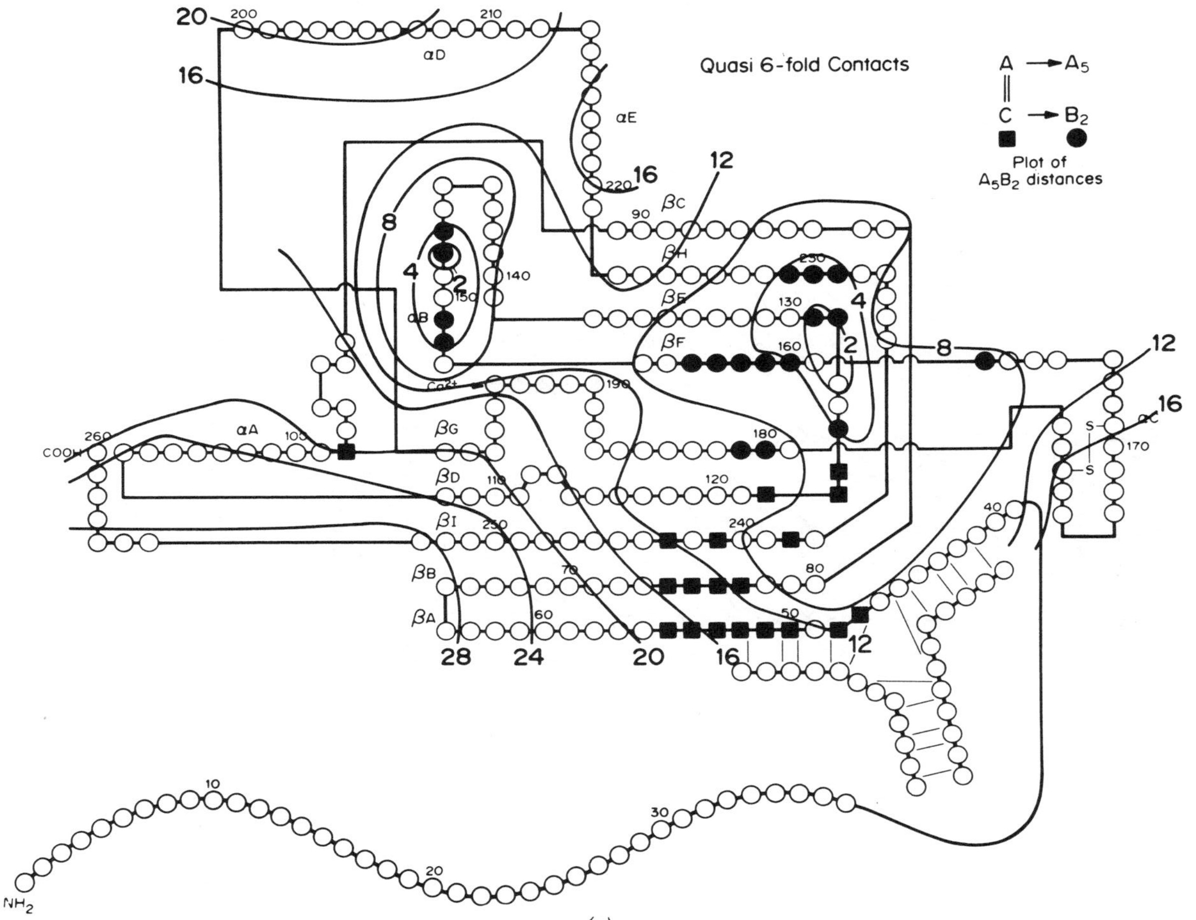

(e)

Figure 14. Continued

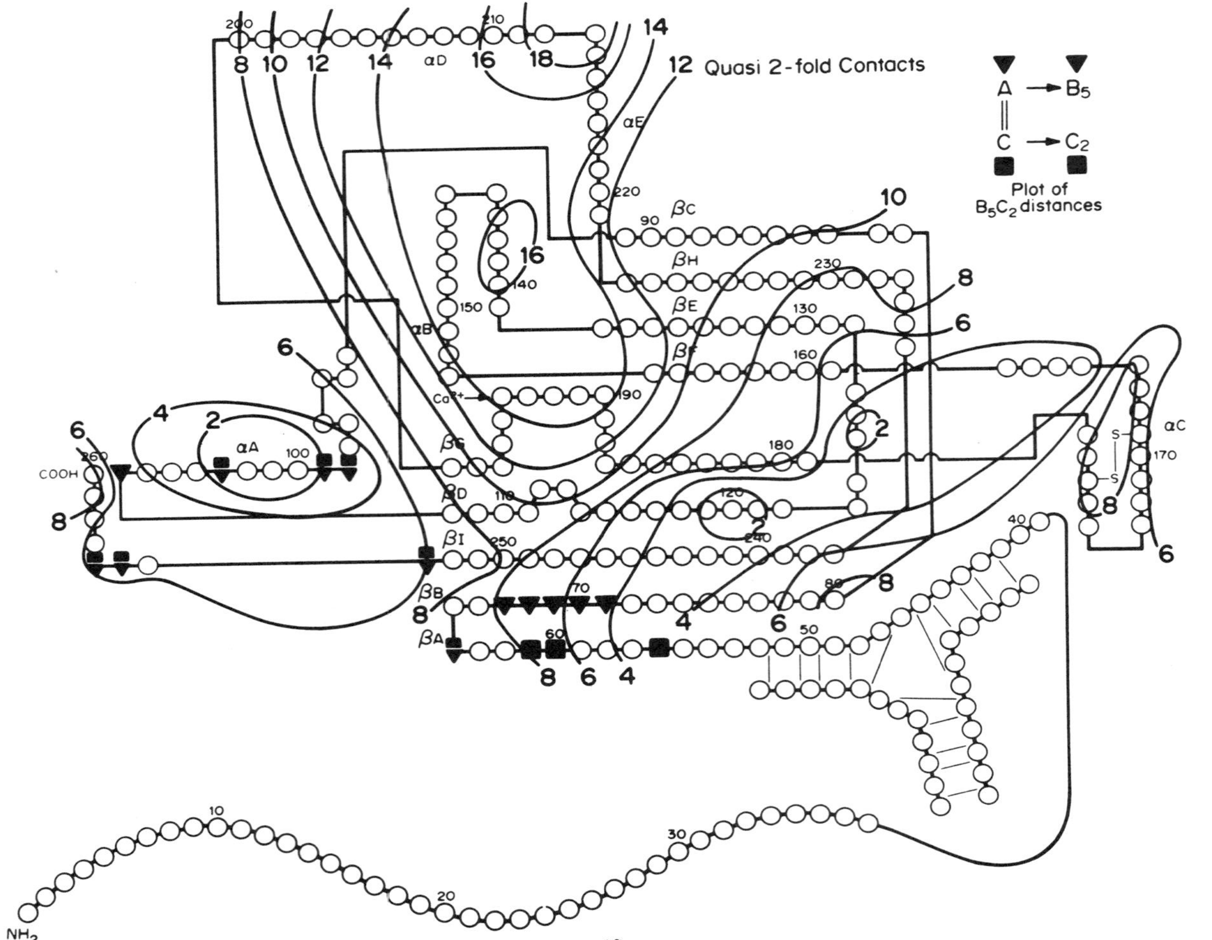

(f)

Figure 14. Continued

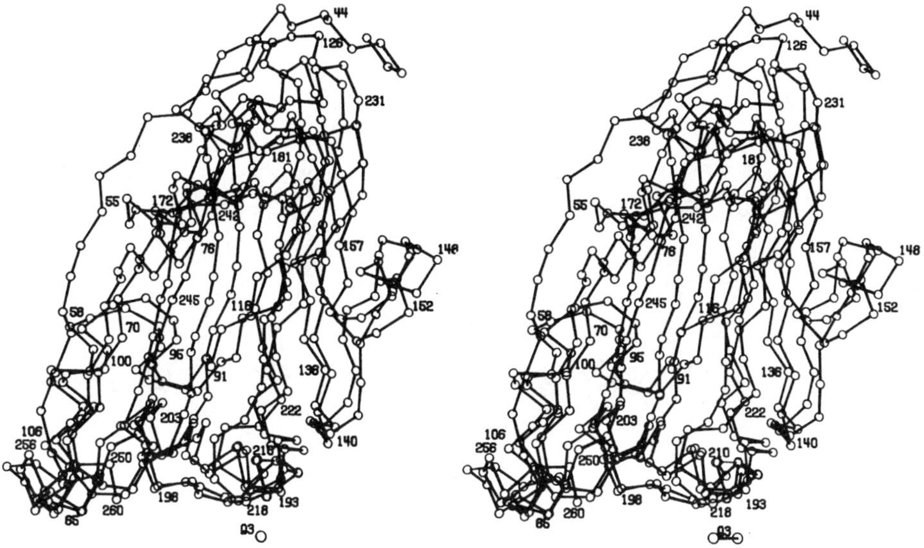

Figure 15. Superposition of the C_α backbone between subunits A and C when subunit B is superimposed on C, corresponding to the results shown in Table 6*b* and Figure 14*b*. The BC contact is different by about 5° to the AB and CA quasi-threefold-generated contacts causing a 4 Å separation of the superimposed subunits at the wedge-shaped end, far away from the contact area.

C_α 148, with S_2–S_4 distances less than 2 Å. These coincide with subunit contact areas. Hence, there is a line (the "vector of rotation") in the subunits which must pass near these two C_α atoms, about which the subunit A_5 in the AA_5 interface must rotate in order to superimpose on B_2 in the CB_2 interface. Furthermore, in light of the small A_5–B_2 distances around C_α 128 and C_α 148, there is little slip along this "vector of rotation" between the subunits. A similar situation pertains to a comparison of the icosahedral and quasi-twofold interactions (Fig. 14*f*) where the atoms C_α 102, C_α 120, and C_α 124 are all close to the "vector of rotation." The amount of rotation for the comparisons of the quasi-sixfold and quasi-twofold axes with their corresponding icosahedral axes are 39° and 36°, respectively. This situation is something far different than the slight deformations proposed by Caspar and Klug (1962).

The conserved interactions along the "vector of rotation" serve as hinges relating the two different environmental states of the protein subunits. Hence, the residues along these lines of contact may be of some

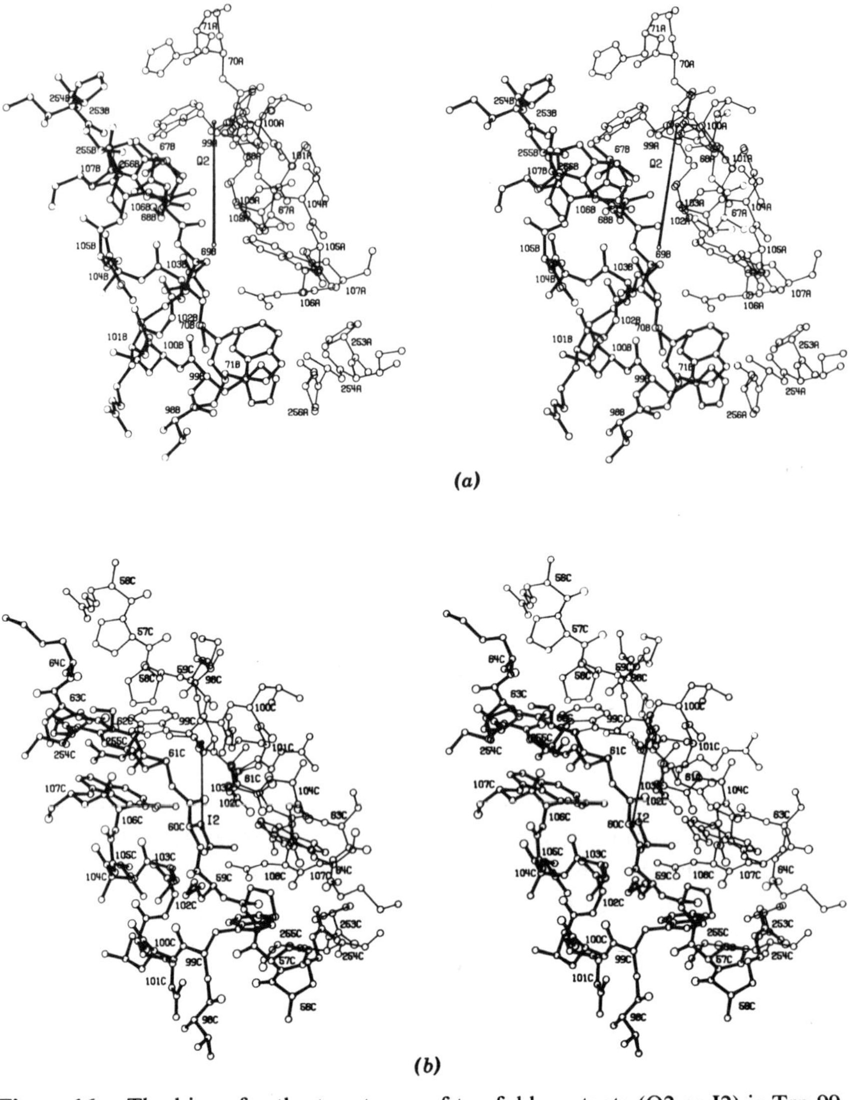

(b)

Figure 16. The hinge for the two types of twofold contacts (Q2 or I2) is Trp 99 inserted into a hydrophobic pocket. This residue is seen interacting with Trp 107 in the neighboring subunit in (*a*) the quasi-twofold interaction Q2 and (*b*) the icosahedral twofold interaction I2.

importance to viral assembly. The twofold interactions generate the hinge by inserting Trp 99 into a hydrophobic pocket of the interacting subunit (Fig. 16). The hydrophobic pocket is lined by residues Trp 107, Pro 253, Ile 254, and Ala 255. The two tryptophan residues 99 and 107 in neighboring subunits are situated to form π electron interactions. There are two points of interaction along the "vector of rotation" relating AA_5 to CB_2: helix αB in one subunit associates with the strand βD in the other and also the neighboring strands βH and βE contact the same strands across the subunit boundary along the "vector of rotation." These contacts provide a complex set of hydrophobic and hydrogen bonding interactions. It could be significant that α-helices are involved in all the hinge regions. Possibly, it is easier for an α-helix to permit some strain (in the Caspar–Klug sense) than when sheets interact. This possibility is further emphasized by observing the substantial involvement of αE in the quasi-threefold interactions where the quasi-equivalence is also accommodated by strain. In contrast, the interaction of β-sheets requires the fitting of residues in one sheet with holes in another (Chothia and Janin, 1981).

6. HEAVY ATOMS AS PROBES FOR QUASI-SYMMETRY

The heavy atom substitutions used in solving the crystallographic structure frequently lacked quasi-symmetry (Table 7). This was of some concern (Abad-Zapatero et al., 1981) while determining the structure as it was contrary to expectation. However, now that an analysis of the structure is available, many of these anomalous substitutions are readily explained. Some of the heavy atom sites are well buried and often at the protein–RNA interface. It remains a mystery how the reagents were able to penetrate the viral shell. The Ⓐ site, although far below the mean viral surface, is accessible, but that does not seem to be the case for the Ⓗ, Ⓕ, Ⓒ, Ⓣ, Ⓛ, Ⓞ, and Ⓟ sites.

The Ⓐ site depends on the substitution of Cys 72. Substitution in the C subunit is significantly poorer than in the A and B subunits, presumably due to steric obstruction by Arg 55 on the βA arm. In addition, the Ⓐ site in the A and B subunits is modified by Asn 152 due to the AA_5 or BC_5 contacts. Apparently, the polar environment is essential for the UO_2F_5, K_2PtCl_4, and $HAuCl_4$ species to fully substitute Cys 72. Hence,

Table 7. Quasi-Symmetry Probed by Heavy Atom Substitutions

			Occupancy in Each Subunit[a]			Subunit Interactions[b]		
Site	Residue	Derivative	A	B	C	A	B	C
Ⓐ	Cys 72	PHMBS	110	99	87	AA_5	BC_{-5}	
		$K_3UO_2F_5$	87					
		K_2PtCl_4	69	44				
		$HAuCl_4$	89	62	53			
Ⓑ	Cys 123	PHMBS	73	61	76	AA_5	BC_{-5}	CB_2
Ⓗ	Cys(169)– Cys(176)	PHMBS	26	37		AA_5	BC_{-5}	
		$HAuCl_4$	27	21				
Ⓕ	Met 59	PHMBS			26	R	R	
		K_2PtCl_4			33			
Ⓘ	Met 139	K_2PtCl_4			27		BC	
Ⓠ	Glu 194 Ca^{2+}	$K_3UO_2F_5$	(121)	(121)	(121)	AB	BC	CA
Ⓔ	Val 145	$K_3UO_2F_5$	30	62	36			
Ⓔ	Glu 229	K_2PtCl_4	20			AA_{-5}		CB_5
Ⓒ	Tyr OH 117	$HAuCl_4$	40		20	Internal cavity		
Ⓣ	Asn 172	PHMBS			26	AA_5	BC_{-5}	
Ⓛ	Glu 77 Asn 238	$K_3UO_2F_5$			37			CB_{-2}
Ⓞ	Thr 179	$K_3UO_2F_5$			26	AA_{-5}		CB_5
Ⓟ	Ser 43	$K_3UO_2F_5$			10	R	R	
Ⓖ	Gln 165	K_2PtCl_4		6	6	AA_{-5}	CB_5	

[a] The occupancies are on a scale roughly proportional to electrons.
[b] "Subunit interactions" shows the nature of the environment whenever a heavy atom substitution is in contact with a neighboring subunit.

the C subunit has only a small or no substitution for all but the PHMBS derivative. No such modification exists to the environment of Cys 123 at site Ⓑ and it is, therefore, only substituted by PHMBS.

The disulfide bond between Cys 169 and Cys 176 is reduced by PHMBS and $HAuCl_4$ at site Ⓗ. However, this occurs only in subunits A and B where the S–S bridge is close to Ser 180 of the neighboring subunit generated by the AA_5 or BC_5 contacts. Possibly, the deeper immersion of αC of subunit C in RNA sterically prevents the reaction of the heavy metal reagent.

Methionine 59 and 139 are both substituted by K_2PtCl_4 and represent the Ⓕ and Ⓘ sites in subunit C, respectively. The substitution of methionine by platinum compounds has been observed frequently in other proteins (Blundell and Johnson, 1976). Site Ⓕ occurs only in subunit C because residue 59 is disordered in subunits A and B. Site Ⓘ interacts with the carboxy-terminal hinge which is very different in subunit C. This might hinder the substitution in subunit B, but it is not obvious why only the C subunit is substituted.

The Ⓠ site is produced by the replacement of Ca^{2+} (Abdel-Meguid et al., 1981) on the quasi-threefold axis by the positively charged uranyl ion.

The © site is caused by the reaction of the hydroxyl of Tyr 117 with $HAuCl_4$ in the interior of the subunits. It is, therefore, surprising that any difference should be detected in subunit B relative to A and C.

7. CHARGE CLUSTERS

The subunit contacts generated by the quasi-threefold axes are mediated by a number of charge interactions (Table 8). The most extensive cluster (Fig. 11) is in the vicinity of the Ca^{2+} binding site on the quasi-threefold axis (Abdel-Meguid et al., 1981). The first Ca^{2+} ion itself is liganded by Glu 194 from all three subunits A, B, and C. In addition, Arg 196, Arg 249, and Glu 252 are juxtaposed to Asp 188 and Glu 191 of the neighboring subunit in the vicinity of this Ca^{2+} site. A second cluster occurs on the viral surface where helix αD (roughly perpendicular to the quasi-threefold axis) and helix αE (almost parallel to the quasi-threefold axis) interact. Here, Lys 200 and Asp 204 on one subunit interact with Asp 138 and Asp 214 on the neighboring subunit. However, Asp 138 and Asp 141 are ligands to the second Ca^{2+} site. The third ionic interaction site, between quasi-threefold-related subunits, is between Asp 65 and Lys 154, also far from the wedge-shaped end of the subunit. The nondirectional ionic interactions are, perhaps, a partial mechanism, reinforcing the large hydrophobic contact areas, which permits strain without substantial loss of binding force. Also, as previously mentioned, removal of the Ca^{2+} ions permits swelling which, in TBSV, also controls orientation of the P domains (Robinson and Harrison, 1982). In SBMV, the second Ca^{2+} site determines the vestigial hinge orientation.

Table 8. Comparison of Intersubunit Charge Clusters in the Cowpea and Bean Strains of SBMV[a]

	Residue Number	Subunit	SBMV (Cowpea)	SBMV (Bean)
Quasi-threefold interactions				
First Ca^{2+} binding site (Fig. 11)				
	194	A, B, C	Glu	Lys
Second Ca^{2+} binding site				
	138	A, B, C	Asp	?
	141		Asp	?
	199	B, C, A	Main chain carbonyls	—
	259			
First charge cluster (Fig. 11)				
	196	A	Arg	Trp
	249	A	Arg	Gln
	252	A	Glu	Glu
	188	C	Asp	Asp
	191	C	Glu	Glu
Second charge cluster				
	200	A	Lys	Lys
	204	A	Asp	Asp
	138	C	Asp	?
	214	C	Asn	Asn
Third charge cluster				
	65	A	Asp	?
	154	C	Lys	?
Icosahedral fivefold interactions				
First charge cluster				
	181	C	Arg	Asn
	181	B_5	Arg	Asn
	163	B_5	Glu	?
Second charge cluster				
	132	C	His	?
	229	C	Glu	Leu
	241	B_5	Arg	Arg
Quasi-sixfold interactions (B_2C) with interposed arm				
First charge cluster				
	132	B_2	His	?
	229	B_2	Glu	Leu
	77	C	Glu	?

[a] Note the close interrelationship of the two Ca^{2+} sites and the first two charge clusters. For instance, Asp 138 is common to the second Ca^{2+} binding site and the second charge cluster (Fig. 11).

Two charge clusters also exist in the icosahedral fivefold contacts A_5A and the corresponding quasi-sixfold interactions CB_5. One cluster (Arg 181, Glu 163) is close to the "vector of rotation" and, hence, is similar (but not identical) in the CB_5 and B_2C quasi-sixfold interactions. The other cluster (His 132, Glu 229 in one subunit and Glu 77, Arg 241 in the other) is completely different in CB_5 compared to B_2C. It is of particular interest that a histidine is in this cluster which can act as a physiological switch because of its near neutral pK.

8. ASSEMBLY

One of the major problems in the understanding of viral function is how it assembles from its newly synthesized components either in the cytoplasm or while leaving the host cell. A similar but converse problem occurs on entering a host cell. In the case of spherical viruses, it will be necessary to appreciate how the assembly generates a, say, $T = 3$ particle rather than any other quasi-symmetrical arrangement. The size of the RNA is likely to be a controlling factor, but even here it is necessary to guide the assembly into, say, a $T = 3$ envelope as in SBMV rather than into a cylindrical bacillus-shaped envelope as in AMV, or other various possibilities.

An important controlling factor is the amino-terminal arm. Removal of this arm almost invariably directs the assembly into empty $T = 1$ capsids of size similar to STNV; examples are SBMV (Sehgal et al., 1979b; Tremaine and Ronald, 1978; Erickson and Rossmann, 1982), turnip crinkle virus (Leberman and Finch, 1970; Crowther and Amos, 1972), AMV (Bol et al., 1974), and brome mosaic virus (Pfeiffer and Hirth, 1975). The structures of SBMV and TBSV show that the arm not only helps to immobilize the RNA, but is required to form quasi-sixfold axes, an essential component of structures with $T > 1$. Nevertheless, it is not clear what induces the arm to become ordered and form the β-annulus and why the assembly is then guided into the correct surface lattice. Indeed, the presence of an arm does not necessarily inhibit the formation of a $T = 1$ capsid, as for example in STNV, SBMV (Savithri and Erickson, 1983), or turnip crinkle virus.

Savithri and Erickson (1983) have shown that the assembly of $T = 3$ particles from SBMV protein and RNA requires Ca^{2+} at alkaline pH. $T =$

1 particles are formed at acid pH in the presence of low molecular weight RNA components. Removal of the amino-terminal residues 1 to 30 or 1 to 61 prevents the formation of $T = 3$ particles. These observations support the view that some ionizable carboxyl groups are switches to induce the formation of the B_2C interaction and, hence, $T = 3$ capsid formation.

It is unknown whether the native protein in aqueous solution has an ordered βA arm. The circular dichroism studies on SBMV and turnip rosette virus of Hull and co-workers (Denloye et al., 1978; Odumosu et al., 1981) do not give a definitive answer. However, in light of the absence of many polar interactions of βA with its own subunit, its lack of acidic and aromatic groups, the presence of three basic groups and its preference for prolines, it would seem probable that it is disordered (Huber, 1979) prior to assembly. Hence, the assembly is likely to direct the ordering of βA in subunit C, rather than the disordering of βA in subunits A and B.

Rossmann et al. (1983c) have formulated an hypothesis on the sequence of events during the self-assembly of the $T = 3$ native virus particle and $T = 1$ empty capsids. No doubt other hypotheses could be made, but its consideration provides at least a testable model. The strengths of the charge interactions and the larger number of interactions on the fivefold subunit contact surfaces suggest that, in the presence of RNA, there may

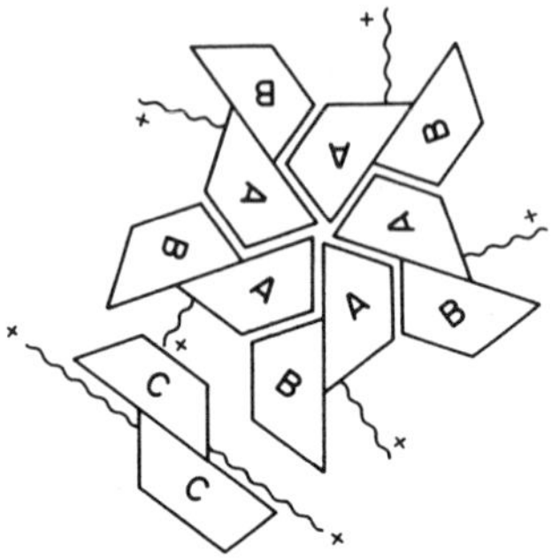

Figure 17. A 10-mer $T = 1$ cap composed of five AB_5 dimers might assemble initially. All the amino-terminal arms are disordered. Addition of Ca^{2+} on the quasi-threefold axes at alkaline pH may throw a switch, charging His 132, and hence altering the properties of the cap's periphery. Additional dimers must then "tense" in order to bind, leaving a hydrophobic channel between subunits C and B. The βA arm will then be induced to fold into this cavity and cause the formation of the β-annulus.

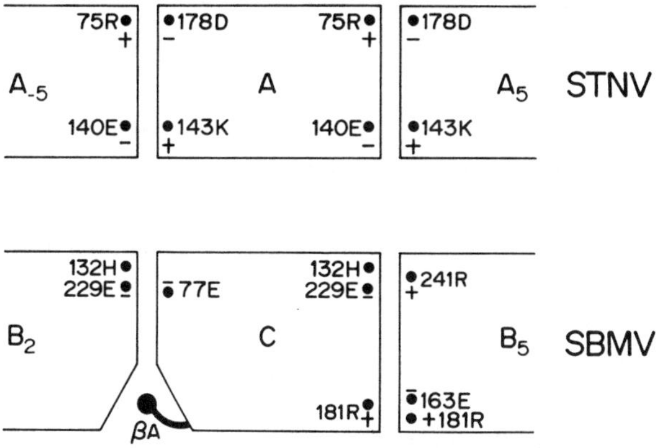

Figure 18. Diagrammatic representation showing a comparison of the spatial distribution of charge clusters in the pentamer and hexamer interfaces of STNV and SBMV, respectively.

occur an initial 10-mer nucleation center composed of five "relaxed" AB_5 dimers (Fig. 17). Addition of further AB_5 dimers can occur in two ways depending on the state of protonation or Ca^{2+} binding of the carboxyl clusters in the subunit contact regions. In the low charge state, the "relaxed" AB_5 dimers will nucleate a $T = 1$ capsid. However, in the high charge state, the periphery of the 10-mer would require the binding of "tensed" CC_2 dimers, creating a hydrophobic cleft between the B and C subunits. This would induce the ordering of the βA arm and formation of the β-annulus, thus leading to the assembly of a $T = 3$ capsid. Comparison of the spatial distribution of charge clusters in SBMV and STNV (Fig. 18) also suggests an involvement of His 132 as a mediator in the switch for the two binding states of the 10-mer nucleation center.

In summary, there are two distinct structural states of SBMV dimers which may be considered as being "relaxed" and "tensed." The possibility of multistate subunits was not envisaged by Caspar and Klug (1962), but might well explain the breakdown of the quasi-equivalence hypothesis in the polyoma structure (Li et al., 1983). The principles of quasi-equivalence were considered only in terms of two-dimensional surfaces by Caspar and Klug. In reality, the protein coat is 30–40 Å thick for SBMV and similar viruses. It appears that the two-dimensional analysis is insufficient for three-dimensional viral coats.

9. EVOLUTION

The author, together with co-workers, has developed an algorithm for the comparison of similar protein folds (Rao and Rossmann, 1973; Rossmann and Argos, 1975, 1976), which determines not only the amino acid alignment of similar structures but also provides a crude measure of similarity. In general, it has been found that the greater the structural similarity, the greater is the homology of amino acids or other measures of evolutionary relationships. These techniques have been applied for studying the evolution of TBSV, SBMV and STNV (Table 9; Fig. 19; Rossmann et al., 1983b). The amino acid sequence of STNV coat protein is known (Ysebaert et al., 1980) and the TBSV and STNV C_α coordinates were supplied by Drs. S. C. Harrison and B. Strandberg, respectively.

There is no clear indication of homologous amino acid sequences between SBMV and STNV. However, there is some conservation of the following functional groups:

1. Threonines and serines at the hexagonal-pentagonal wedge-shaped end of the subunit.
2. Lysines and arginines at the protein–RNA interface.
3. Hydrophobic residues in the cavity within the antiparallel β-barrel.
4. An aspartic acid near a site which binds Ca^{2+} in TBSV.
5. Ionic interactions in the contacts between fivefold-related subunits.

A variety of measures can be chosen to estimate structural similarity (Rossmann and Argos, 1977). Three such measures, the number of equivalenced residues, the percentage of equivalenced residues, and the rms

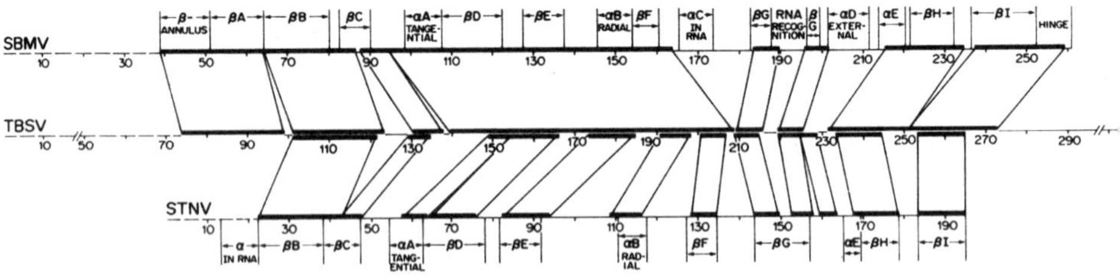

Figure 19. Alignment of TBSV, SBMV, and STNV polypeptide chains.

Table 9. Results of Pairwise Comparison of C_α Positions[a]

Molecule 1	Molecule 2	Number of Equivalences	rms (Å)	MBC/C[b]
TBSV(C)	SBMV(C)	179	2.15	—
STNV	TBSV(C)	109	3.44	—
STNV	SBMV(A)	104	3.74	1.40
STNV	SBMV(C)	103	3.49	1.43

[a] TBSV(C) refers to the C subunit of TBSV, and so on.
[b] MBC/C signifies minimum base change per codon. Random polypeptides align with a mean MBC/C of about 1.45.

deviation between C_α atoms of equivalenced residues, are shown in Table 10. The descending order shown in Table 10 is based on the percentage of equivalenced residues in the smaller domain. It is evident that the degree of similarity between the S domains of TBSV and SBMV is of the same order as the α and β domains of hemoglobin. The demanding case for evolutionary divergence between the genes for the α and β chains of hemoglobin was immediately recognized from their structural similarity (Perutz et al., 1960). It will, therefore, be of interest, when the amino acid sequence of TBSV is known, whether the rate of change of amino acids between these viruses is also comparable to the structural change. The similarity of TBSV or SBMV to STNV is of the same order as that between the NAD binding domains of dehydrogenases (Rossmann et al., 1974, 1975) and certainly greater than the similarity of hen egg white and phage lysozyme (Rossmann and Argos, 1976). The structural and functional similarity in both these examples makes it probable that they represent cases of divergence from a common ancestor (Artymiuk et al., 1981; Matthews et al., 1981). The principal known function of viral coat proteins is to envelop the nucleic acid. It was shown above that there is considerable conservation of charge distribution relating to the subunit contact areas, and conservation of basic residues at the protein–RNA interface. However, these conservations relate only to three-dimensional spatial positions, not to topological equivalence along the polypeptide. Nevertheless, the topological similarity of the folds of TBSV, SBMV, and STNV is so compelling in relation to the data shown in Table 10 that the probability for divergence would seem greater than that for convergence. If that is the case, then the functionally important protein–protein and protein–

Table 10. Structural Comparisons

Molecule 1	Molecule 2	Number of Residues Molecule 1	Number of Residues Molecule 2	Number of Equivalences	Percentage of Equivalences Molecule 1	Percentage of Equivalences Molecule 2	rms (Å)	Reference
Hb β	Hb α	146	141	139	95	99	1.9	Argos and Rossmann (1979)
SBMV(C)	TBSV(C)	221	198	179	81	90	2.2	Rossmann et al. (1983b)
GAPDH(NAD)	LDH(NAD)	148	144	96	65	67	2.9	Rossmann and Argos (1977)
TBSV(S)[a]	STNV[b]	163	173	109	67	63	3.4	Rossmann et al. (1983b)
TBSV(S)[a]	TBSV(P)	167	110	69	41	63	3.8	Argos et al. (1980)
SBMV(A)[a]	STNV[b]	187	173	104	56	60	3.7	Rossmann et al. (1983b)
T4 lysozyme	HEW lysozyme	164	129	78	48	60	4.1	Rossmann and Argos (1977)
LDH(NAD)	Flavodoxin	144	138	39	27	28	2.4	Eventoff and Rossmann (1975)

[a] Without βA arm or P domain.
[b] Without amino-terminal helix.

RNA surfaces undergo relatively rapid change, whereas in contrast the active sites of enzymes are especially well conserved.

Having established probable evolutionary divergence among the spherical plant viruses of known coat protein structure, the next logical step is an attempt to establish a possible sequence of events in the development of these viruses. For this purpose, the following assumptions will be made:

1. The amount of structural difference is proportional to the rate of acceptance of mutational events and that these in turn are proportional to elapsed time. Hence, the S domains of TBSV and SBMV have more recently diverged from each other than they each have from STNV and presumably, AMV.

2. The rate of acceptance of mutational events in the P domain is the same as in the S domain, although it could be that the different functional requirements of the P domain might permit it to change faster than the S domain. Hence, the S and P domains have diverged from each other at a time preceding the divergence of the S domains in the various viruses.

3. The histonelike R domains, present in many plant viruses, whose function is to bind and fold RNA (Bol et al., 1974), all had a common ancestor. The alternative hypothesis that the R domain of each virus coat protein had a different origin seems less likely as it is improbable that the essential R to S gene fusion event could occur frequently. Nevertheless, there is little evidence in the amino acid sequence of a common origin for the various R domains.

Application of the above assumptions yields the possible evolutionary sequence shown in Figure 20. This is probably the simplest set of events that satisfy the above assumptions and, hence, presents the tree of greatest parsimony. The rate of change for RNA controlled genes far exceeds that for DNA genes (Holland et al., 1982). Hence, the time of the events in Figure 20 might be far less than such a tree based on even the fastest changing protein structure derived from a DNA gene.

In considering the evolution of viruses, it is essential to recognize their host specificity for they must have had a common host at the time they separated into recognizably distinct viruses. However, all the viruses shown in Figure 20 infect many dicotyledons. Only SBMV has its host range restricted to legumes.

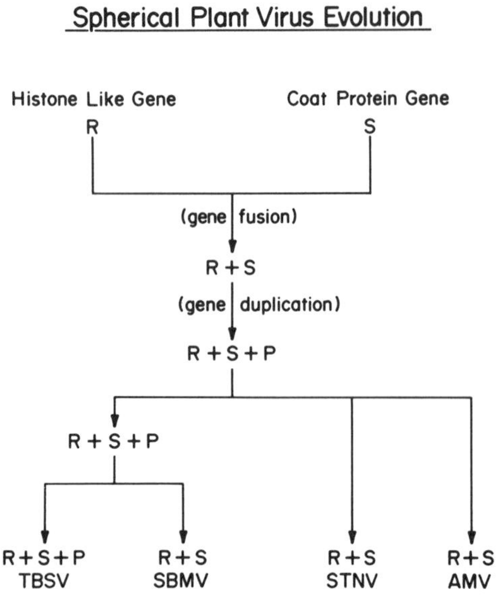

Figure 20. The most probable series of events in a divergent evolutionary process of virus evolution consistent with the assumptions given in the text.

10. CONCLUSIONS

The remarkable similarity of subunit fold of SBMV, TBSV, STNV, and possibly also of the bacillus-shaped virus AMV (Fukuyama et al., 1983) must determine many basic properties of these and, presumably, many other plant viruses. It is, therefore, useful to differentiate those viruses which have a protein shell surrounding a nucleic acid core as opposed to tobacco mosaic virus-like viruses where the RNA has a distinct structure intertwined with the protein. It will be of interest whether the structural motif in the plant virus shells will appear in some simple animal viruses such as picornaviruses (Hogle, 1982; Erickson et al., 1983). Similarities among viruses will give guidance to their general functional requirements whereas the detailed structural variations relate to their differentiation. Such observations will be independent of whether the similarities are due to evolutionary divergence from a common viral precursor or convergence dictated by common functional necessities.

Many animal and bacterial viruses recognize their host cells (Lonberg-Holm and Philipson, 1974) by virtue of a specific polysaccharide. The latter may be a part of a virus [cf. influenza virus (Klenk et al., 1975; Lazarowitz and Choppin, 1975)] or on the host cell surface [cf. adenovirus (Philipson et al., 1976)]. In the case of plant cells, the infective process is less well understood. The usual mechanism of infection is said to involve the mechanical insertion of the virus through the cell wall and membrane by a vector (cf. Matthews, 1970). Proposals have also been made (cf. Hughes, 1976; Gaard and De Zoeten, 1979) that plant viruses recognize host cells with the aid of specific polysaccharides in a manner similar to animal and bacterial viruses. It is, therefore, tempting to speculate that the difference in the viral surfaces between SBMV and TBSV generated by the additional P domain in TBSV might be related to host specificity. Possibly, there is some binding of the virus to a cell membrane component to which the viral protein can bind during assembly or disassembly. However, no such receptor has yet been identified for simple plant viruses. The observation that concanavalin A has a structure similar to the S or P domains of plant viruses (Argos et al., 1980) is of interest as concanavalin A can compete with human rhinovirus and poliovirus for receptors on HeLa cells (Lonberg-Holm, 1975).

The occurrence of a basic, histonelike arm in many plant virus coat proteins is another remarkable similarity. It has already been observed that the positive charge on the polypeptide roughly cancels the charge on the nucleic acid. Where such histone domains are absent, as in turnip yellow mosaic virus, the virus has a substantial complement of polyamines such as spermidine (Cohen and McCormick, 1979). The complementary folding of the R domain with protein during assembly lacks detailed information at this time. Is there a unique RNA structure associated with a unique set of R domain conformations? To what extent does the basic surface on the interior of the viral coat participate in the RNA folding? How is specificity of the coat protein binding to RNA achieved in AMV and possibly other viruses?

The process of self-assembly is also not well understood. Certainly, removal of the R domain inhibits formation of particles larger than $T = 1$. However, the presence of the R domain does not necessarily determine the geometry of the viral coat. The formation of a β-annulus about quasi-sixfold axes is not sufficient for determining the surface lattice, and the size of the RNA is not a completely controlling factor. Why, for example, does AMV form bacillus-shaped particles rather than a series of spherical particles? Fukuyama et al. (1981, 1983) and Abdel-Meguid et al. (1982)

have proposed a two-state association of dimers as an important switch in the assembly of AMV. In AMV, there must be two types of dimer environments in the cylindrical portion of a particle (Fig. 21). Those dimers situated along the lengthwise ridges of the virus (A in Fig. 21) would provide curvature and require less space in the viral interior, whereas those sitting on the faces (B in Fig. 21) of the virus would consume more space in the interior. The icosahedral ends would require "A type" dimers (Fig. 21). The switch between A and B type dimers might be the insertion of an ordered arm analogous to the situation in SBMV and TBSV.

Another aspect of the similarity among the known viral structures is the requirement of cations as a glue between subunits. The binding of Ca^{2+} to the vestigial hinge in SBMV and the conservation of this position in TBSV suggests that these sites also control conformation. Thus, their conservation and functional importance may be a controlling factor in the

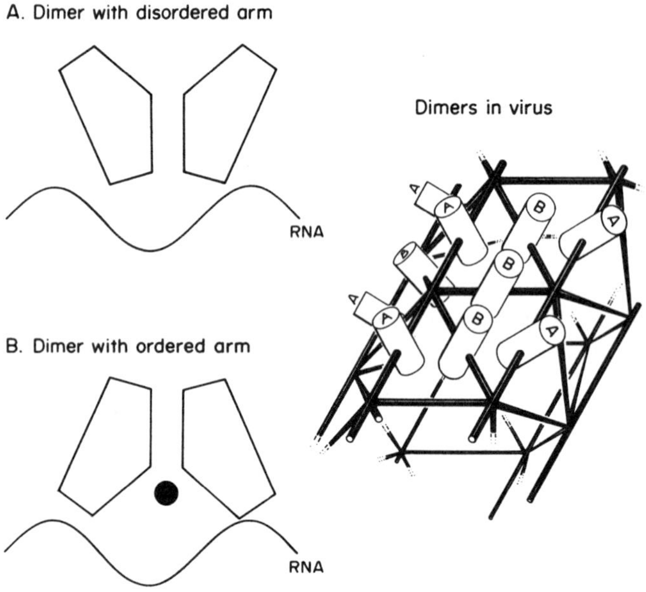

Figure 21. Two types of dimer environments must occur in the cylindrical portion of AMV. The A type dimers have less space in the interior. The B type dimers have greater space and may, therefore, have ordered amino-terminal arms in the subunit interface. The icosahedral virus ends would contain only A type dimers. The large hole at the sixfold axes excludes the possibility of forming a β-annulus type structure around these axes.

evolution of viral structure in much the same way as the most conservative part of enzyme structure is usually the active center.

Similarly, the occurrence of a large number of basic residues on the protein–RNA interfaces in SBMV, TBSV, and STNV represents an important conserved function. This requirement would inhibit large structural changes during evolution once the property has been established as beneficial to viral function.

It is apparent that the subunit fold found in SBMV, TBSV, and STNV, as well as possibly AMV, exhibits many features useful to viral structure. Not least, the wedge-shaped character of the subunit is well adapted for the formation of protein shells. Given, therefore, all manner of constraints of polypeptide folding, the observed subunit in plant viruses may be one of only a very few structures capable of performing all aspects in which it participates during a viral life cycle.

Acknowledgments. Many parts of this manuscript have been derived from earlier papers. In particular, the manuscript borrows heavily from three papers (Rossmann et al., 1983a,b,c) still in the refereeing process at the time of writing. I wish to thank all my co-workers for the education and stimulation they have given me and, in particular, I wish to thank Drs. Cele Abad-Zapatero and John Erickson in my laboratory and Drs. Lars Liljas, Alwyn Jones, and Bror Strandberg at the Wallenberg Laboratory of the University of Uppsala in Sweden who are co-authors on the three papers mentioned above. I also wish to thank Dr. Jack Johnson for numerous stimulating encouragements and Sharon Wilder for much help in the preparation of this manuscript. The work was supported by a National Institutes of Health grant (no. AI 11219) and a National Science Foundation grant (no. PCM78-16584) to M.G.R. and a United States Department of Agriculture grant (no. 182117620) to M.G.R. and John Erickson.

REFERENCES

Abad-Zapatero, C., Abdel-Meguid, S. S., Johnson, J. E., Leslie, A. G. W., Rayment, I., Rossmann, M. G., Suck, D., and Tsukihara, T. (1980). *Nature (London)* **286,** 33–39.

Abad-Zapatero, C., Abdel-Meguid, S. S., Johnson, J. E., Leslie, A. G. W., Rayment, I., Rossmann, M. G., Suck, D., and Tsukihara, T. (1981). *Acta Crystallogr.* **B37,** 2002–2018.

Abdel-Meguid, S. S., Yamane, T., Fukuyama, K., and Rossmann, M. G. (1981). *Virology* **114,** 81–85.

Abdel-Meguid, S. S., Fukuyama, K., and Rossmann, M. G. (1982). *Acta Crystallogr.* **B38,** 2004–2008.

Akimoto, T., Wagner, M. A., Johnson, J. E., and Rossmann, M. G. (1975). *J. Ultrastruct. Res.* **53,** 306–318.

Argos, P. (1981). *Virology* **110,** 55–62.

Argos, P., and Rossmann, M. G. (1979). *Biochemistry* **18,** 4951–4960.

Argos, P., Tsukihara, T., and Rossmann, M. G. (1980). *J. Mol. Evol.* **15,** 169–179.

Arndt, U. W., and Wonacott, A. J. (1977). *The Rotation Method in Crystallography.* North-Holland, Amsterdam.

Arnott, S., Smith, P. J. C., and Chandrasekaran, R. (1976). In *Handbook of Biochemistry and Molecular Biology. Nucleic Acids*, G. D. Fasman, Ed., 3rd ed., Vol. 2. CRC Press, Cleveland, Ohio, pp. 411–422.

Artymiuk, P. J., Blake, C. C. F., and Sippel, A. E. (1981). *Nature (London)* **290,** 287–288.

Bancroft, J. B., Hills, G. J., and Markham, R. (1967). *Virology* **31,** 354–379.

Blundell, T. L., and Johnson, L. N. (1976). *Protein Crystallography.* Academic Press, New York.

Bol, J. F., Kraal, B., and Brederode, F. Th. (1974). *Virology* **58,** 101–110.

Bricogne, G. (1976). *Acta Crystallogr.* **A32,** 832–847.

Butler, P. J. G., and Klug, A. (1978). *Sci. Am.* **239(5),** 62–69.

Caspar, D. L. D., and Klug, A. (1962). *Cold Spring Harbor Symp. Quant. Biol.* **27,** 1–24.

Chothia, C., and Janin, J. (1981). *Proc. Natl. Acad. Sci. U.S.A.* **78,** 4146–4150.

Cohen, S. S., and McCormick, F. P. (1979). *Adv. Virus Res.* **24,** 331–387.

Crowther, R. A., and Amos, L. A. (1972). *Cold Spring Harbor Symp. Quant. Biol.* **36,** 489–494.

Denloye, A. O., Homer, R. B., and Hull, R. (1978). *J. Gen. Virol.* **41,** 77–85.

Dietrich, J. B., Peter, R., Das, B. C., Peter, C., and Collot, D. (1981). *Arch. Biochem. Biophys.* **210,** 794–796.

Durham, A. C. H. (1977). *Nature (London)* **267,** 375–376.

Durham, A. C. H., Hendry, D. A., and von Wechmar, M. B. (1977). *Virology* **77,** 524–533.

Erickson, J. W., and Rossmann, M. G. (1982). *Virology* **116,** 128–136.

Erickson, J. W., Frankenberger, E. A., Rossmann, M. G., Fout, G. S., Medappa, K. C., and Rueckert, R. R. (1983). *Proc. Natl. Acad. Sci. U.S.A.* **80,** 931–934.

Eventoff, W., and Rossmann, M. G. (1975). In *CRC Critical Reviews in Biochemistry*, G. D. Fasman, Ed., Vol. 3. CRC Press, Cleveland, Ohio, pp. 111–140.

Fukuyama, K., Abdel-Meguid, S. S., and Rossmann, M. G. (1981). *J. Mol. Biol.* **150,** 33–41.

Fukuyama, K., Abdel-Meguid, S. S., Johnson, J. E., and Rossmann, M. G. (1983). *J. Mol. Biol.* **167,** 873–894.

Fuller, W., Hutchinson, F., Spencer, M., and Wilkins, M. H. F. (1967). *J. Mol. Biol.* **27,** 507–524.

Gaard, G., and De Zoeten, G. A. (1979). *Virology* **96,** 21–31.

Harrison, S. C. (1980). *Biophys. J.* **32,** 139–153.

Harrison, S. C., Olson, A. J., Schutt, C. E., Winkler, F. K., and Bricogne, G. (1978). *Nature (London)* **276,** 368–373.

Hermodson, M. A., Abad-Zapatero, C., Abdel-Meguid, S. S., Pundak, S., Rossmann, M. G., and Tremaine, J. H. (1982). *Virology* **119,** 133–149.

Hogle, J. M. (1982). *J. Mol. Biol.* **160,** 663–668.

Holland, J., Spindler, K., Horodyski, F., Grabau, E., Nichol, S., and VandePol, S. (1982). *Science* **215,** 1577–1585.

Holmes, K. C. (1980). *Trends Biochem. Sci.* **5,** 4–7.

Houwing, C. J., and Jaspars, E. M. J. (1980). *Biochemistry* **19,** 5261–5264.

Hsu, C. H., Sehgal, O. P., and Pickett, E. E. (1976). *Virology* **69,** 587–595.

Hsu, C. H., White, J. A., and Sehgal, O. P. (1977). *Virology* **81,** 471–475.

Huber, R. (1979). *Trends Biochem. Sci.* **4,** 271–276.

Hughes, C. (1976). In *Specificity in Plant Diseases*, R. K. S. Wood and A. Graniti, Eds. Plenum Press, New York, pp. 77–99.

Hull, R. (1977a). *J. Gen. Virol.* **36,** 289–295.

Hull, R. (1977b). *Virology* **79,** 50–57.

Hull, R. (1977c). *Virology* **79,** 58–66.

Hull, R. (1978). *Virology* **89,** 418–422.

Johnson, J. E. (1978). *Acta Crystallogr.* **B34,** 576–577.

Johnson, J. E., Rossmann, M. G., Smiley, I. E., and Wagner, M. A. (1974). *J. Ultrastruct. Res.* **46,** 441–451.

Johnson, J. E., Akimoto, T., Suck, D., Rayment, I., and Rossmann, M. G. (1976). *Virology* **75,** 394–400.

Jonard, G., Richards, K. E., Guilley, H., and Hirth, L. (1977). *Cell* **11,** 483–493.

Jones, T. A. (1978). *J. Appl. Crystallogr.* **11,** 268–272.

Kan, J. H., Andree, P. J., Kouijzer, L. C., and Mellema, J. E. (1982). *Eur. J. Biochem.* **126,** 29–33.

Klenk, H. D., Rott, R., Orlich, M., and Blödorn, J. (1975). *Virology* **68,** 426–439.

Koper-Zwarthoff, E. C., and Bol, J. F. (1980). *Nucl. Acids Res.* **8,** 3307–3318.

Krüse, J., Timmins, P. A., and Witz, J. (1982). *Virology* **119,** 42–50.

Langridge, R., Billeter, M. A., Borst, P., Burdon, R. H., and Weissmann, C. (1964). *Proc. Natl. Acad. Sci. U.S.A.* **52,** 114–119.

Lazarowitz, S. G., and Choppin, P. W. (1975). *Virology* **68,** 440–454.

Leberman, R., and Finch, J. T. (1970). *J. Mol. Biol.* **50,** 209–213.

Leonard, B. R., Jr., Anderegg, J. W., Shulman, S., Kaesberg, P., and Beeman, W. W. (1953). *Biochim. Biophys. Acta* **12,** 499–507.

Leslie, A. G. W., and Tsukihara, T. (1980). *J. Appl. Crystallogr.* **13,** 304–305.

Li, J., Fricks, C., Caspar, D. L. D., and Rayment, I. (1983), *Virology*, in press.

Lifson, S., and Sander, C. (1979). *Nature (London)* **282,** 109–111.

Liljas, L., Unge, T., Jones, T. A., Fridborg, K., Lövgren, S., Skoglund, U., and Strandberg, B. (1982). *J. Mol. Biol.* **159,** 93–108.

Lilley, D. M. J., Howarth, O. W., Clark, V. M., Pardon, J. F., and Richards, B. M. (1976). *FEBS Lett.* **62,** 7–10.

Lonberg-Holm, K. (1975). *J. Gen. Virol.* **28,** 313–327.

Lonberg-Holm, K., and Philipson, L. (1974). In *Monographs in Virology*, J. L. Melnick, Ed., Vol. 9. Karger, Basel.

McCain, D. C., Virudachalam, R., Markley, J. L., Abdel-Meguid, S. S., and Rossmann, M. G. (1982a). *Virology* **117,** 501–503.

McCain, D. C., Virudachalam, R., Santini, R. E., Abdel-Meguid, S. S., and Markley, J. L. (1982b). *Biochemistry* **21,** 5390–5397.

Magdoff, B. S. (1960). *Nature (London)* **185,** 673–674.

Mang, K. Q., Ghosh, A., and Kaesberg, P. (1982). *Virology* **116,** 264–274.

Matthews, B. W., Remington, S. J., Grütter, M. G., and Anderson, W. F. (1981). *J. Mol. Biol.* **147,** 545–558.

Matthews, R. E. F. (1970). *Plant Virology*. Academic Press, New York.

Miller, G. L., and Price, W. L. (1946). *Arch. Biochem.* **10,** 467–477.

Moss, T., Cary, P. D., Crane-Robinson, C., and Bradbury, E. M. (1976). *Biochemistry* **15,** 2261–2267.

Munowitz, M. G., Dobson, C. M., Griffin, R. G., and Harrison, S. C. (1980). *J. Mol. Biol.* **141,** 327–333.

Odumosu, A. O., Homer, R. B., and Hull, R. (1981). *J. Gen. Virol.* **53,** 193–196.

Perutz, M. F., Rossmann, M. G., Cullis, A. F., Muirhead, H., Will, G., and North, A. C. T. (1960). *Nature (London)* **185,** 416–422.

Pfeiffer, P., and Hirth, L. (1975). *FEBS Lett.* **56,** 144–148.

Philipson, L., Everitt, E., and Lonberg-Holm, K. (1976). In *Cell Membrane Receptors for Viruses, Antigens and Antibodies, Polypeptide Hormones and Small Molecules*, R. F. Beers, Jr., and E. G. Bassett, Eds. Raven Press, New York, pp. 203–216.

Randles, J. W., Davies, C., Hatta, T., Gould, A. R., and Francki, R. I. B. (1981). *Virology* **108,** 111–122.

Rao, S. T., and Rossmann, M. G. (1973). *J. Mol. Biol.* **76,** 241–256.

Rayment, I., Johnson, J. E., Suck, D., Akimoto, T., and Rossmann, M. G. (1978). *Acta Crystallogr.* **B34,** 567–578.

Rayment, I., Johnson, J. E., and Rossmann, M. G. (1979). *J. Biol. Chem.* **254,** 5243–5245.

Rayment, I., Baker, T. S., Caspar, D. L. D., and Murakami, W. T. (1982). *Nature (London)* **295,** 110–115.

Robinson, I. K., and Harrison, S. C. (1982). *Nature (London)* **297,** 563–568.

Rosenberg, J. M., and Greene, P. (1982). *DNA* **1,** 117–124.

Rossmann, M. G. (1979). *J. Appl. Crystallogr.* **12,** 225–238.

Rossmann, M. G., and Argos, P. (1975). *J. Biol. Chem.* **250,** 7525–7532.

Rossmann, M. G., and Argos, P. (1976). *J. Mol. Biol.* **105,** 75–95.

Rossmann, M. G., and Argos, P. (1977). *J. Mol. Biol.* **109,** 99–129.

Rossmann, M. G., and Blow, D. M. (1962). *Acta Crystallogr.* **15,** 24–31.

Rossmann, M. G., Moras, D., and Olsen, K. W. (1974). *Nature (London)* **250,** 194–199.

Rossmann, M. G., Liljas, A., Brändén, C. I., and Banaszak, L. J. (1975). In *The Enzymes*, P. D. Boyer, Ed., 3rd ed., Vol. 11. Academic Press, New York, pp. 61–102.

Rossmann, M. G., Leslie, A. G. W., Abdel-Meguid, S. S., and Tsukihara, T. (1979). *J. Appl. Crystallogr.* **12,** 570–581.

Rossmann, M. G., Chandrasekaran, R., Abad-Zapatero, C., Erickson, J. W., and Arnott, S. (1983a). *J. Mol. Biol.* **166,** 73–80.

Rossmann, M. G., Abad-Zapatero, C., Murthy, M. R. N., Liljas, L., Jones, T. A., and Strandberg, B. (1983b). *J. Mol. Biol.* **165,** 711–736.

Rossmann, M. G., Abad-Zapatero, C., Hermodson, M. A., and Erickson, J. W. (1983c). *J. Mol. Biol.* **166,** 37–83.

Rutgers, T., Salerno-Rife, T., and Kaesberg, P. (1980). *Virology* **104,** 506–509.

Savithri, H. S., and Erickson, J. W. (1983). *Virology* **126,** 328–335.

Sehgal, O. P., and Hsu, C. H. (1977). *Virology* **77,** 1–11.

Sehgal, O. P., Van, M., and White, J. A. (1979a). *Virology* **94,** 479–483.

Sehgal, O. P., Hsu, C. H., White, J. A., and Van, M. (1979b). *Phytopathol. Z.* **95,** 167–177.

Stubbs, G., Warren, S., and Mandelkow, E. (1979). *J. Supramolec. Struct.* **12,** 177–183.

Thomas, J. O. (1978). In *International Review of Biochemistry: Biochemistry of Nucleic Acids II*, B. F. C. Clark, Ed., Vol. 17. Univ. Park Press, Baltimore, Maryland, pp. 181–232.

Tomita, K. I., and Rich, A. (1964). *Nature (London)* **201,** 1160–1163.

Tremaine, J. H., and Ronald, W. P. (1977). *Can. J. Bot.* **55,** 2274–2277.

Tremaine, J. H., and Ronald, W. P. (1978). *Virology* **91,** 164–172.

Tremaine, J. H., Ronald, W. P., and Kelly, E. M. (1981). *Can. J. Microbiol.* **27,** 654–663.

Tremaine, J. H., Ronald, W. P., and Kelly, E. M. (1982). *Virology* **118,** 35–44.

Vriend, G., Hemminga, M. A., Verduin, B. J. M., De Wit, J. L., and Schaafsma, T. J. (1981). *FEBS Lett.* **134,** 167–171.

Warrant, R. W., and Kim, S. H. (1978). *Nature* (*London*) **271,** 130–135.

Weber, K. A., and Sehgal, O. P. (1982). *Phytopathology* **72,** 909–913.

Williams, R. J. P. (1977). In *Calcium-Binding Proteins and Calcium Function*, R. H. Wasserman, R. A. Corradino, E. Carafoli, R. H. Kretsinger, D. H. MacLennan, and F. L. Siegel, Eds. North-Holland, New York, pp. 3–12.

Yphantis, D. A. (1964). *Biochemistry* **3,** 297–317.

Ysebaert, M., van Emmelo, J., and Fiers, W. (1980). *J. Mol. Biol.* **143,** 273–287.

Zaumeyer, W. J., and Harter, L. L. (1942). *Phytopathology* **32,** 438–439.

Zimmern, D. (1977). *Cell* **11,** 463–482.

3

The Structure of Satellite Tobacco Necrosis Virus

LARS LILJAS
BROR STRANDBERG
Institute of Molecular Biology
University of Uppsala
Uppsala, Sweden

CONTENTS

1. PROPERTIES OF SATELLITE TOBACCO NECROSIS VIRUS

Satellite tobacco necrosis virus (STNV) is a small spherical RNA virus that can grow on many types of plants. It can propagate, however, only in the presence of another virus, tobacco necrosis virus (TNV) (Kassanis and Nixon, 1961; Kassanis, 1962). When coinfected with TNV, STNV is under certain conditions produced in excess of its helper virus. *In vivo* it is transmitted by the soil fungus *Olpidium brassicae* to the root cells of the plant (Kassanis and MacFarlene, 1968).

The STNV particle is very stable. It resists both low and high salt concentrations, is stable between pH 2 and pH 11, and it is not digested by either nucleases or proteases. The particle can, however, be disrupted in sodium dodecyl sulfate or 6 *M* urea (Unge and Strandberg, 1979) and is sensitive to treatment with ethylenediaminetetraacetic acid (EDTA) at pH above 7.0.

STNV has a single-stranded RNA, which is the messenger for the coat protein. It consists of 1239 bases and has been sequenced (Ysebaert et al., 1980). The protein for which it codes has 195 amino acids. Most of the protein sequence obtained from the RNA sequence has been verified by direct amino acid sequencing (Henriksson et al., 1981).

No evidence for a sequence homology between STNV and TNV has as yet been found, but it is still reasonable to believe that STNV has evolved from either its helper virus or another virus (Mossop and Francki, 1979). STNV is the only known plant virus with this type of dependence on a helper virus, but a similar type of satellite relationship has been observed for the parvovirus adeno-associated virus (AAV) and adenovirus.

2. NONCRYSTALLOGRAPHIC STRUCTURAL STUDIES

From electron micrographs STNV particles were observed to be 170 Å in diameter and to have icosahedral symmetry (Höglund, 1968; Kassanis and Woods, 1968). But it was not until 1971 that it was clearly shown with analysis of X-ray data from STNV crystals (Klug, 1971; Åkervall et al., 1971) that STNV had a $T = 1$ (Caspar and Klug, 1962) icosahedral symmetry.

The STNV particle has been studied by solution scattering experiments, using both neutrons and X-rays. Chauvin et al. (1977) found in their neutron work on STNV evidence for a model with little intermixing of protein and nucleic acid. In this model most of the protein was between a 56 and 88 Å radius, and the RNA was mainly inside a 60 Å radius. The measurements also indicated the presence of a central hole with a radius less than 30 Å.

In an X-ray solution scattering experiment (Sjöberg, 1977), the observed data suggested a model in which the RNA was mainly distributed between a 14 and 56 Å radius, with a lower concentration in a shell between 28 and 42 Å. The protein was found within a shell of 56–84 Å radius.

STNV crystallizes easily and is actually observed to crystallize *in vivo* (Kassanis et al., 1970). STNV crystals were analyzed with X-rays as early as 1945 (Crowfoot and Schmidt, 1945). The X-ray structure determination of STNV in our laboratory started in 1964 and has reached moderately high resolution. A 3.0 Å resolution study has been published (Liljas et al., 1982), and a 2.5 Å resolution refinement of the protein portion is in progress (Jones and Liljas, 1983; Jones and Liljas, to be published).

STNV crystals have also been studied with neutron diffraction at 32 Å resolution, using contrast matching with different D_2O concentrations in the crystals to match out the scattering from either the protein or the nucleic acid (G. Bentley, personal communication). Maps in projections based on amplitudes from neutron diffraction and phases obtained by refinement of a model of the protein have shown a distribution of protein and nucleic acid similar to that obtained with other methods, although the low resolution precludes any detailed interpretation.

3. THE SATELLITE TOBACCO NECROSIS VIRUS STRUCTURE

3.1. Techniques Employed in the Structure Determination

STNV crystallizes in space group C2 with one virus particle in the asymmetric unit of the crystallographic unit cell. Because the icosahedral particle is built up of 60 chemically identical protein subunits, there exists a 60-fold noncrystallographic symmetry, which was of great use in the structure determination. Two heavy atom derivatives, one with I_2 and one with $PtCl_4^{2-}$, have been used in the structure determination. Both derivatives were first used in a 10 Å study, after the heavy atom sites were found by Patterson search technique (Lentz et al., 1976). Later, a 4.0 Å structure was determined based on isomorphous replacement with the iodine derivative and phase refinement with noncrystallographic symmetry (Unge et al., 1980). In this study the boundary of the protein subunit could be delineated. The electron density maps also showed a helix protruding into the RNA from every subunit.

The high resolution data were collected using oscillation cameras with rotating anode X-ray generators as sources. A set of about 500,000 independent reflections was collected, which represents 85% of the total number of reflections to 2.5 Å resolution (Liljas et al., 1982). Native STNV crystals diffract well beyond the resolution limit used in the data collection, but there is a considerable intensity decline with increasing resolution. For some of the crystals the reflections between 2.8 and 2.5 Å resolution were rather weak or suffered from absorption of the reflected beam in the crystal. They were therefore omitted in the data processing, resulting in an abbreviated data set near 2.5 Å.

Data from crystals of one derivative, $PtCl_4^{2-}$, were utilized for the initial phase determination. A total of 230,000 independent reflections to 3.0 Å resolution were collected. Three platinum sites per subunit were found in a difference Fourier map using 4.0 Å phases from earlier work (Unge et al., 1980). After refinement of the occupancy of the 180 platinum sites in the asymmetric unit, phases were calculated to give a first electron density map. This map was based on single isomorphous replacement phases for 55% of the reflections between 5.5 and 3.0 Å resolution, and phases from the earlier work between 15 and 5.5 Å resolution. Phase refinement using the noncrystallographic symmetry was carried out using the programs written by Gerard Bricogne (1974, 1976). The electron density in the particle was 60-fold averaged inside an envelope that was chosen to contain as closely as possible the whole virus particle. The internal RNA-containing region of the virus particle was averaged as well. The region outside the envelope, between the particles, was leveled to zero. Accurate determination of the orientation of the particle relative to the crystal axes (needed for the phase refinement) was obtained using the rotation function (Rossmann and Blow, 1962).

In spite of the limited phase information from the single derivative, this starting model resulted in an interpretable map after 21 cycles of averaging (Liljas et al., 1982). In the corresponding structure determination of two other spherical viruses, tomato bushy stunt virus (TBSV; Harrison et al., 1978) and southern bean mosaic virus (SBMV; Abad-Zapatero et al., 1980), two and four heavy atom derivatives were used, respectively. The convergence of the phase refinement in STNV was probably enhanced by the high noncrystallographic symmetry. However, plots of R-factors and correlation coefficients as a function of resolution (Liljas et al., 1982) indicated that the phase refinement produced good phase angles only to about 3.7 Å resolution. The low quality of the phases between 3.7 and 3.0 Å resolution has also been verified by comparisons with phases from the refinement to 2.5 Å resolution (Jones and Liljas, 1983). The refined phases to 3.7 Å resolution have very small phase errors, and the final electron density is therefore clear and easy to interpret. This is not normally the case for electron density maps of proteins at 3.7 Å resolution.

Only the protein part of the virus was seen in the electron density map. A model for the protein subunit was built in a graphics display (Jones, 1982). This first model was then used for phase extension to 2.5 Å resolution by alternating cycles of phase refinement with symmetry averaging and real space refinement employing a procedure developed by T. A. Jones. The details of the subunit interactions described below have become clear only as a result of this refinement.

3.2. Polypeptide Chain and Secondary Structure of the STNV Protein

The polypeptide chain of the STNV protein subunit can be followed in the electron density map from residue 12 to the C terminus, residue 195. Residues 12–24 form a radial helix which lies between 43 Å and 58 Å radius in the particle. The remainder of the polypeptide chain forms a β-barrel between 58 and 85 Å radius and is composed of two four-stranded antiparallel β-sheets (Fig. 1). The β-strands are denoted β B to β I (after Hermodson et al., 1982). Stereo drawings of the α-carbon atoms of the main chain are shown in Figures 2 and 3. The BIDG sheet is quite regular and has a twist of about 5° per strand. At one end, near the fivefold axis of the icosahedron, the strands of the two sheets are connected with four short loops. At the other end, the two sheets are bridged by two long loops. A third long loop, which contains two short helices, connects the two strands β E and β F of the CHEF sheet. The particular topology of this β-barrel has been called "β-roll" or "Swiss roll," because it can be thought of as formed by rolling two long, antiparallel β-strands into a right-handed helix (Fig. 4). Figure 5 is a plot of the secondary structure.

3.3. Tertiary Structure of the Virus Protein

The tertiary structure of the protein subunit is stabilized by hydrophobic interaction of several bulky side chains between the two sheets, and be-

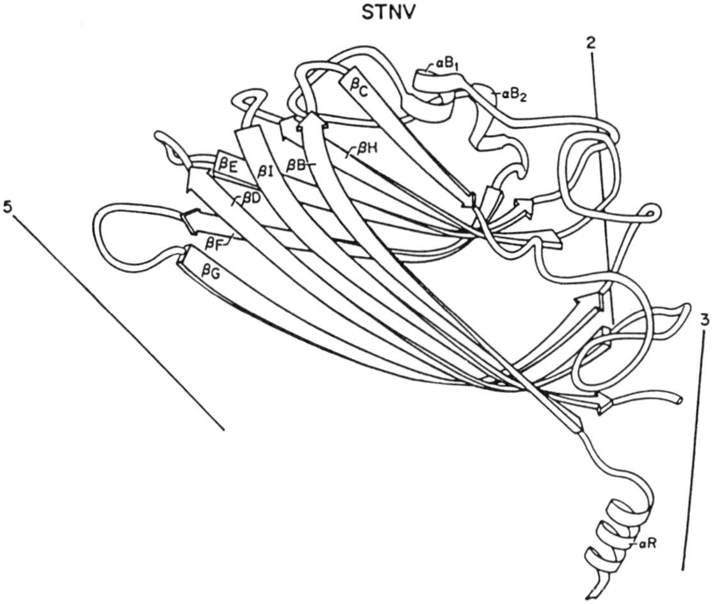

Figure 1. Schematic drawing of the STNV protein subunit. The view is chosen to simplify the comparison with the corresponding drawings of TBSV and SBMV.

tween the CHEF sheet and the EF loop (Fig. 6). Near the ends of the sheets and in the connecting loops the structure is rich in polar residues and several internal hydrogen bonds are found there. Four internal salt bridges are found on the surface of the subunit, between Arg 60 and Asp 63, Arg 86 and Asp 130, Arg 28 and Glu 189, and between His 73 and

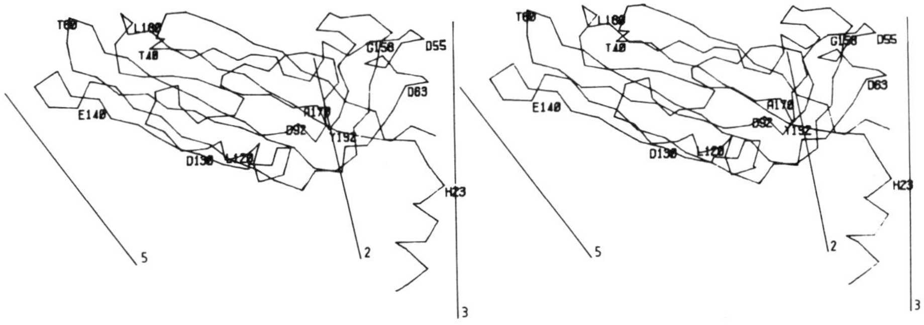

Figure 2. Stereo drawing of the α-carbons of STNV. View perpendicular to the threefold axis.

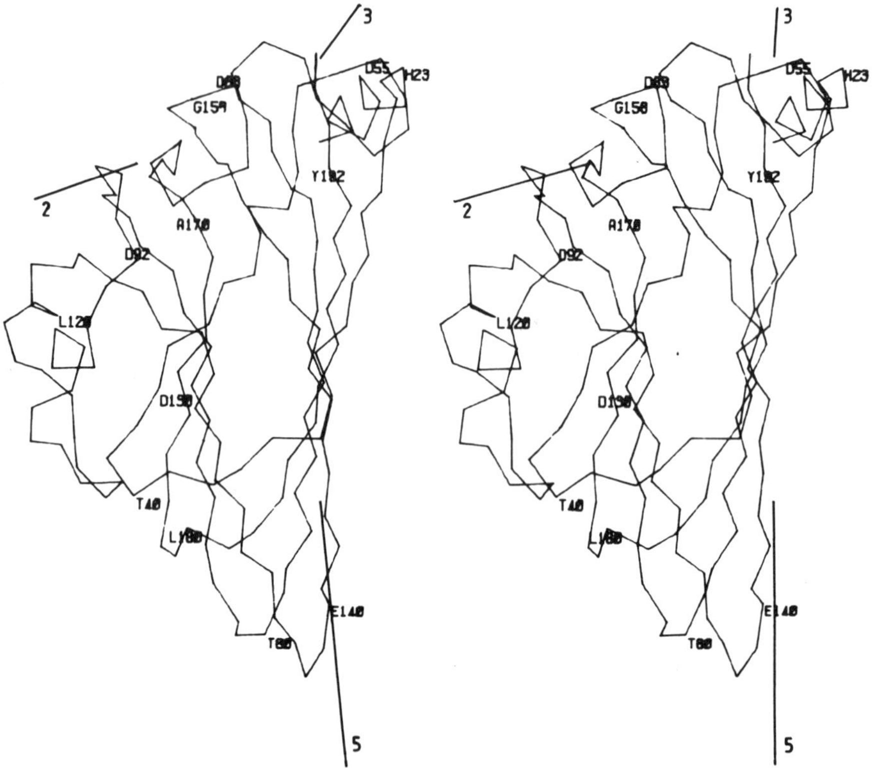

Figure 3. Stereo drawing of the α-carbons of STNV. View along threefold axis toward particle center.

Asp 144 (Fig. 7) The first one of these stabilizes the long CD loop. The other three bridge neighboring strands in the β-sheets. They are all between strands in the antiparallel hairpin that forms the β-roll (Fig. 4).

The two β-sheets contain four β-bulges. At the inside of the protein shell the D and G strands make bulges (residues Val 68, Ser 69, Pro 150,

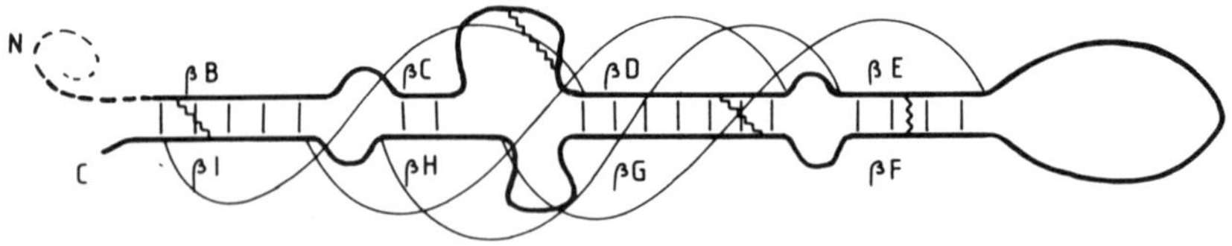

Figure 4. The topology of the STNV protein subunit. An antiparallel beta hairpin forms a right-handed helix. The positions of the four intrasubunit salt links are indicated with wavy lines.

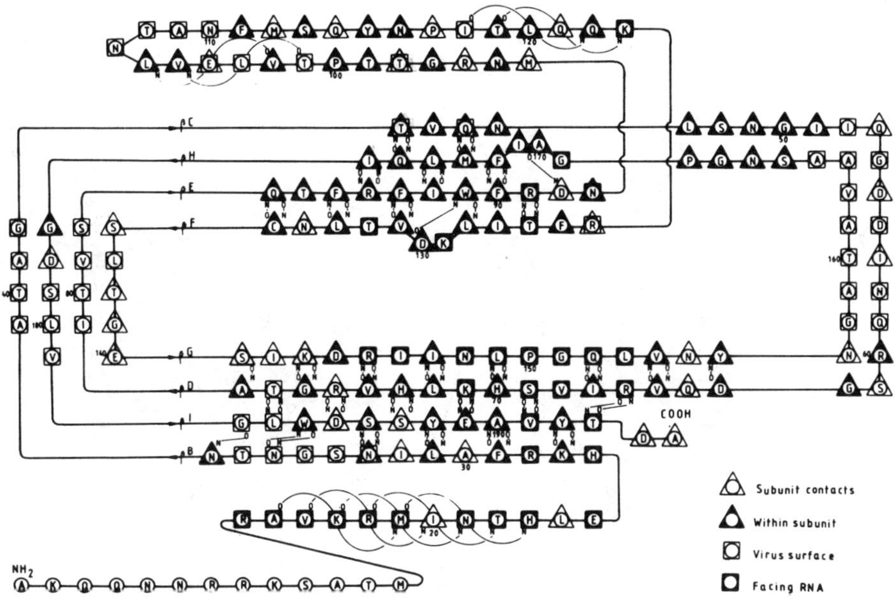

Figure 5. The secondary structure of the STNV protein subunit. This drawing also shows the type of interactions in which the different side chains are involved. Slightly modified from Liljas et al. (1982).

and Gly 151; Fig. 2). Another bulge is found near the inner surface of the CHEF sheet (Leu 128 and Lys 129). In all these cases the main chain protrudes slightly into the interior of the particle. A fourth exception from perfect β-sheet hydrogen bonding is found on strand B, where Ser 34 and Gly 35 bulge out. These residues are on the surface of the virus particle. The three bulges on the BIDG sheet are at the outermost (34,35) and innermost (68,69; 150,151) corner of the sheet. They might be structurally necessary to increase the twist of the sheet in these corners to prevent them from protruding excessively.

3.4. Subunit Packing in the Virus

The packing of identical protein subunits in a $T = 1$ icosahedral arrangement involves three types of subunit interactions, one for each of the two-, three- and five-fold axes of the icosahedron. A general view of how the STNV subunits pack around these axes is shown in Figure 8. The greatest area of contact is between subunits related by the fivefold axis, principally

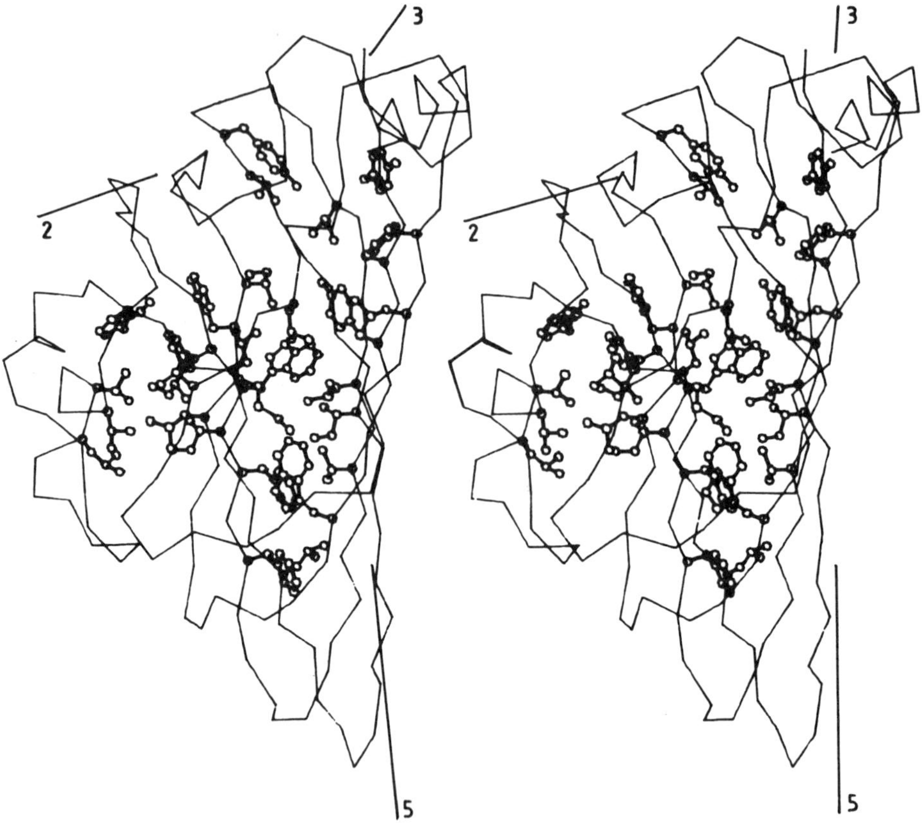

Figure 6. Stereo drawing showing several hydrophobic side chains in the interior of the STNV subunit.

where the EF loop and a small part of the CHEF sheet pack against the BIDG sheet of another subunit. Of the short connecting loops near the fivefold axis only the FG loop makes subunit–subunit contact. The twofold contact is mainly due to another part of the EF loop and part of the GH loop. The threefold packing consists of the N-terminal helix, the C terminus, and the CD loop.

The thickness of the protein shell is roughly 20 Å, but at the fivefold axis it is less, about 10 Å. The protein shell has no obvious holes or channels. The two primary heavy atom sites for the two derivatives used in the X-ray work, however, were Met 19 ($PtCl_4^{2-}$) and His 23 ($PtCl_4^{2-}$

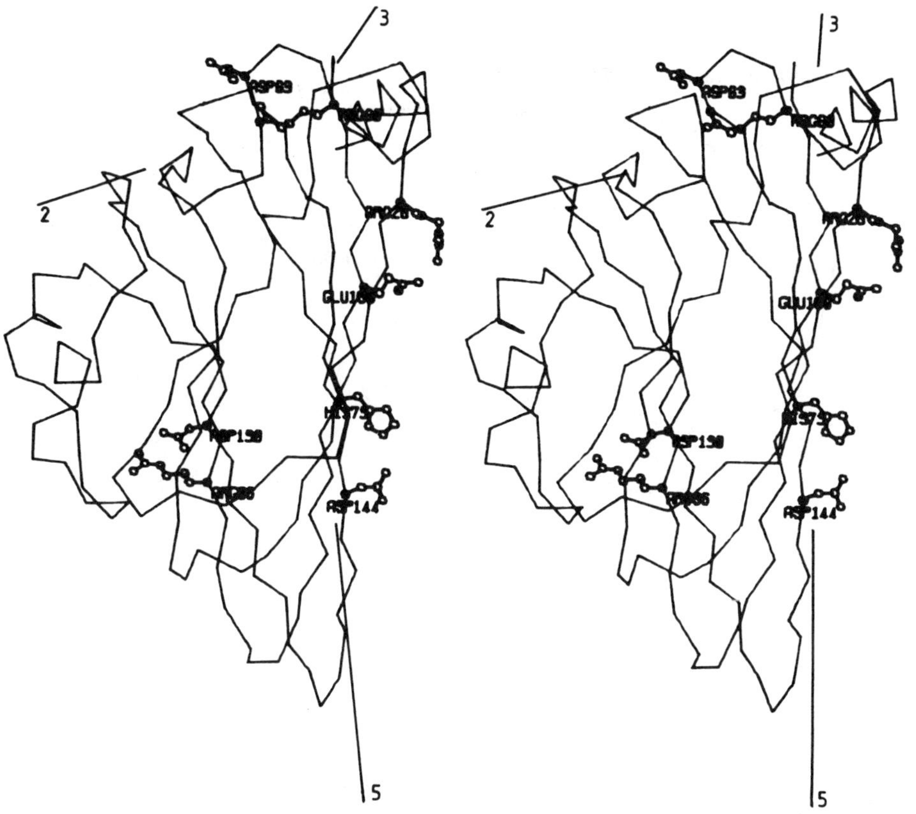

Figure 7. Stereo drawing of the four internal salt bridges in STNV protein subunit.

and I_2). Both these residues are found inside the shell on the N-terminal helix, 30 and 25 Å, respectively, from the virus surface.

3.4.1. Threefold Contacts

The N-terminal helix of the STNV subunit, which protrudes into the interior of the particle, interacts with the symmetry-related helices around the threefold axis. The contact is via hydrophobic side chains (Val 16, Ile 20, Leu 24; see Fig. 9). The hydrophilic side chains, of which several are basic, are exposed at the surface of this helix bundle, that is, toward the RNA. The helices are tilted relative to the threefold axis and constitute

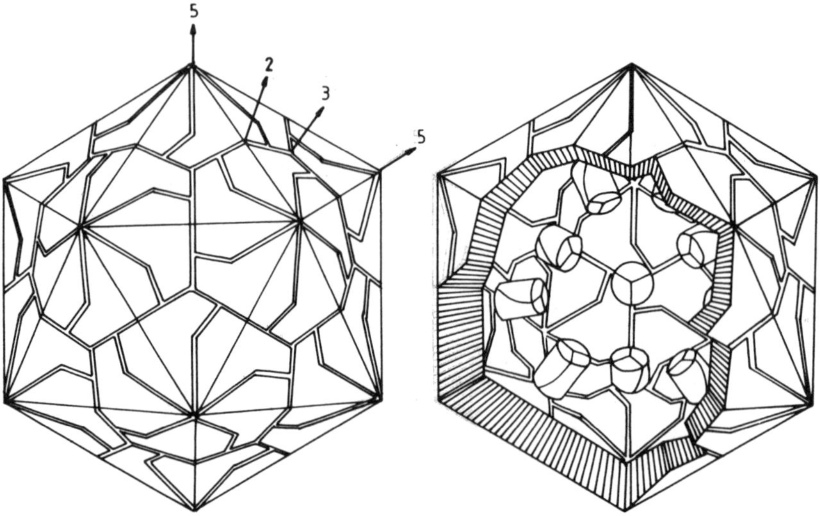

Figure 8. Drawing of the subunit packing in STNV.

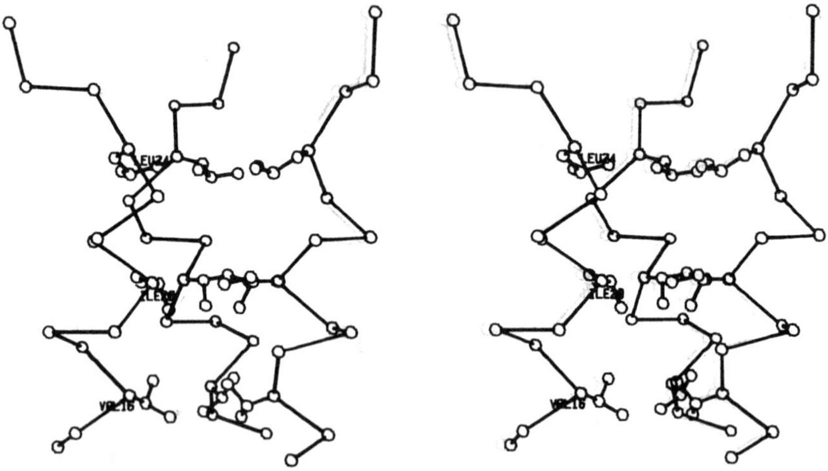

Figure 9. Stereo drawing of the main chain of threefold-related N-terminal helices, and the three different hydrophobic side chains involved in their packing.

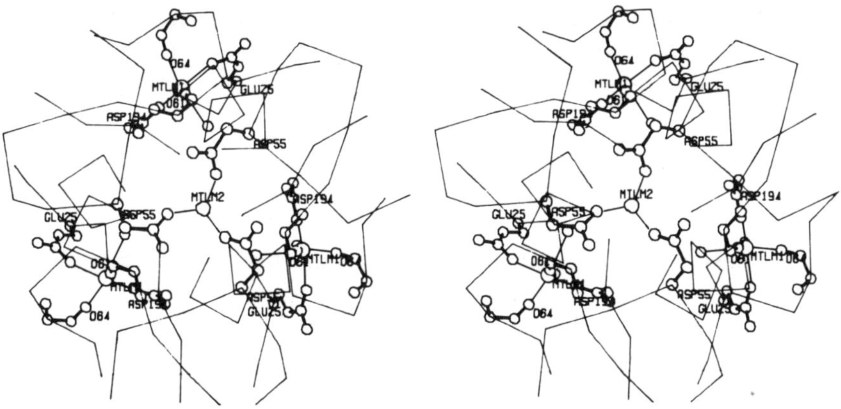

Figure 10. Stereo drawing of the interaction between threefold-related subunits. The main chain is indicated by thin lines connecting the α-carbons. The three acidic side chains involved in liganding the metal ions are shown as well as the main chain atoms of residues 61 and 64, whose carbonyl oxygens bind to one of the metal ions (MTLM1). The two water molecules bound to this metal ion are also shown.

a short right-handed superhelix. Packing of helices from different subunits around a threefold axis is also found in SBMV (Abad-Zapatero et al., 1980) and in influenza virus hemagglutinin (Wilson et al., 1981).

The remainder of the threefold contact, in the shell domain of the protein, is hydrophilic in nature. It is stabilized by two metal ions. Electron density, which most likely is due to a Ca^{2+} ion, is found on the threefold axis at 80.5 Å radius (Fig. 10). It partly neutralizes the negative charges of the three Asp 55 residues, which come close to the threefold axis. This Ca^{2+} ion is accessible from the outside of the virus particle. Another Ca^{2+} ion is in a general position between pairs of threefold-related subunits. It is found at a 6.8 Å distance from the threefold axis at a 72 Å particle radius. It is liganded by Glu 25 from one subunit, two water molecules, Asp 194, and carbonyl oxygens O-61 and O-64 from another subunit. The six ligands are arranged in a slightly distorted octahedral configuration.

3.4.2. Twofold Contacts

The twofold contact involves both hydrophobic and hydrophilic interactions. The side chains of Met 94 from the two subunits pack against each other around the twofold axis (Fig. 11). A salt link is formed between Arg 96 from one subunit and Glu 104 and Asp 92 from the twofold-related one. In addition there are several side chain hydrogen bonds between the subunits.

3.4.3. Fivefold Contacts

The extensive fivefold contact involves some hydrophobic interaction (e.g., Ile 32, Ile 142 of the BIDG sheet of one subunit and Met 112 from the EF loop of another subunit), but mainly polar interactions with several hydrogen bonds and two salt linkages. These are between Arg 75 of one and Asp 178 of the neighboring subunit, and (close to the fivefold axis) between Glu 140 of one and Lys 143 of the other subunit (Fig. 12). In addition, a Ca^{2+} ion is found on the fivefold axis, at 90 Å radius. This metal is coordinated by five surrounding carbonyl oxygens of residue 138, and by two water molecules lying on the fivefold axis, one inside and one outside the metal ion.

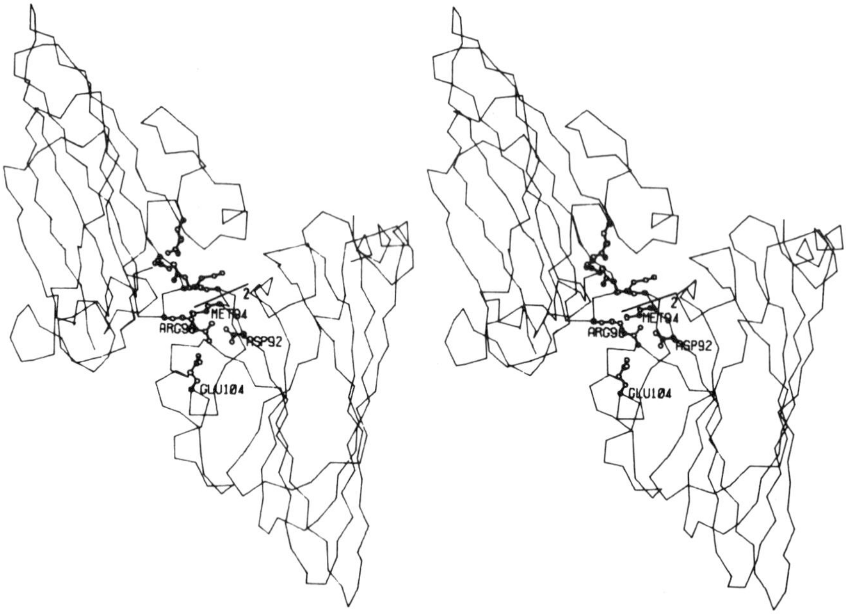

Figure 11. Stereo drawing of the α-carbons of two twofold-related subunits, and a few of the side chains utilized in subunit–subunit interactions.

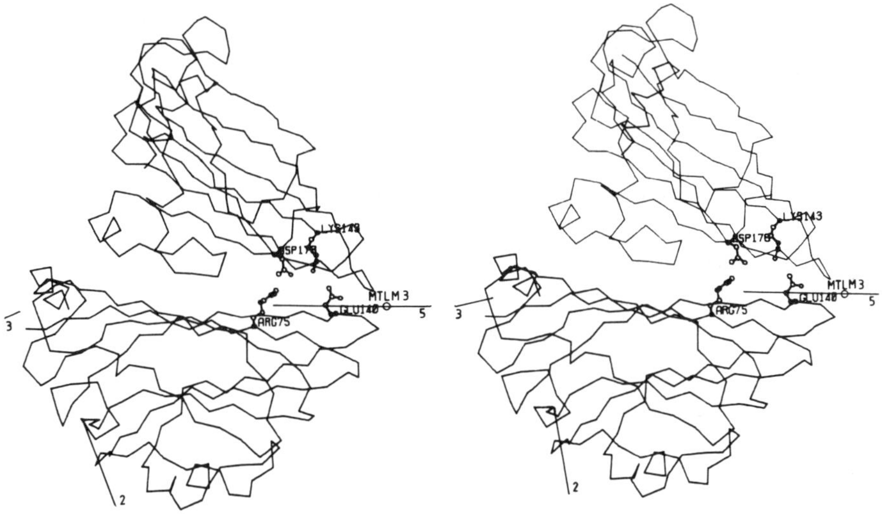

Figure 12. Stereo drawing of α-carbons of two fivefold-related subunits. The metal ion on the fivefold axis and the side chains of those amino acids that are involved in salt links are included.

3.5. Protein-Nucleic Acid Interactions

Assuming a degree of hydration similar to that of tRNA, the RNA of STNV will occupy almost the entire interior of the virus particle. At least part of the protein is, therefore, in direct contact with the nucleic acid. However, the protein subunits do not seem to impose icosahedral symmetry on the RNA, since no density is seen in the RNA region of the icosahedrally averaged map. Instead the parts of the protein that most likely interact with the RNA seem to be disordered. This is also the case in TBSV and SBMV.

The inside of the STNV protein shell confronting the RNA contains several basic residues. On the central part of the subunit six Lys and Arg residues are exposed to the interior. Of these Lys 123, Lys 129, Arg 66, Arg 91, and Arg 145 are not completely ordered, so that only part of the side chains are seen in the map. In the case of Lys 71 the whole side chain is seen, probably because this residue is fixed by additional interactions with neighboring side chains.

On the visible part of the N-terminus three basic residues are found: Arg 14, Lys 17, and Arg 18. For all of these residues no side chain density is observed, whereas the side chains of Val 16 and Met 19, for example, are visible in the map. Four positively charged side chains are located on the first 11 residues of the sequence. Together with the N-terminal amino group this sums to $14 \times 60 = 840$ exposed positive charges on the inside of the protein shell. This should be compared with the total number of negatively charged phosphates on the STNV RNA, which is 1239.

3.6. Correlation with Other Structural Information

The present STNV model confirms the structural results obtained from solution studies with neutrons and X-rays. With both methods the radial distribution of protein and nucleic acid was reasonably well determined. However, no clear evidence was found from scattering studies for the penetration of the N-terminal part of the protein into the nucleic acid (Chauvin et al., 1977; Sjöberg et al., 1977).

The relative inertness of the STNV particle (Unge and Strandberg, 1979) is consistent with the structure. No reactive side chains are accessible to enzymes or bulky chemical reagents approaching from outside of the virus. It is striking that the only useful heavy atom modifications have been obtained on the N-terminal helix inside the protein coat. In these two cases relatively small reagents (I_2 and $PtCl_4^{2-}$) have been used. The iodine derivative was prepared in solution, and a high degree of substitution was achieved after 1 hr.

The essential role of the calcium ions in maintaining the native structure is confirmed by experiments in this laboratory. The number of calcium ions per virus particle has been determined to be 70 ± 15 by chemical analyses (Liljas et al., 1982), and 92 by the structural investigation. The chemical analyses also indicated a content of about 30 magnesium ions per particle. Dialysis of a virus solution against EDTA at pH 8.5 removes the divalent cations and results in a swelling of the particle as determined with analytical ultracentrifugation. The degree of swelling is pH dependent. At pH 7.5 the metal-free virus has a sedimentation coefficient corresponding to a swelling of 10% in diameter, but at pH 4.5 it nearly retains its original S value. At pH 7.5 the swollen virus is sensitive to trypsin treatment, which is not the case for native virus. Also, the nucleic acid becomes accessible to nuclease attack (T. Unge, personal communication). A similar swelling has been observed for SBMV (Hsu et al., 1977) and TBSV, for which a low resolution crystal structure of a swollen state of the virus has been determined (Robinson and Harrison, 1982), as well as for other spherical plant viruses.

3.7. The Relation of the Structure to the Activity of the Virus

The role of the complete virion in the life cycle of a virus is to protect the genetic information in the nucleic acid and to release it by some means when it is able to utilize the replicating and translating machinery of the host cell. The STNV RNA is well protected against degradation inside the very stable protein shell of the native virus. In light of the known structure it is reasonable to suggest that the disassembly process involves removal of the bound Ca^{2+} ions, which are necessary for the stability of the particle at neutral pH. A removal of, for example, the Ca^{2+} ions on and near the threefold axes would cause repulsion of carboxyl groups. Conceivably, the virus might then disassemble as a result of proteolytic degradation of exposed cleavage points on the protein subunits in the swollen virus.

The assembly process of the rod-shaped tobacco mosaic virus (TMV) is well known. The TMV protein subunit can bind to a specific initiation site on the corresponding RNA molecule, and the virus can elongate from this starting point. The protein subunits can in this case unwind the RNA, which probably exhibits some secondary structure when in free form. The assembly of STNV or other spherical viruses is, however, not understood. In these cases the nucleic acid does not reflect the symmetry of the protein shell and the protein subunits do not appear to unwind the RNA. For some of these viruses the protein subunits can assemble into a shell in the absence of RNA. Whether this is the case for STNV is not known. The structure of the protein shell does not, however, show any obvious requirements of nucleic acid for its assembly or stability.

4. SIMILARITIES BETWEEN SPHERICAL PLANT VIRUSES

The topology of the β-barrel in STNV is also found in the two other plant viruses for which a detailed structure is known, TBSV (Harrison et al., 1978) and SBMV (Abad-Zapatero et al., 1980). It is also found in one domain of influenza virus hemagglutinin (Wilson et al., 1981) and in one domain of the catabolite-activating protein (CAP) (McKay and Steitz, 1981). One part of the concanavalin A β-barrel also has the same β-roll topology. Whereas the structural similarity between STNV and the latter three examples seems to be restricted to the topology, the three plant viruses also exhibit great similarities in their tertiary structure, although SBMV and TBSV, which both are $T = 3$ viruses, are clearly more similar to each other than they are to STNV. The β-strands are of similar lengths and conformation. The four short connecting loops near the fivefold axis in STNV correspond to short loops in TBSV and (with one exception) SBMV, but the conformation is different. Of the three long loops, the EF loop is much shorter in TBSV and SBMV than in STNV, and all three loops have completely different conformations. One similarity, however, is in the EF loop where STNV has a short helix (residues 118–123) just before strand F, in a position topologically similar to the radial helix of TBSV and SBMV. The structure of these loops also differ considerably between TBSV and SBMV.

The STNV structure can be superimposed on TBSV and SBMV with an rms deviation of about 3.5 Å for most of the α-carbons in the β-strands (Rossmann et al., 1983). Only very few amino acids in structurally identical positions, however, are found to be identical.

The TBSV and SBMV virus particles are also stabilized by divalent cations in a way similar to STNV. In both viruses acidic side chains from quasi-threefold related subunits are liganded by Ca^{2+} ions. In TBSV two metal ions are found in general positions between pairs of quasi-threefold related subunits (Harrison, 1980). In SBMV a Ca^{2+} is found on the quasi-threefold axis, bound by three Glu residues. However, the ligands are not in the same position relative to the β-barrels of the three viruses. Asp 55 in STNV, which binds the metal on the threefold axis, is found in the CD loop, whereas Glu 194, which in SBMV ligands the Ca^{2+} ion on the quasi-threefold axis, is found in the EF loop. This is due to a considerable difference in packing of the STNV subunits compared to the more structurally similar subunits of TBSV and SBMV (Liljas et al., 1982).

5. INFERENCE OF AN RNA STRUCTURE FROM A SPHERICAL VIRUS

Originally, a principal stimulant for structural investigations of viruses was to obtain information on the conformation of the nucleic acid. Because little information has emerged from these studies, it is worth discussing the likelihood of obtaining RNA or DNA structure from a spherical virus.

Even if the free RNA molecule is stabilized by base pairing and forms stems or groups of stems (Kaempfer et al., 1982; Fiers et al., 1976), it cannot have a compact tertiary structure unless the negatively charged phosphates are neutralized by positive charges. In the STNV, TBSV, and SBMV virus particles, the negative charges have mainly been neutralized by the positive charges of the coat protein. In other cases, for example, bacteriophage R17, the neutralization is at least partly achieved with spermidine molecules (Fukuma and Cohen, 1975). The RNA has been shown to be densely packed, and to have very restricted mobility in TBSV (Munowitz et al., 1980), SBMV (McCain et al., 1982), and cowpea chlorotic mottle virus (CCMV; Vriend et al., 1981).

Is the packing, then, of the RNA inside the protein shell specific, creating a unique RNA conformation, or nonspecific, resulting in a densely packed nucleic acid, but with more or less varying conformations in different virus particles? In the first case the RNA structure can be crystallographically determined provided the crystal packing is such that the nucleic acid obeys the crystal symmetry. This requires that the asymmetry of the RNA can impose an asymmetry on the entire virus particle that can be experienced by other particles in the lattice. Because all spherical viruses have icosahedral symmetry, the virus particles can pack in crystals utilizing some of the two- and threefold axes of the icosahedron as crystallographic axes. In the cases of TBSV and SBMV the space groups are such that one-twelfth and one-sixth, respectively, of the virus particle comprise the crystal asymmetric unit. If the RNA is completely asymmetric, this will result in a 12-fold or sixfold averaging of the RNA density, respectively, even if the noncrystallographic symmetry is not used in the structure determination. In the case of STNV, and also for one crystal form of CCMV (Heuss et al., 1981), the crystal asymmetric unit is built up of a whole virus particle, and the RNA structure could in principle be solved. However, it is doubtful that the crystal packing of STNV depends on an asymmetry of the particle, because no significant deviation from icosahedral symmetry can be seen in the parts of the protein shell exposed on the viral surface.

The TBSV, SBMV, and STNV protein subunits all have flexible, basic domains at their N termini, which interact with the nucleic acid, but the basic amino acid side chains on the inside of the ordered protein shell are disorderd at least in STNV. The 11 N-terminal residues of the STNV protein, which are invisible in the map, can extend to the innermost regions of the nucleic acid and enable dense packing of the RNA by neutralizing the negative charges. In TBSV and SBMV, which both have considerably more nucleic acid and a much larger radius, the disordered N-terminal strand is longer (roughly 75 and 40 amino acids, respectively, in C subunits and another 25 in A and B subunits). The icosahedral protein shell of these viruses is not well suited for specific interactions with the whole of an asymmetric nucleic acid molecule, while the flexible parts of the protein subunits seem to be well designed for nonspecific interaction with an asymmetric RNA molecule.

The closely related brome mosaic virus (Bancroft et al., 1969) and cowpea chlorotic mottle virus (Adolph and Butler, 1977) have both been reassembled with different nucleic acids. In the case of CCMV both TBSV and TMV RNA can be encapsidated. Therefore, the protein–nucleic acid interaction is probably nonspecific for these viruses. The CCMV (Rees and Short, 1982) and BMV (Ahlquist et al., 1981) protein subunits have recently been sequenced. They have a length (about 190 amino acids) and a distribution of positively and negatively charged side chains similar to that of STNV, including a basic N terminus. Although they seem to lack specificity, these viruses are classified as stabilized by protein–nucleic acid interactions by Kaper (1975). Therefore, among viruses with a tertiary structure likely similar to STNV, TBSV, and SBMV, there is a lack of nucleic acid specificity. These viruses could provide a description of the packing of an RNA molecule of appropriate size in a nonspecific way.

A different situation exists for the spherical but anisotropic bacteriophages, for example, MS2. These viruses have in their protein shell in addition to about 180 copies of the normal coat protein, a single protein subunit of a different kind, the A protein, with a molecular weight three times that of the coat protein. The A protein enters the bacterium together with the RNA (Kozak and Nathans, 1971), and interacts with the RNA in a specific way (Shiba and Suzuki; 1981). The coat protein of the very similar bacteriophage R17 has been shown to protect a specific sequence of 59 bases of its own RNA from nuclease attack (Bernardi and Spahr, 1972). Thus, crystallographic studies of these viruses might provide some insight into the interactions between viral proteins and nucleic acid provided the anisotropy of the particle is maintained by the crystal packing. MS2 crystals (Min Jou et al., 1979) analyzed in this laboratory diffract to only very low resolution. This disorder could be due to isotropic packing of the anisotropic particle. To overcome this problem and achieve a unique orientation of the A protein in the MS2 crystals, work is now in progress in Uppsala to utilize monoclonal antibodies directed against specific determinants on thc A protein, thereby enhancing the anisotropy of the MS2 particle.

Acknowledgment. We thank Drs. Alwyn Jones, Anders Liljas, and Torsten Unge for helpful discussions. This work was supported by the Swedish Natural Science Research Council (Grant no. 2142-112), the Magnus Bergvall Foundation, and the Knut and Alice Wallenberg Foundation.

REFERENCES

Abad-Zapatero, C., Abdel-Meguid, S. S., Johnson, J. E., Leslie, A. G. W., Rayment, I., Rossmann, M. G., Suck, D., and Tsukihara, T. (1980). *Nature (London)* **286,** 33–39.

Adolph, K. W., and Butler, P. J. G. (1977). *J. Mol. Biol.* **109,** 345–357.

Ahlquist, P., Luckow, V., and Kaesberg, P. (1981). *J. Mol. Biol.* **153,** 23–38.

Åkervall, K. et al. (1971). *Cold Spring Harbor Symp. Quant. Biol.* **36,** 469–488.

Bancroft, J. B., Hiebert, E., and Bracker, C. E. (1969). *Virology* **39,** 924–930.

Bernardi, A., and Spahr, P. F. (1972). *Proc. Natl. Acad. Sci. U.S.A.* **69,** 3033–3037.

Bricogne, G. (1974). *Acta Crystallogr. A* **30,** 395–405.

Bricogne, G. (1976). *Acta Crystallogr. A* **32,** 832–847.

Caspar, D. L., and Klug, A. (1962). *Cold Spring Harbor Symp. Quant. Biol* **27,** 1–24.

Chauvin, C., Jacrot, B., and Witz, J. (1977). *Virology* **83,** 479–481.

Crowfoot, D., and Schmidt, G. M. J. (1945). *Nature (London)* **155,** 504–505.

Fiers, W., Contreras, R., Haegeman, G., Iserentant, D., Merregaert, J., Min Jou, W., Molemans, F., Raeymaekers, A., Vandenberghe, A., Volckaert, G., and Ysebaert, M. (1976). *Nature (London)* **260,** 500–507.

Fukuma, I., and Cohen, S. S. (1975). *J. Virol.* **16,** 222–227.

Harrison, S. C. (1980). *Biophys. J.* **32,** 139–153.

Harrison, S. C., Olson, A. J., Shutt, C. E., Winkler, F. K., and Bricogne, G. (1978). *Nature (London)* **276,** 368–373.

Henriksson, D., Tanis, R. J., Tashian, R. E., and Nyman, P. O. (1981). *J. Mol. Biol.* **152,** 171–179.

Hermodson, M. A., Abad-Zapatero, C., Abdel-Meguid, S. S., Pundak, S., Rossmann, M. G., and Tremaine, J. A. (1982). *Virology* **119,** 133–149.

Heuss, K. L., Rao, J. K. M., and Argos, P. (1981). *J. Mol. Biol.* **146,** 635–640.

Höglund, S. (1968). *J. Gen. Virol.* **2,** 427–436.

Hsu, C. H., White, J. A., and Sehgal, O. P. (1977). *Virology* **81,** 471–475.

Jones, T. A. (1982). In *Computional Crystallography*, D. Sayre, Ed. Oxford University Press, New York. pp. 307–317.

Jones, T. A., and Liljas, L. (1983). *Acta Crystallogr.*, in press.

Kaempfer, R., van Emmelo, K., and Fiers, W. (1981). *Proc. Natl. Acad. Sci. U.S.A.* **78,** 1542–1546.

Kaper, J. M. (1975). *The Chemical Basis of Virus Structure, Dissociation and Assembly*. North Holland, Amsterdam.

Kassanis, B. (1962). *J. Gen. Microbiol.* **27,** 477–488.

Kassanis, B., and MacFarlane, I. (1968). *J. Gen. Virol.* **3,** 227–232.

Kassanis, B., and Nixon, H. L. (1961). *J. Gen. Microbiol.* **25,** 459–471.

Kassanis, B., and Woods, R. D. (1968). *J. Gen. Virol.* **2,** 395–398.

Kassanis, B., Vince, P. A., and Woods, R. D. (1970). *J. Gen. Virol.* **7,** 143–151.

Klug, A. (1971). *Cold Spring Harbor Symp. Quant. Biol.* **36,** 483–487.

Kozak, M., and Nathans, D. (1971). *Nature (London) New Biol.* **234,** 209–211.

Lentz, P. J., Jr., Strandberg, B., Unge, T., Vaara, I., Borell, A., and Petef, G. (1976). *Acta Crystallog. B* **32,** 2979–2983.

Liljas, L., Unge, T., Jones, T. A., Fridborg, K., Lövgren, S., Skoglund, U., and Strandberg, B. (1982). *J. Mol. Biol.* **159,** 93–108.

McCain, D. C., Virudachalam, R., Markley, J. L., Abdel-Meguid, S. S., and Rossmann, M. G. (1982). *Virology* **117,** 501–503.

McKay, D. B., and Steitz, T. A. (1981). *Nature (London)* **290,** 744–749.

Mossop, B., and Francki, R. I. B. (1979). *Virology* **94,** 243–253.

Min Jou, W., Raeymaekers, A., and Fiers, W. (1979). *Eur. J. Biochem.* **102,** 589–594.

Munowitz, M. G., Dobson, C. M., Griffin, R. G., and Harrison, S. C. (1980). *J. Mol. Biol.* **141,** 327–333.

Rees, M. W., and Short, M. N. (1982). *Virology* **119,** 500–503.

Robinson, I. K., and Harrison, S. C. (1982). *Nature (London)* **297,** 563–568.

Rossmann, M. G., and Blow, D. M. (1962). *Acta Crystallogr.* **15,** 24–31.

Rossmann, M. G., Abad-Zapatero, C., Murthy, M. R. N., Liljas, L., Jones, T. A., and Strandberg, B. (1983). *J. Mol. Biol.* **165,** 711–736.

Shiba, T., and Suzuki, Y. (1981). *Biochim. Biophys. Acta* **654,** 249–255.

Sjöberg, B. (1977). *Eur. J. Biochem.* **81,** 277–283.

Unge, T., and Strandberg, B. (1979). *Virology* **96,** 80–87.

Unge, T., Liljas, L., Strandberg, B., Vaara, I., Kannan, K. K., Fridborg, K., Nordman, C. E., and Lentz, P. J., Jr. (1980). *Nature* (*London*) **285,** 373–377.

Wilson, I. A., Skehel, J. J., and Wiley, D. C. (1981). *Nature* (*London*) **289,** 366–373.

Vriend, G., Hemminga, M. A., Verduin, B. J. M., de Wit, J. L., and Schaafsma, T. J. (1981). *FEBS Lett.* **134,** 167–171.

Ysebaert, M., van Emmelo, J., and Fiers, W. (1980). *J. Mol. Biol.* **143,** 273–287.

4

The Structure Determination of Tobacco Mosaic Virus

KENNETH C. HOLMES
Max-Planck-Institut für Medizinische Forschung
Heidelberg, Federal Republic of Germany

CONTENTS

1. INTRODUCTION

Tobacco mosaic virus (TMV), a common plant virus with a wide host range, has been extensively studied by techniques from the differing fields of biology, chemistry, and physics. The singular historical position of TMV may account for the magnitude of this research. Ivanovsky (1892) first discovered a filterable infectious agent from the sap of plants infected with TMV, and Beijerink (1898) first realized that this agent was something distinct from bacteria and called it a "contagious living fluid." Stanley (1935) was able to isolate the infectious agent from plant sap and about the same time Bawden and Pirie produced liquid crystalline samples of the virus which were examined by X-ray diffraction methods by Bernal and Fankuchen (Bawden et al., 1936). TMV was, therefore, the first virus discovered, and later the first material which could be considered living to be investigated and characterized by physicochemical methods. Bawden and Pirie showed that TMV, in addition to protein, contained a small percentage of RNA.

In 1946 Stanley was awarded part of the Nobel prize in chemistry for the purification of TMV. Stanley also prepared crystals of the virus. Unfortunately, these crystals proved to be paracrystals and true single crystals of the virus have yet to be demonstrated. However, the protein component can be isolated (Schramm, 1947) and was later crystallized as double disks (Macleod et al., 1963; Finch et al., 1966; see also Chapter 5).

TMV particles are rodlike, 300 nm long, and 18 nm in diameter (Klug and Caspar, 1960). The particles consist of 94% protein and 6% nucleic acid. A particle contains 2140 identical protein subunits, each of molecular weight 17,420 (158 residues) (Anderer, 1963), arranged on a helix of pitch 2.3 nm with $16\frac{1}{3}$ subunits per turn. Winding through this helix is a single strand of RNA 6400 nucleotides in length. TMV has a central hole of diameter 4.0 nm. Concentrated solutions of TMV show an unusual property, namely, they form spontaneously birefringent tactoids that may be easily persuaded to fuse. Thus, a capillary tube may be obtained containing essentially one highly ordered, oriented gel (see Gregory and Holmes, 1965, for a description of the phenomenon). This property has made it possible to obtain X-ray fiber diagrams of unsurpassed quality. These may be analyzed by an extension of the methods of protein crystallography to fiber diagrams.

The pioneering work of Bernal and Fankuchen (1941) on oriented gels of TMV showed that the rodlike virus particles formed a highly regular two-dimensional hexagonal array in which the axes of the virus particles were parallel. The ordered virus gel was paracrystalline rather than truly crystalline since the arrangement of virus particles was not regular along the direction of the orientation axis. Moreover the orientation of each virus particle about its axis was random. The diffraction pattern obtained from these gels (Fig. 1) consisted of a set of layer lines at right angles to the particle axis with a spacing of 6.9 nm, thus showing that

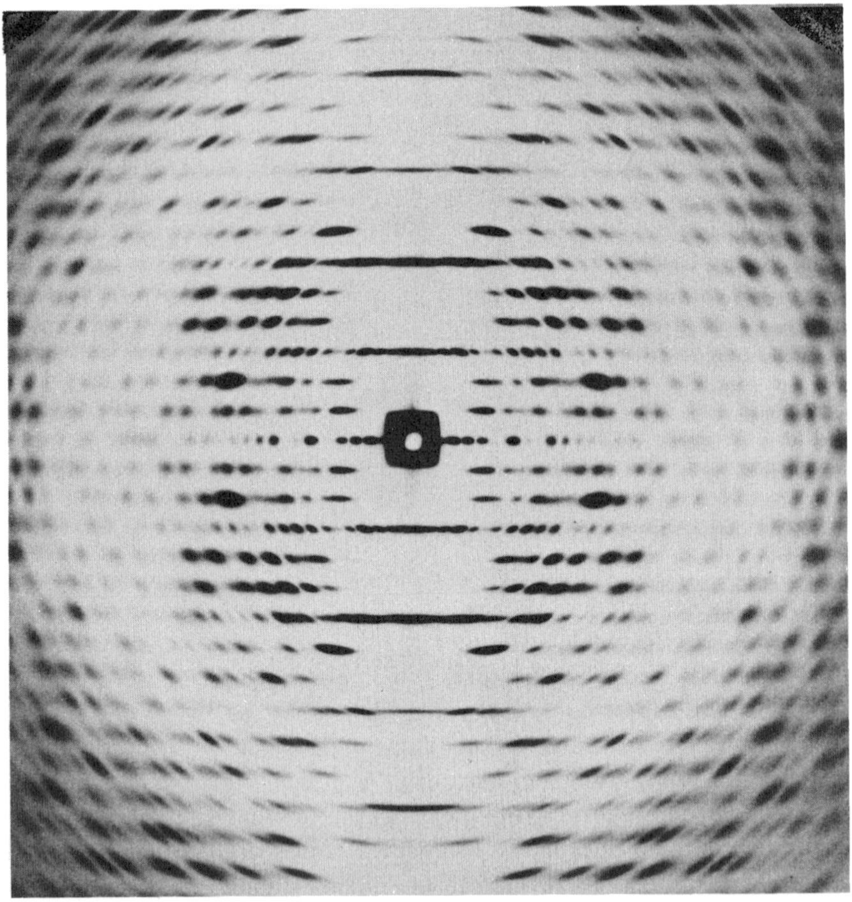

Figure 1. An X-ray diffraction fiber diagram of an oriented gel of TMV (Holmes et al., 1975) obtained with the apparatus depicted in Figure 5 and an Elliott (London) GX6 fine-focus rotating anode tube (λ = 0.154 nm). Note the layer lines corresponding to an axial spacing of 6.9 nm. The third-order near-meridional layer lines with a repeat distance of 2.3 nm arise from the genetic helix of the virus. The ring of maxima particularly clear on the second and seventh layer lines have a Bragg spacing of about 1.0 nm. This is the limit of the region where only one order of Bessel function contributes. The weak meridional maxima come from the "stacked disk" form of the protein which is present as a minor impurity.

there was a regular repeat along the axis. Bernal and Fankuchen were able to show that the diffraction pattern consisted of two distinct regions: on the equator at spacings of less than 10 nm one or two reflexions occurred which arose from the interparticle ordering (the positions of these reflexions varied with the concentration of virus); all other parts of the diffraction pattern were virtually unaffected by changes in concentration, and could therefore arise only from the intramolecular structure of each particle. Bernal and Fankuchen further suggested that each TMV particle was more like a microcrystalline array of protein molecules than like a giant molecule, a concept that has proved to be of considerable significance.

Bernal and Fankuchen recognized correctly the similarity of the diffraction pattern given by TMV gels to that given by a collection of crystallites oriented randomly about a common long axis. Such diffraction patterns arise from fibrous specimens such as wool and are referred to as fiber diagrams. A fiber diagram can conceptually be thought of as arising by rotating a single crystallite about its long axis during an X-ray exposure. Thus, each Bragg reflexion is spread out into a circle in reciprocal space. In general, the recorded diffraction pattern consists of a section through reciprocal space, that is a section through the cylindrically averaged square of the transform of the individual scattering unit. If the scattering unit is a crystallite then the transform is constrained to Bragg reflections which show up as sharp maxima arranged in layer lines. If, however, the sample should not be crystalline then the scattering is still constrained to layer lines but the intensity distribution along each layer line is *continuous*. This turns out to be the case for TMV (Fig. 1). Bernal and Fankuchen attempted to interpret the TMV fiber diagram as arising from an assembly of crystallites; each virus particle was taken to be a microcrystal having a three-dimensional lattice. They recognized that the small diameter of the virus would lead to an imperfect expression of the Bragg relationship and indeed that there are many more "reflections" than could possibly be accounted for by a particle having the size and shape that TMV was known to have. They attempted to index the TMV fiber diffraction pattern in terms of a rhombohedral unit cell having a threefold screw axis but allowing extra half-order Bragg reflexions to take account of the small crystallite size. The connection between the postulate of a threefold screw axis and the helical form given by later work is that the trigonal screw axis generates equivalent planes in differing orientations at heights of $\frac{1}{3}c$, $\frac{2}{3}c$, and so on. This situation is also approximated at low resolution by a helical structure of small pitch which repeats in three turns.

Intact TMV does not have crystal symmetry as was suggested by Bernal and Fankuchen but has rather the symmetry of a helix. Thus, it was necessary to wait for the theoretical elucidation of the structure factor of a helical array of subunits (Cochran, et al., 1952) before it was possible to proceed with the analysis of the diffraction of TMV gels. The form of the structure factor from a helix shows that each layer line is created by the addition of Bessel functions (one for each atom in the repeating unit) and the orders of the Bessel functions which may occur are determined by the symmetry of the helix (helix selection rule). By analyzing the reciprocal space radial positions of the Bessel function maxima on various layer lines, and noting the presence of the pseudo-repeat every three turns, Watson (1954) was able to show that the structure of TMV was helical with $3N + 1$ subunits repeating in three turns every 6.9 nm. Watson's best value for N was 10.

2. THEORY OF DIFFRACTION FROM ORIENTED GELS OF HELICAL ASSEMBLIES

The following theory follows the development of Klug et al. (1958). Because TMV particles are helical the structure factor has the form shown in Eq. 1:

$$F\left(R, \psi, \frac{l}{c}\right) = \sum_n \sum_j f_j J_n(2\pi R r_j) \cdot \exp i(-n\phi_j + 2\pi l z_j/c) \exp in(\psi + \pi/2) \quad (1)$$

where R, ψ, l/c are the reciprocal space cylindrical coordinates and r_j, ϕ_j z_j are the real space cylindrical coordinates of the jth atom of the repeating subunit. We equate complex number G_{nl} to that part of the structure factor not dependent on ψ,

$$G_{nl}(r) = \sum_j f_j J_n(2\pi R r_j) \exp i(-n\phi_j + 2\pi l z_j/c) \quad (1a)$$

so that

$$F(R, \psi, l/c) = \sum_n G_{nl}(R) \exp in(\psi + \pi/2) \quad (1b)$$

On averaging the square of the modulus of the structure factor over ψ, the observed intensity is

$$I = \langle FF^* \rangle_\psi = \sum_n G_{nl}(R)^2 \tag{2}$$

on any given layer-line l.

If we express each G_{nl} in terms of its real and imaginary parts A_{nl} and B_{nl} we obtain

$$I = \sum_n A_{nl}^2 + \sum_n B_{nl}^2 \tag{2a}$$

If we consider the simplest case of a single atom helix, then in place of each G_{nl} we would have a Bessel function,

$$I = \langle FF^* \rangle_\psi = \sum_n J_n^2\,(2\pi Rr) \tag{3}$$

Thus, the cylindrically averaged squared structure factor is a sum of Bessel functions. Because of the helical symmetry of TMV not every Bessel function can occur on every layer line. The permitted Bessel functions of order n on the layer line l are given by the helix selection rule:

$$l = tn + um \tag{4}$$

where $t = 3$ and $u = 49$ for TMV. m is a positive or negative integer or zero. Because the value of a Bessel function is essentially zero until the value of the argument is equal to or greater than the order, diffraction from atoms at a large radius will occur nearer the center of the diagram than will diffraction from atoms at a small radius. Furthermore, diffraction from low-order Bessel functions will occur at a small reciprocal space radius. Because the maximum diameter of TMV is about 9 nm one finds that to a resolution of about 1.0 nm^{-1} essentially only one Bessel function can contribute to the diffraction pattern. Thus, to this resolution no information has been lost by cylindrical averaging, for we may equate the modulus of the lowest order G_{nl} with the square root of the observed intensity:

$$\sqrt{I} = |F| = |G_{nl}| \tag{5}$$

where n takes the lowest order values. If it is possible to recover the phases of the G_{nl} then knowledge of the helical symmetry provides the ψ variation and by use of Eq. 1b, it is possible to reconstruct the three-dimensional structure factor. By a Fourier inversion of the structure factor the three-dimensional electron density could then be obtained. In fact one proceeds without explicitly calculating the three-dimensional structure factor because the terms (g_{nl}) in the series expansion of the electron density ρ in ϕ and z may be directly calculated from the corresponding G_{nl} by a Fourier–Bessel inversion:

$$g_{nl}(r) = \int_{R=0}^{\infty} G_{nl}(R)\, J_n(2\pi Rr)\, 2\pi R dR \tag{6}$$

$$\rho(r, \phi, z) = \sum_n \sum_l g_{nl} \exp i(n\phi - 2\pi lz/c) \tag{7}$$

The problem, therefore, reduces to finding the phases of the G_{nl} and this may be done by the method of isomorphous replacement. The contribution of a single (heavy) atom to the term G_{nl} is

$$J_n(2\pi Rr_h) \exp i(-n\phi_h + 2\pi lz_h/c) \tag{8}$$

where the angular expression $(-n\phi_h + 2\pi lz_h/c)$ is constant for a given Bessel function term on a particular layer line (Φl). If the coordinates of the heavy atom are known, then the heavy atom structure factor $J_h\,(2\pi Rr_j)$ is calculable. The moduli of the virus structure factor (G^v) and the virus heavy-atom structure factor (G^{vh}) are experimentally available (Eq. 5) at low resolution. Therefore, by an application of the Harker construction (1956) it is possible to calculate the real and imaginary parts of the G_{nl}. The essential problem therefore, in common with normal protein crystallography is to determine the coordinates of the heavy atoms. For the case of TMV, because of the low resolving power and high symmetry, this problem is particularly tiresome. In the end six heavy atoms were used to provide enough cross-checks. However, the availability of this large number of derivatives eventually made it possible to separate overlapping G_{nl} terms so that the resolution could be extended to 0.4 nm.

3. LOW RESOLUTION ANALYSIS OF THE X-RAY FIBER PATTERN

Franklin (1955) attempted to establish the helical parameters of TMV by calculating a cylindrically averaged Patterson map. Franklin suggested that there were 37 subunits in 3 turns ($N = 12$).

Franklin and Klug (1955) noticed that the third layer line at low angles is composed primarily of a rapidly fluctuating series of intensity maxima which are subjected to a slow spatial modulation. This was shown to arise from a helical groove following the flat genetic helix of the virus which arises from the tapering of the subunits toward the outside.

Further progress necessitated the use of a systematic method of analysis. Emboldened, therefore, by the success with hemoglobin (Green et al., 1954) Franklin and Caspar independently started to employ the method of isomorphous replacement. It was quickly established that the virus had 49 subunits in three turns ($N = 16$) (Franklin and Holmes, 1958), that the virus was hollow, and that the nucleic acid was located at a radius of 4.0 nm (Franklin, 1956; Caspar, 1956). These results were obtained by analysis of the low resolution part of the zero layer line ($<1.0\ \text{nm}^{-1}$) where the data are effectively the diffraction pattern from a centrosymmetric structure (the scattering arises purely from J_0 Bessel functions). The radial density distribution can be obtained by a Fourier–Bessel inversion of the amplitudes derived from the diffracted intensity (Figs. 2 and 3). The signs of the peaks on the inner part of the zero layer line were determined independently by Caspar (1955) and by Franklin and Holmes (1958).

With the help of two heavy atom derivatives, methyl mercury (Franklin and Holmes, 1958) and osmic acid, Franklin proceeded to analyze the third layer line in order to arrive at a low resolution projection down the flat genetic helix of the virus. This showed the inclination of the subunit from the proximal to distal end which gives the end of the virus a recognizable polarity in electron micrograph images. The trace of the RNA chain could also be seen. This phase of the work (see Klug and Caspar, 1960), which is summarized in Caspar's drawing (1963) (Fig. 4), was interrupted by the premature death of Rosalind Franklin.

With the establishment of Klug's group in Cambridge, England, in 1962 the necessary technical milieu became available to continue and extend Franklin's earlier work. In an unpublished study Klug was able to demonstrate that the RNA was single-stranded by comparing the virus diffraction pattern with that given by the helical form of the repolymerized protein *without* RNA. Changes were visible only on the near meridional third order layer lines, which implies that the RNA must follow the track of the single start helix which gives rise to these layer lines. Klug and Holmes continued the structural analysis of the oriented gels by means of the method of isomorphous replacement. Six independent heavy atom

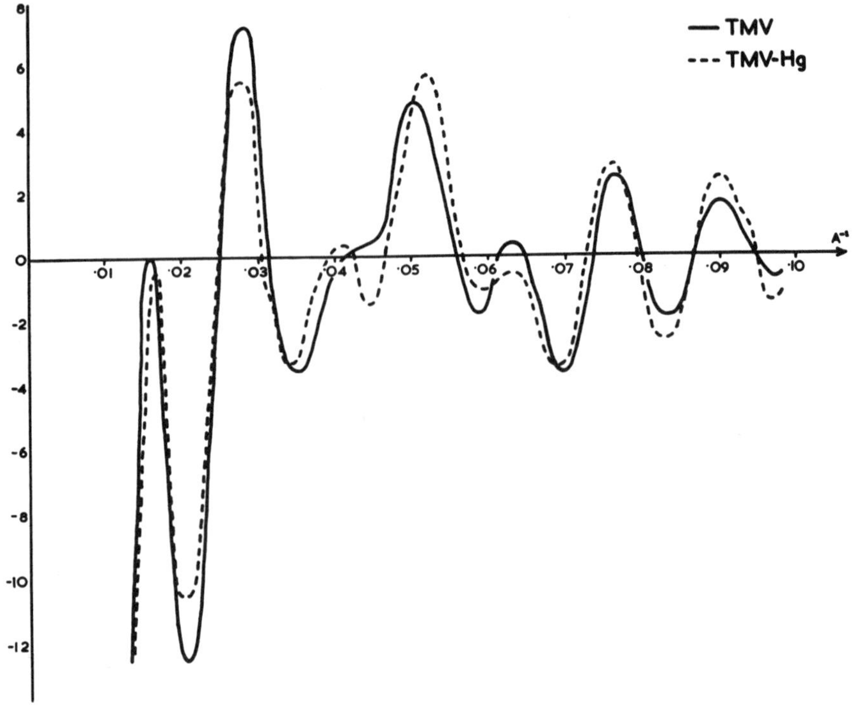
— TMV
--- TMV-Hg

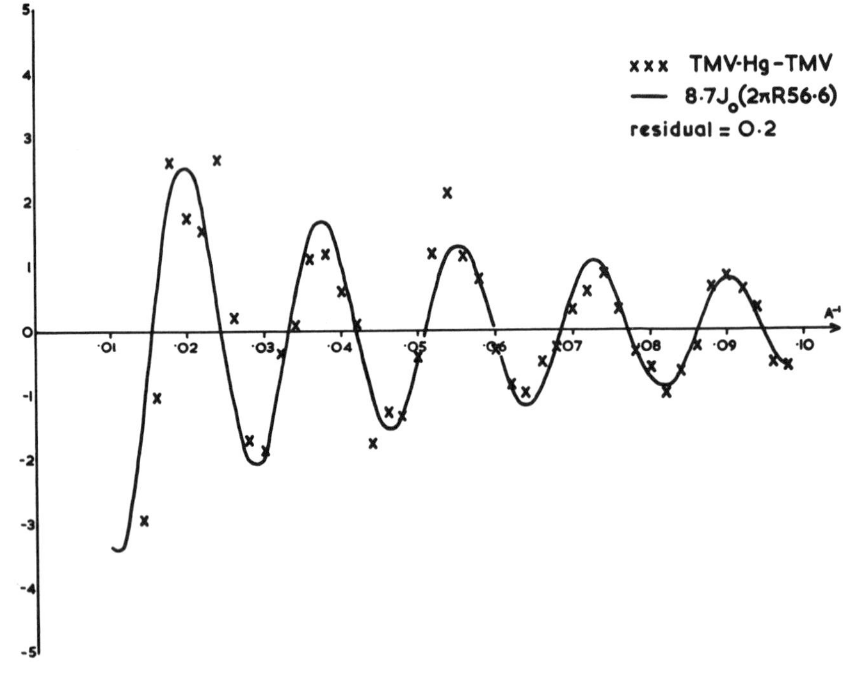
xxx TMV-Hg – TMV
— 8·7J$_0$(2πR56·6)
residual = 0·2

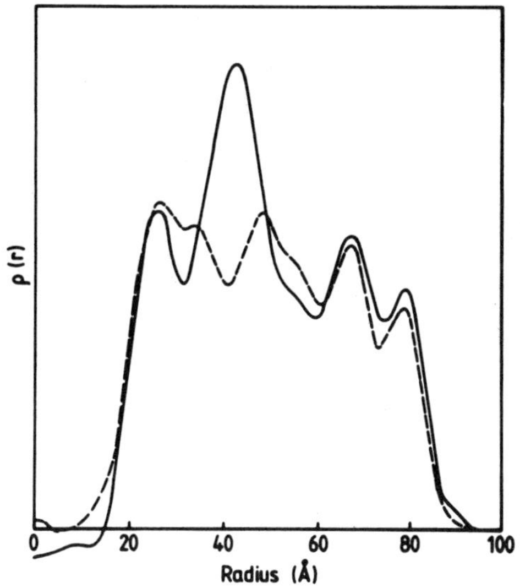

Figure 3. The radial density distribution, computed by a Fourier–Bessel transformation of the zero layer line amplitudes (Fig. 2) of TMV (solid line) and reagregated TMV protein in the helical form (broken line). The difference shows the radial position of the RNA (Franklin, 1956; Caspar, 1956).

derivatives were characterized and the first low resolution electron density map was calculated (Holmes et al., 1966). In 1968 Holmes moved to Heidelberg, Germany, where work on the virus fiber diagram was continued. After the publication of the 0.4 nm resolution electron density map of the virus in 1977 (Stubbs et al.) active work on the oriented virus moved (with Stubbs) to Brandeis University (Caspar's laboratory) and then to Vanderbilt University (Stubbs's laboratory) where it continues. The studies of the crystalline double disks were continued in Cambridge under Klug's direction.

4. DATA COLLECTION AND REDUCTION

The fiber diffraction data from TMV gels are very reproducible and free from systematic errors. However, effort must be expended to reduce the

---

Figure 2. (*Opposite*) The scattering amplitude on the zero layer line of TMV (solid line) and TMV plus methyl mercury (broken line). The difference gives the scattering amplitude of the mercury and corresponds to a single-site substitution with radius 5.6 nm (Franklin and Holmes, 1958).

Figure 4. Drawing (Caspar, 1963) of about one-twentieth of the length of a TMV particle. The 49 protein subunits repeat every three turns of the virus in 6.9 nm. A single strand of RNA winds between the subunits at a radius of 4.0 nm. with three bases per subunit.

background scattering as much as possible. Franklin's first attempts with a micropinhole camera and microfocus X-ray tube of the kind used for the DNA studies were plagued by the problem of radiation damage, a consequence of the high specific loading of a small portion of the specimen. Success was obtained with an evacuated Guinier camera using an asymmetrically cut bent quartz monochromator (Franklin and Holmes, 1958) using slit geometry. Later attempts to extend the data to 0.3 nm neccesitated use of a doubly focusing camera (point focus) in order to confine the layer lines to as small an area as possible so that the background could still be measured between layer lines. A suitable camera was constructed by Witz (1969) (Fig. 5). The use of focusing monochromators necessitated the development of a fine-focus rotating anode

X-ray tube (Longley and Holmes, unpublished; Fig. 6), which was the prototype of the Elliott GX6.

For data collection computer-driven flatbed film scanners were developed (Barrington Leigh, 1970; Mandelkow, 1973). These have since been replaced by systems involving drum scanners. The appropriate disorientation correction (Holmes and Barrington Leigh, 1974) allowed the intensity (integrated across the layer line) to be evaluated from scans along Debye–Scherrer arcs or from traces along layer lines (Stubbs, 1974). The

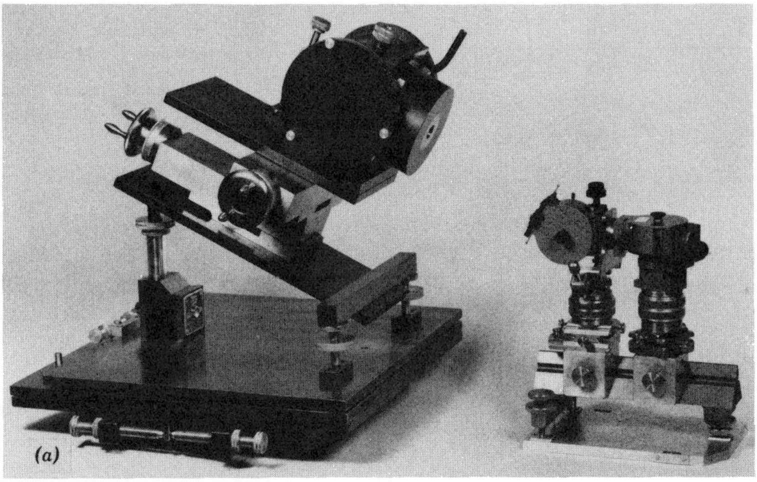

Figure 5. (*a*) The "ski jump" point-focus camera with double monochromators (see Witz, 1969) designed for photographing TMV. On the right is a short optical bench arranged to hold two CGR (Compagnie Generale de Radiologie, Paris) asymmetrically cut (Guinier geometry) bent quartz monochromators at right angles. The first monochromator (extreme right) is mounted adjacent to the X-ray tube. The focused X-ray beam leaves the second monochromator at an angle of 26° to the horizontal. The second monochromator is mounted in a cradle so that the axes of the two bent monochromators can be adjusted to be exactly at right angles. On the left is an evacuable 12 cm focusing camera (Rowland focusing) with the specimen mounted on the circumference of the focusing circle. For alignment a microscope fitted with a CsI crystal is mounted behind the camera. The camera has a port along the optic axis. On page 133: (*b*) Typical dimensions used in the ski jump camera. In this diagram the X-ray source is on the left. σ is the angle between the reflecting Bragg planes of the quartz monochromator and the surface of the monochromator. R is the radius of curvature of the monochromator. The focal point (extreme right) is arranged to coincide with the back wall of the focusing camera.

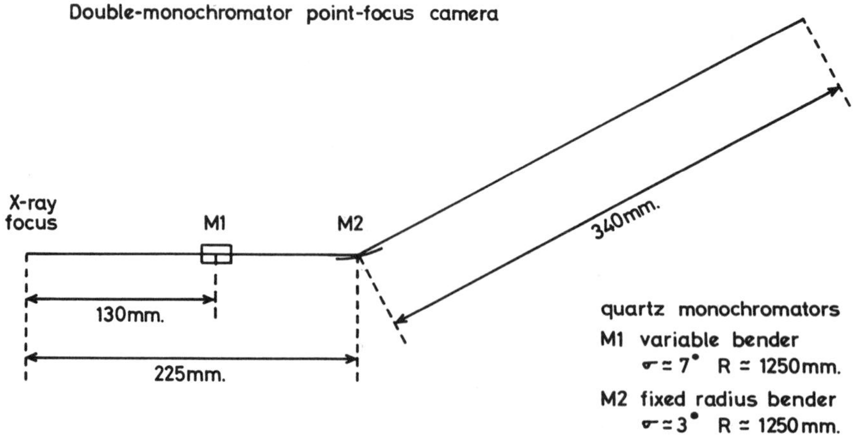

(b)

Figure 5. Continued

problem of background fitting is severe and the initial processing, including the data for the 0.4 nm electron density map. (Stubbs et al., 1977), was done using an on-line interactive display which allowed the operator to draw the background under the layer line trace. Background fitting can now be performed entirely automatically as part of the total problem of determining the layer line splitting (Stubbs and Makowski, 1982).

5. LOCATING HEAVY ATOMS FROM FIBER DIAGRAMS

By taking the difference between the structure factor of the virus and virus plus heavy atom on the inner part of the zero layer line the radii of attachment of heavy atoms may be established (Fig. 2). The difference should correspond to a single zero-order Bessel function or a small number of Bessel functions. In this way the radii of attachment of methyl mercury, lead, and osmium, each of which bind at single sites, were established (Barrett et al., 1971). Single site substitutions were chosen for further investigation because initially the problem of locating multiple site heavy atoms in three dimensions could not be solved.

The major difficulty is fixing the ϕ and z coordinates of the heavy atoms. Initially, one was restricted to a resolution limit of 0.1 nm^{-1} because outside this region more than one Bessel function term contributes to the diffraction pattern, a complication that cannot be accommodated by the simple methods first used to determine the heavy atom coordinates. With the restriction of low resolution, on any one layer-line the angle Φ_l is constant. If Φ_l can be determined by some means, then by combining results from a number of layer lines the relative ϕ_h and z_h may be determined by a graphic method (Barrett et al., 1971). It is necessary to choose one heavy atom as origin because the helical symmetry does not define an origin for ϕ and z. Unfortunately, on any one layer line, comparing any

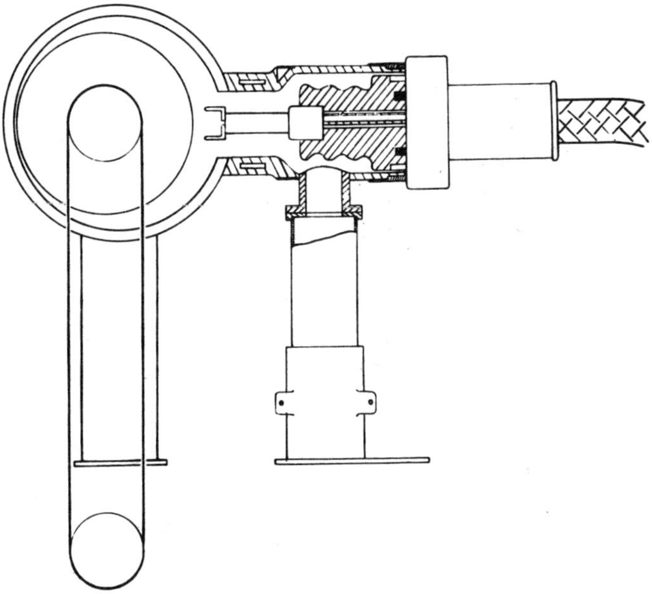

Figure 6. The fine focus rotating anode X-ray tube designed for work on viruses (Holmes and Longley, unpublished, with technical assistance of L. G. Hayward, D. Hart, and A. Woollard). On the left is a 10 cm diameter water-cooled rotating copper anode (after A. Broad). On the right is a cathode capable of delivering 20 mA at 40 kV into a line focus 1 × 0.07 mm. This cathode is a modification of a commercial design (Beaudouin, Paris). The line focus is arranged to be horizontal so as to appear as a small intense point when viewed at 4°. A source of this intensity is necessary for operating the double quartz focusing monochromator. The small size is necessary because the monochromators can only accept radiation from a very small area (<0.1 × 0.1 mm). A commercial X-ray tube derived from this design (Elliot GX6) was used for later work.

two heavy atoms there is always an ambiguity in the sign of Φ_l. Fortunately the uncertainties introduced by this ambiguity may be resolved by studying the appropriate double derivative (Holmes et al., 1972). Barrett et al. (1971) were able to calculate a three-dimensional electron density map of the virus with a nominal resolution of 1.0 nm. This map correctly showed the general appearance of the nucleic acid running circumferentially at 4.0 nm radius and the overall appearance of the subunit, including strong radial features which were later shown to be α-helices. The structure of the virus extends proximally to a radius of 2.2 nm, in sharp contrast to the situation in the disk.

6. HIGH RESOLUTION STUDIES OF THE FIBER DIFFRACTION PATTERN

At high resolution two, three, or more Bessel function terms fall on top of each other. Each measured intensity, therefore, consists of the sum of squares of the real and imaginary parts of a number of terms which must be separated (Eq. 2). Holmes (1959) had shown that it is possible in principle to separate Bessel functions by means of heavy atom derivatives and the method was analyzed in detail by Stubbs (1975). The problem of separating Bessel functions by the method of isomorphous replacement is a multidimensional analog of the method of determining phases by the Harker construction (1956) and can formally be solved by constructing phase circles in a hyperspace which has two dimensions per Bessel function (i.e., a real and imaginary part for each G_{nl}). Thus, at low resolution (only one Bessel function active $<1.0\ \text{nm}^{-1}$), the problem is two-dimensional and appears quite conventional. However, at 0.4 nm resolution at least three Bessel functions must be taken into account so the problem becomes six-dimensional and requires at least six heavy atoms for a solution. For the normal crystallographic two-dimensional case, a simple search procedure suffices to establish the best solution to the phase. However, locating the best intersection of the Harker hyperspheres by a six-dimensional search for every point in reciprocal space is a very time-consuming computing problem. An algebraic solution was developed by Stubbs and Diamond (1975) which evaluates the real and imaginary parts of each G_{nl} and this was successfully employed to compute first a 0.7 nm and then a 0.4 nm Fourier–Bessel map of TMV (Holmes et al., 1975;

Stubbs et al., 1977). The limiting factor was the availability of enough well-defined single heavy atom derivatives of the virus. Both genetic and chemical modifications to the virus were used to introduce new SH groups (Wittmann 1964; Mandelkow and Holmes, 1974; Gallwitz et al., 1974; Perham and Thomas, 1971) which were then tagged with methyl mercury.

In the high resolution study (0.4 nm) reported by Stubbs et al. (1977), the refinement of the heavy atom coordinates was carried out by a modification of the method of Dickerson et al. (1968): a determination of the G_{nl} was carried out as described above; a search was made for the heavy atom parameters which minimized the lack of closure on the Argand diagram for each point and the process was repeated.

7. THE USE OF CONTINUOUS DATA

Noncrystallographic symmetry is sometimes employed in protein structure determinations as an aid to the isomorphous replacement determination of phase angles (see Crowther, 1969; Jack, 1973; Bricogne, 1976; Makowski, 1982). Indeed such methods were essential for the solution of the structure of the TMV protein disk. Two kinds of noncrystallographic symmetry operator need to be considered: rotation and translation. Crowther's analysis showed that rotation is by far the more powerful constraint. A 49-fold noncrystallographic rotational operator exists in TMV but its effects have been exhausted by building the information into the algebra of the structure factor plus the helix selection rule. Translation operators essentially allow a glimpse of the molecular transform *between* reciprocal lattice points. Crowther showed that this extra information is not a strong constraint unless the molecular transform can be sampled very frequently. TMV should be an excellent case on which to try out this method because the continuous molecular transform is experimentally available.

Such methods find their simplest expression in the "minimum wavelength principle" (Bragg and Perutz, 1952). Indeed the zero layer line of TMV can be phased purely by means of the minimum wavelength principle (Caspar, 1955). On other layer lines it has proved useful to constrain the rate of change of the G_{nl}. An example of how this may be done on an Argand diagram is given in Figure 7. The values of $G_{-16},1$ (Bessel function order -16 on the first layer line) derived from isomorphous re-

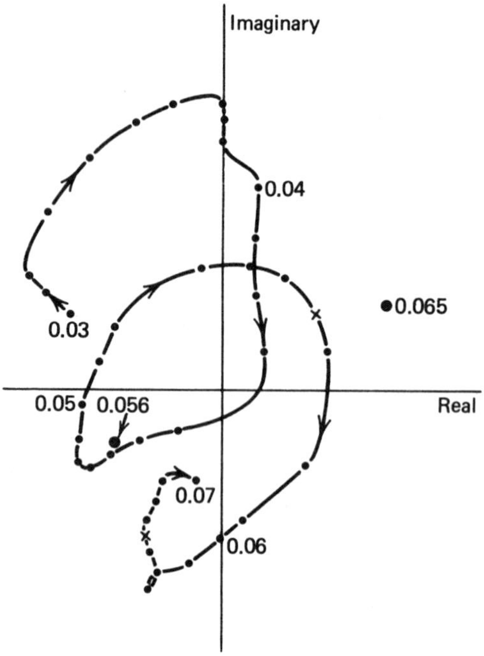

Figure 7. Continuity of the transform of TMV, for layer line l reciprocal space radius (R) values between 0.3 and 0.7 nm^{-1} (0.03 and 0.07 $Å^{-1}$). Each point represents the value of the most probable $G_{-16,l}$ on the Argand diagram as phased by isomorphous replacement. The line connects points adjacent in R. Circled points do not fit in with the general continuity. The crosses represent minor probability maxima, which, however, on grounds of continuity are accepted rather than the major solution (Holmes et al., 1975; Klug, unpublished)

placement are shown plotted as a function of reciprocal space radius. Certain "wild points" are easily recognizable (e.g., $R = 0.065$ $Å^{-1}$), because they do not lie on the structure factor trajectory, and may be corrected by interpolation. A graphic method of ensuring continuity using an interactive computer terminal was developed by Mandelkow and Holmes (1974). Stubbs (1975) sought an algorithmic realization of this principle and found that it was sufficient to take as the best solution that with the minimum curvature when plotted on an Argand diagram.

The procedure most often used for computing Fourier equations from phase angles derived from isomorphous replacement is to use the centroid of the probability distribution of the phase angle of each reflection to produce the "best" Fourier (Blow and Crick, 1959). This is the "best" procedure if the phase determinations are uncorrelated. However, for TMV neighboring points are highly correlated so that all Fourier summations have been calculated using instead the most probable phase angles produced by Stubbs' algorithm, which produces the "most probable" Fourier in the parlance of Blow and Crick.

8. THE STRUCTURE OF THE VIRUS

The density map at 0.67 nm resolution (Holmes et al., 1975) showed that the most important features of the subunit were four approximately radial rods, two of which continued to a low radius. Between these two, running circumferentially at a radius of 4.0 nm between subunits was the single-stranded RNA. Both the C and N termini were correctly located at the outside surface of the virus. In 1976 Champness et al. solved the structure of the double disk to a resolution of 0.5 nm and were able to see that the structure of the subunit in the virus was similar to that in the disk except for the missing RNA and the proximal region of the disk (the region at low radius from the RNA binding site) where the protein density was singularly absent. They were able to identify the radial features of Holmes et al. (1975) as α-helices and to show that the subunit boundaries chosen by Holmes et al were incorrect (i.e., the radial features were not correctly joined together).

In 1977 Stubbs et al. produced the 0.4 nm map of the virus. The low radius region of the map appears to be well resolved, showing details that allowed the orientation of the α-helices and the large side chains to be determined. The handedness of the α-helices was seen to be correct, which was gratifying because the chirality of TMV had been determined by electron microscopy (Finch, 1972). The detailed interpretation of the map was based on the results of Champness et al. (1976) for the disk, so that in discussing the results it is helpful to use the terminology for the parts of the molecule introduced by Champness et al. This distinguishes between two kinds of α-helix, radial (R) and slew (S) with the prefix left (L) or right (R). The identification of these features (LR, RR, LS, and RS) in the 0.4 nm electron density map is shown in Fig. 8. Outside 6.0 nm radius the use of only three Bessel functions in the Fourier–Bessel synthesis leads to a drop of resolving power so that an unambiguous

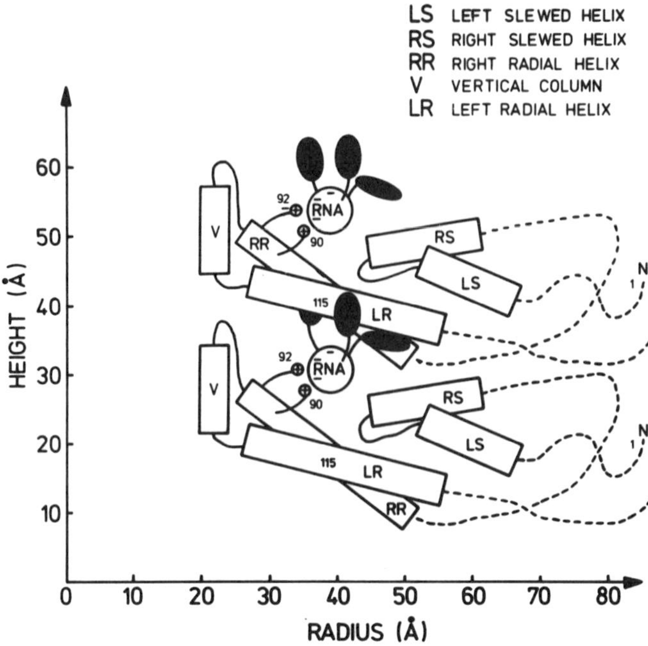

Figure 8. Schematic view of the low radius part of TMV as interpreted by Stubbs et al. (1977) from their 0.4 nm resolution electron density map. The names of the α-helices, LR, RR, LS, and RS, are taken from Champness et al. (1976). The three bases of the RNA form a claw around the hydrophobic LR helix. Arginine residues 90, 92, and 41 were suggested as elements of the phosphate binding site. The LR and RR helices extend to a low radius compared with the structure of the double disk.

interpretation of the electron density was no longer possible although a general similarity to the electron density of the disk was apparent. Proximally from 4.0 nm radius, where the density in the disk fades out, one sees the LR and RR helices *extending* to a low radius (2.5 nm) (Fig. 9) where they are joined together by a strong vertical column of density (the V column) which was identified by Stubbs et al. (1977) as an α-helix although more recent studies show this identification to be somewhat uncertain. Stubbs et al. point out that the sequence comprising the extension to the LR and RR helices and the V column contains a large concentration of ionizable side chains, particularly carboxyl groups. It is suggested that the spatial concentration of these groups brought about by the polypeptide chain changing direction in this region accounts for the abnormally low carboxyl pK seen in TMV (Caspar, 1963).

Fortunately, the nucleic acid and its environment could be clearly identified. Three nucleotides are associated with each protein subunit. Subsequent studies by Mandelkow et al. (1981) on the repolymerized helical

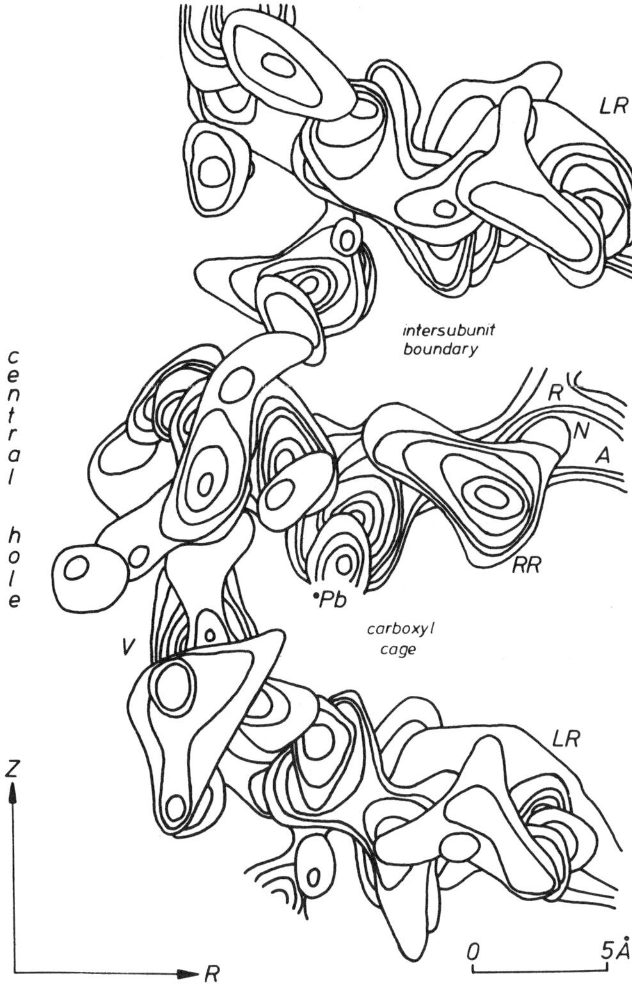

Figure 9. The very low radius part of the electron density of TMV (Stubb et al., 1977). Note the continuous high density linking the extended LR helix with the extended RR helix. This density is not visible in the double disk (Bloomer et al., 1978).

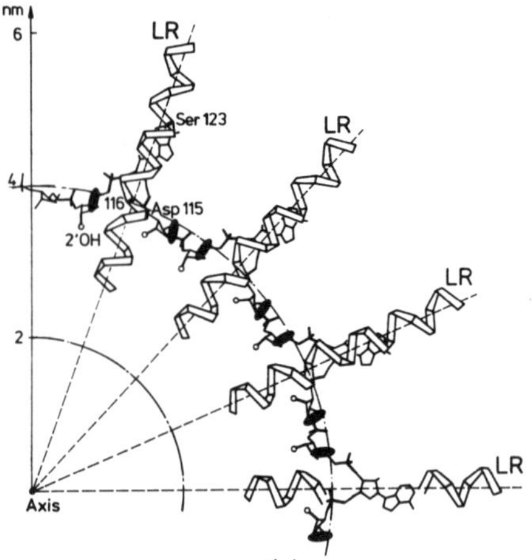

(a)

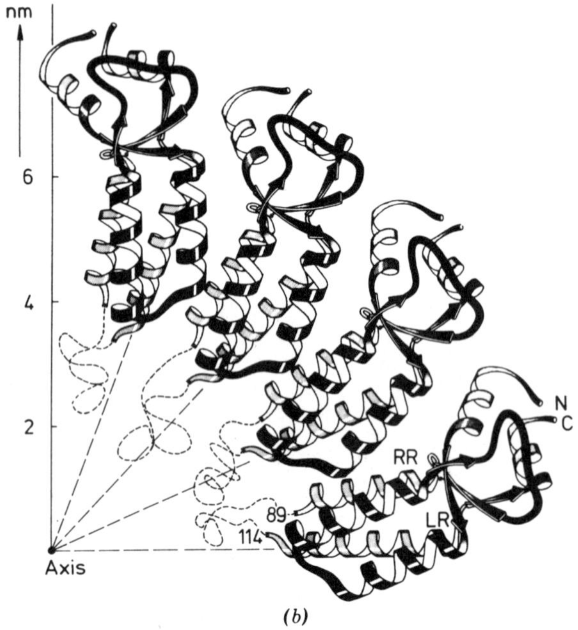

(b)

Figure 10. (*a*) The geometry of the RNA and its relationship to the LR helix (Stubbs and Stauffacher, 1981). The sugar–phosphate backbone runs under and at right angles to the LR helix. Two of the three bases (black ellipses) stick up from the plane of the sugar–phosphate chain to lie along both sides of the LR helix. A third base extends to a large radius where it may hydrogen bond to Ser 123. The OH groups of two of the sugars may make hydrogen bonds to the aspartic acid residues 115 and 116 (Holmes, 1979). (*b*) The structure of the double disk shown on the same scale for comparison (after J. Richardson, 1981). The flexible part of the chain is shown with a broken line.

form of the coat protein without RNA substantiated the identification of the electron density associated with the RNA. A least-squares refinement of an RNA model into this density was made by Stubbs and Stauffacher (1981). The phosphates are clustered around a radius of 4.0 nm and appear to bond to the adjoining RS and RR helices from the neighboring subunit in the next turn down. Thus, the RNA binding site lies between the rings of a disk, in agreement with the model for assembly proposed by Butler et al. (1976). The three bases nestle round the LR helix (Fig. 10), forming a kind of claw with two at about the same radius as the phosphates and one at a larger radius. Details of this interaction are discussed in Chapter 5.

9. THE RELATIONSHIP BETWEEN THE DISK AND VIRUS

In 1978 the details of the structure of the double disk were revealed by the calculation of a 0.28 nm electron density map (Bloomer et al., 1978). It is clearly desirable to combine the resulting detailed model with the virus map. Bloomer et al. (1981) have analyzed the relationship between the A disk, the B disk, and the virus and have shown that, in addition to a small radial movement, the subunits must be rotated by 10° (A disk) or 20° (B Disk) about an axis perpendicular to the common disk or helix axis in order to bring them into the same orientation as the virus. The rotation and translation necessary to bring the A and B disk density into coincidence with the virus density has been determined by comparing heavy atom positions and by a comparison of the electron density maps, which also shows that the disk and helix structures are approximately related by a rigid body transformation. Applying this transformation to the A ring of the disk results in the composite structure made by combining the disk model with the RNA of Stubbs and Stauffacher shown in Figure 11. The missing density in the disk is depicted with a broken line to show the apparent continuation of the LR and RR helices for two turns to be joined together by the V column. Comparing the resulting structure with the virus electron density indicates that some revisions will be necessary to take account of local distortions (Bloomer et al., 1981). Comparing the model so produced with the model fitted to the helix electron density by Stubbs et al. shows that the disk structure is one turn further out (3.6 residues) along each α-helix (LR and RR) than the Stubbs et al. model

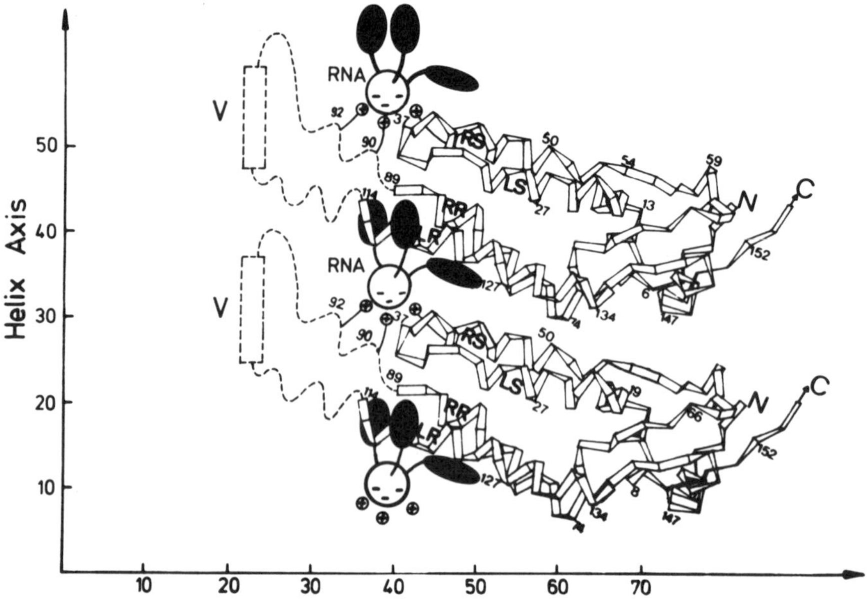

Figure 11. The structure of TMV produced by combining the RNA structure deduced from the virus electron density map (Stubbs et al. 1977) with the atomic model obtained from the crystalline double disk (Bloomer et al., 1978) using the coordinate transformation deduced by Bloomer et al. (1981). The LR and RR helices have been extrapolated to low radius and are shown joined by the V column whose exact geometry is uncertain. The bases are shown as black ellipses.

(Bloomer et al., 1978). This means that about eight fewer residues have to be accommodated in the V column than was thought by Stubbs et al. to be the case.

The general appearance of the subunit, as deduced from the high resolution studies of Bloomer et al. (1978), is shown in Figure 10*b*. Both the C and N termini are located distally. This is described in detail in Chapter 5.

10. SEGMENTAL FLEXIBILITY IN THE DISK

DeWitt et al. (1978), on the basis of NMR spectroscopy, first pointed out that TMV coat protein shows regions of unusual flexibility. A fairly unambiguous identification of the flexible region was made by Jardetzky *et al.* (1978) by the use of proton NMR. The 360 MHz spectrum of the 4S aggregate (a trimer) shows numerous sharp lines, which is not to be expected from an object of such high molecular weight. The aromatic region is ill resolved, in line with the dense packing of the hydrophobic girdle, whereas the aliphatic region shows several sharp peaks. The relative sharpness of these lines persists as the effective molecular weight of the protein is increased to the 8S aggregate and to the 20S double disk. Some signs of these sharp lines can still be seen in the RNA-free helix but they are much attenuated. In the intact virus they completely vanish. It is estimated that *all* the amino acids in the flexible loop (i.e., Asp 88–Arg 112) contribute to the observed spectrum, demonstrating that this segment must be very mobile in the disk. Therefore, the absence of electron density in the lumen of the disk seems to be due to movement rather than static disorder. In contrast, the virus shows no NMR spectrum and the inner region of the virus has a well-defined ordered structure. The significance of this change from disorder to order during virus assembly is discussed in Chapter 5.

11. USE OF ISOMORPHOUS REPLACEMENT WITH LAYER LINE SPLITTING

Franklin and Klug (1955) first noted that the layer lines in TMV are slightly split. This comes about because the TMV helix is not exactly integral, there being 49.01 subunits in three turns of the helix, whereby the small fractional part varies with conditions and the strain of TMV. Bessel functions of different order on the same layer line (such as the terms $G_{-16},1$ and $G_{-33},1$ on the first layer line) in fact do not occur exactly on the same layer line. Although it had long been realized that this splitting represented an additional source of information about the correct way of distributing the observed intensity between overlapping Bessel functions, because of the difficulties involved, the method has only recently been put to a practical test. Stubbs and Makowski (1982) have shown that the splitting of the Bessel functions can be measured and used as a source of information which seems to be worth about three heavy atoms at 0.67 nm resolution. The layer line splitting is considerably less than the disorientation so that the arcs arising from neighboring Bessel functions merge to form a Gaussian-like profile which must be deconvoluted to yield the contributing G_{nl}

values. In addition, considerable care is needed with the background fitting. However, the success of this trial leads one to expect that the resolution of the virus electron density map may be extended toward 0.3 nm which should facilitate an understanding of the processes involved in the disk–helix transition.

REFERENCES

Anderer, F. A. (1963). *Adv. Protein Chem.* **18,** 1–35.

Barrett, A. N., Barrington Leigh, J., Holmes, K. C., Leberman, R., Mandelkow, E., Von Sengbusch, P., and Klug, A. (1971). Cold Spring Harbor Symp. Exp. Biol. **36,** 433–448.

Barrington Leigh, J. (1970). Thesis, Univ. Cambridge.

Bawden, F. C., Pirie, N. W., Bernal, J. D., and Fankuchen, I. (1936). *Nature* (*London*) **138,** 1051–1055.

Beijerink, M. W. (1898). *Ver. Akad. Wet. Amsterdam* **65,** 3–10.

Bernal, J. D., and Fankuchen, I. (1941). *J. Gen. Physiol.* **25,** 111–165.

Bloomer, A. C., Champness, J. N., Brigogne, G., Staden, R., and Klug, A. (1978). *Nature* (*London*) **276,** 362–368.

Bloomer, A. C., Graham, J., Hovmoeller, S., Butler, P. J. G., and Klug, A. (1981). In *Structural Aspects of Recognition in Biological Macromolecules*, M. Balaban, J. L. Sussmann, W. Traub, and A. Yonath, Eds. Balaban ISS, Rehovoth, pp. 851–864.

Blow, D. M., and Crick, F. H. C. (1959). *Acta Crystallogr.* **12,** 794–802.

Bragg, W. L. and Perutz, M. F. (1952). *Proc. R. Soc. London, Ser. A* **213,** 425–435.

Bricogne, G. (1976). *Acta Crystallogr. A* **32,** 832–847.

Butler, P. J. G., Bloomer, A. C., Bricogne, G., Champness, J. N., Graham, J., Guilley, H., Klug, A., and Zimmern, D. (1976). In *Structure-Function Relationships in Proteins*, R. Markham, and R. W. Horne, Eds., 3rd John Innes Symposium. North Holland Amsterdam, pp. 101–110.

Caspar, D. L. D. (1955). Thesis, Yale Univ., New Haven, Connecticut.

Caspar, D. L. D. (1956). *Nature* (*London*) **177,** 928–929.

Caspar, D. L. D. (1963). *Adv. Protein Chem.* **18,** 37–121.

Champness, J. N., Bloomer, A. C., Bricogne, G., Butler, P. J. G., and Klug, A. (1976). *Nature* (*London*) **259,** 20–24.

Cochran, W., Crick, F. H. C., and Vand, V. (1952). *Acta Crystallgr.* **5,** 581–586.

Crowther, R. A. (1969). *Acta Crystallogr. B* **25,** 2571–2580.

de Witt, J. L., Hemminger, M. A., and Schaefomer, T. J. (1978). *J. Mag. Res.* **31,** 97–107.

Dickerson, R. E., Weinzierl, J. E., and Palmer, R. A. (1968), *Acta Crystallgr. B* **24,** 997–1003.

Finch, J. T. (1972). *J. Mol. Biol.* **66,** 291–294.

Finch, J. T., Leberman, R., Chang, Y. S., and Klug, A. (1966). *Nature (London)* **212,** 349–350.

Franklin, R. E. (1955). *Nature (London)* **175,** 379–383.

Franklin, R. E. (1956). *Nature (London)* **177,** 929–930.

Franklin, R. E., and Holmes, K. C. (1958). **11,** 213–220.

Franklin, R. E., and Klug, A. (1955). *Acta Crystallogr.* **8,** 777–783.

Franklin, R. E., and Klug, A. (1956). *Biochim. Biophys. Acta* **19,** 403–411.

Gallwitz, U., King, L., and Perham, R. N. (1974). *J. Mol. Biol.* **87,** 257–264.

Green, D. W., Ingram, V. M., and Perutz, M. F. (1954). *Proc. R. Soc. London, Ser. A* **225** 287–308.

Gregory, J., and Holmes, K. C. (1965). *J. Mol. Biol.* **13,** 796–801.

Harker, D. (1956). *Acta Crystallogr* **9,** 1–9.

Holmes, K. C. (1959). Thesis, Univ. London.

Holmes, K. C. (1979). *J. Supramol. Struct.* **12,** 305–320.

Holmes, K. C., and Barrington Leigh, J. (1974). *Acta Crystallogr. A* **30,** 635–638.

Holmes, K. C., Mandelkow, E., and Barrington Leigh, J. (1972). *Naturwissenschaften* **59,** 247–254.

Holmes, K. C., Klug, A., Leberman, R., and von Sengbusch, P. (1966). Abstr. 2nd Int. Biophys. Congress Vienna, No. 205.

Holmes, K. C., Stubbs, G., Mandelkow, E., and Gallwitz, U. (1975). *Nature (London)* **254,** 192–196.

Ivanowsky, D. (1892). *Bull. Akad. Sci. St. Petersburg* **35,** 67–71.

Jack, A. (1973). *Acta Crystallogr. A* **29,** 545–554.

Jardetzky, O., Akasaka, K., Vogel, D., Morris, S., and Holmes, K. C. (1978). *Nature (London)* **273,** 564–566.

Klug, A., and Caspar, D. L. D. (1960). *Adv. Virus Res.* **7,** 225–325.

Klug, A., Crick, F. H. C., and Wyckoff, H. W. (1958). *Acta Crystallogr.* **11,** 199–213.

Macleod, R., Hills, G. J., and Markham, R. (1963). *Nature (London)* **200,** 932–934.

Makowski, L. (1982). *J. Appl. Crystallogr.* **15,** 546–557.

Mandelkow, E. (1973). Thesis, Univ. Heidelberg.

Mandelkow, E., and Holmes, K. C. (1974). *J. Mol. Biol.* **87,** 265–273.

Mandelkow, E., Stubbs, G., and Warren, S. (1981). *J. Mol. Biol.* **152,** 375–386.

Perham, R. N., and Thomas, J. O. (1971). *J. Mol. Biol.* **62,** 415–418.

Richardson, J. (1981). *Adv. Protein Chem.* **34,** 167–339.

Schramm, G. (1947). *Z. Naturforsch.* **2b,** 112–118, 249–257.

Stanley, W. M. (1935). *Science* **81,** 644–648.

Stubbs, G. (1974). *Acta Crystallogr. A* **30,** 639–645.

Stubbs, G. (1975). *Acta Crystallogr. A* **31,** 709–715.

Stubbs, G., and Diamond, R. (1975). *Acta Crystallogr. A* **31,** 715–718.

Stubbs, G., and Makowski, L. (1982). *Acta Crystallogr A* **38,** 417–425.

Stubbs, G., and Stauffacher, C. V. (1981). *J. Mol. Biol.* **152,** 387–396.

Stubbs, G., Warren, S., and Holmes, K. C. (1977). *Nature (London)* **267,** 216–221.

Watson, J. D. (1954). *Biochim. Biophys. Acta* **13,** 10–19.

Wittmann, H. G. (1964). *Z. Vererbungslehre* **95,** 333–344.

Witz, J. (1969). *Acta Crystallogr. A* **25,** 30–42.

5

Macromolecular Interactions in Tobacco Mosaic Virus

GERALD STUBBS
Vanderbilt University
Nashville, Tennessee

CONTENTS

1. INTRODUCTION

1.1. The Virus Particle

Tobacco mosaic virus (TMV) is probably the most commonly occurring plant virus, reaching high concentrations in a wide variety of host plants and presenting significant problems for tobacco and tomato crops (Matthews, 1981). It was the first virus to be discovered (Iwanowski, 1892), and has been widely studied ever since. Recent work has concentrated particularly on its assembly and molecular biology, much of which has been described in reviews by Butler and Durham (1977), Okada (1980), and Hirth and Richards (1981).

Early work on the structure and assembly of TMV was reviewed by Caspar (1963), see also Chapter 4. The virus is rod-shaped, 3000 Å long, and 180 Å in diameter, containing one molecule of RNA, MW $\sim 2 \times 10^6$ and approximately 2,100 identical coat protein subunits, MW 17,500. The protein subunits form a right-handed helix having 49 subunits in three turns, with the RNA following the basic helix at a radius of 40 Å. There are three nucleotides bound to each protein subunit. There is a hole down the axis of the virus about 40 Å in diameter, so that the RNA is about 20 Å from the inner surface of the virus, completely surrounded by protein. These features are illustrated in Figure 4 in Chaper 4.

This review will examine the protein–nucleic acid interactions in TMV, especially from a structural point of view. However, to appreciate the significance of these interactions, it is necessary to first consider the assembly and disassembly of the virus.

1.2. Dissociation

The intact virion of TMV is easily dissociated into its constituent protein and RNA. Bawden and Pirie (1937) used both glacial acetic acid and heat treatments to extract the RNA, but in a rather degraded state. Cohen and Stanley (1942) were more successful with heat treatment, but the most effective methods for dissociation of TMV were treatment by alkaline degradation (Loring, 1939; Schramm, 1943, 1947 a,b) and sodium dodecyl sulfate (SDS) (Sreenivasnya and Pirie, 1938; Fraenkel-Conrat and Singer, 1954). Although degradation of products, especially RNA, had been a problem until this point, in 1955 Fraenkel-Conrat and Williams, using the SDS method, were able to dissociate TMV sufficiently free of degradation to be able to reconstitute infectious virus from the resulting protein and RNA. This experiment had far-reaching consequences for the understanding of viral assembly, which will be discussed further in Section 1.4.

The dissociation of TMV in alkali, detergent and other systems starts at one end, with protein subunits being progressively lost (Hart, 1955; Harrington and Schachman, 1956; May and Knight, 1965). In most strains of TMV [the cowpea strain is an exception (Beachy et al., 1976)], subunits are first lost from the 5′ end of the RNA under a wide variety of degrading conditions, including alkali and SDS (Perham and Wilson, 1976; Wilson et al., 1976; Ohno and Okada, 1977; Wilson et al., 1981; Blowers and Wilson, 1982). The common strain virus then goes through a succession of stable intermediates, although these have not been found for the closely related U2 strain (Perham, 1969). There is a particularly stable form one-sixth the length of the original particle (Schramm et al., 1955; Symington, 1969; Perham, 1969). Five distinct intermediates were described by Wilson et al. (1978), who noted that all but the shortest two were capable of reassembly. The existence of stable intermediates suggests that the strength of the protein–RNA interaction must vary along the rod, as a consequence of the nucleotide sequence of the RNA. This is supported (Perham and Wilson, 1978) by the fact that the nucleotide sequence that initiates protein binding and assembly (see Section 1.4.2) is located about 1000 nucleotides from the 3′ end of the RNA (Zimmern and Wilson, 1976), just at the site responsible for the one-sixth length stable fragment.

A feature of stripping observed by several groups is the "stable fraction" of TMV (not to be confused with the "stable fragment"): 25–30% of the virus particles resist degradation (Schramm et al., 1955; Perham, 1969). This fraction does not differ from the degradable fraction in RNA sequence, and it does not contain significant amounts of any extra protein (Perham and Wilson, 1978). Powell (1975) has shown that if metal ions are excluded from the virus preparation, the stable fraction disappears, so it seems that binding of these ions probably protects the virus from stripping.

1.3. The Coat Protein

1.3.1. Polymorphic Forms

Before considering the assembly of TMV, it is necessary to examine the aggregates of the coat protein, several of which play important parts in the assembly process. There is a wide variety of aggregates, forming under different solution conditions, including pH, ionic strength, and temperature. Butler and Durham (1977) have reviewed work on these aggregates, their interconversions, and their involvement in virus assembly. The best characterized forms and some of the conditions under which they are stable are summarized in Figure 1 (a revised version of a figure from Durham et al., 1971), although Vogel (1982) has shown that the "phase boundaries" of this figure are somewhat dependent on the experimental conditions, especially the type of salt present. Most of the known aggregates are similar to one of the three basic forms: the A protein, the 34-subunit disk, or the protein helix.

1.3.2. The A Protein

The protein monomer exists at significant levels only at extreme dilution (Ansevin and Lauffer, 1959), or at very high (Anderer, 1959) or low (Ba-

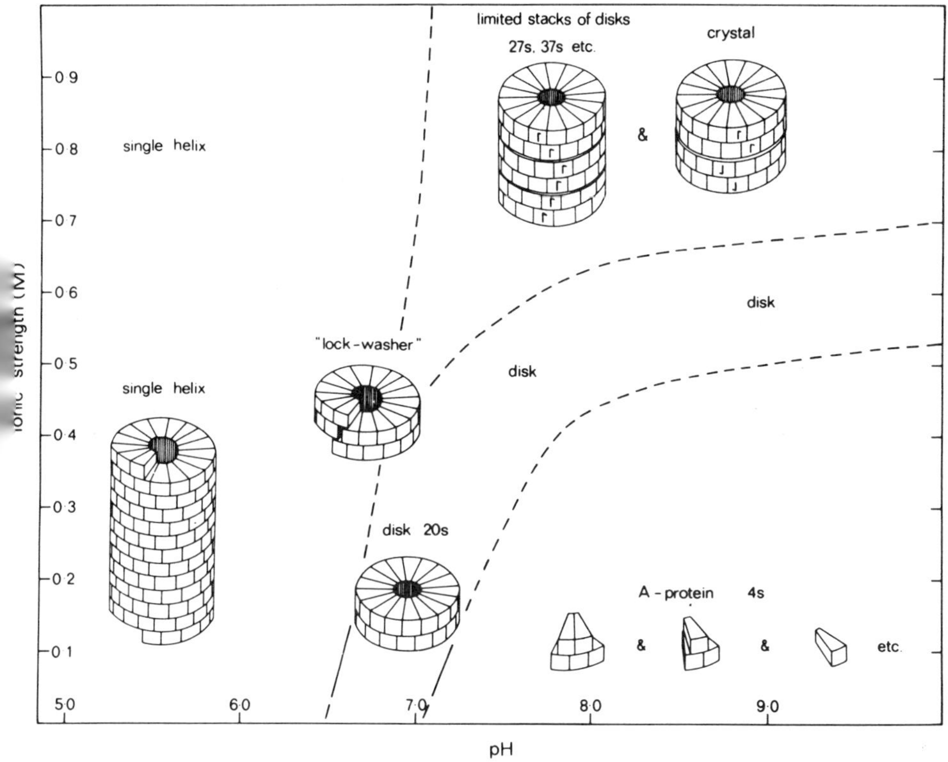

Figure 1. The aggregates of TMV protein present at 20°C, protein concentration 5 mg/ml, as pH and ionic strength are varied. The smaller aggregates are not necessarily absent under the conditions where the larger aggregates form. This is a revised version by P. J. G. Butler of a figure from Durham et al. (1971). I thank Dr. Butler for making this figure available.

nerjee and Lauffer, 1966) pH. At more usual concentrations, at low ionic strength and moderately alkaline or neutral pH, a mixture of low molecular weight oligomers called A protein (Schramm and Zillig, 1955) is present. This mixture often has a major component with a sedimentation constant of about 4S, but the term includes other oligomers with sedimentation constants lower than 20S, that is, anything smaller than the "disk" in Figure 1.

The size distributions and equilibria present in A protein have been studied by Vogel et al. (1979). The 4S A protein has on average four or five subunits arranged in two layers (Durham and Klug, 1971, 1972). A distinct 8S species has been observed (Durham, 1972a; Lonchampt et al., 1972) and although there is no direct evidence for its structure, it may have three layers (Durham, 1972a; Vogel et al., 1977), perhaps being the cyclic heptamer proposed by Caspar (1963).

The A protein has been shown by CD studies (Vogel, 1973; Vogel and Jaenicke, 1974) to undergo a rapid conformational change preceding the slow change in aggregation state occurring on changing pH, ionic strength, or temperature, forming a transient species called A′, which has a CD spectrum identical to that of the disk.

1.3.3. The Disk

Central to the variety of TMV protein forms is the disk, a two-layered aggregate of 34 subunits which sediments at about 20S. It plays a vital role in the assembly of TMV, which will be discussed further in Section 1.4. The disk is the dominant species at neutral pH or at higher pH and moderate ionic strength, although A protein is always present to some degree in disk preparations (Durham, 1972a). The two layers of the disk face the same way, and a pairing interaction between the layers prevents further polymerization into long viruslike rods (Durham et al., 1971). This pairing is in the form of a distinct tilting of the outer end of each subunit toward the opposite layer (Champness et al., 1976). Pairing is also observed in the Dahlemense strain of the intact virus (Caspar and Holmes, 1969) and sometimes, to a very slight degree, in the native and U2 strains and in the helical protein aggregate. However, this pairing has a different nature from that in the disk, because the bonding between layers in the disk (Champness et al., 1976) is quite different from that in the virus (Holmes et al., 1975) (Section 2.4.2).

The disk has been crystallized (Finch et al., 1966), and its molecular structure is known at 2.8 Å resolution (Bloomer et al., 1978). This will be discussed further in Section 2.3. Under conditions similar to the conditions of crystallization, long rods of stacked disks have been shown to form, but this process is irreversible, and Durham (1972b) showed that the subunits in these rods had undergone proteolytic cleavage.

Several aggregates resembling the disk have been reported, sedimenting between 19 and 26S. Vogel et al. (1977) reported a transient 26S species which appeared during formation of disks from A protein at 0.1 *M* ionic strength and pH below 7.1. They suggested that this might be a three-layer aggregate. However, Shire et al. (1979a) have used a mass-balance analysis of the conversion of a 24.4S aggregate to a two-layer aggregate to show that the 24.4S boundary contains a three-layer aggregate, which suggests that the species observed by Vogel et al. is somewhat larger. Shire et al. suggested that the 24.4S species is the helix of approximately three turns predicted by Caspar (1963). Other short helices (Vogel et al., 1977; Shire et al., 1979a; Schuster et al., 1979) and limited stacks of disks (Durham, 1972a; Vogel et al., 1977) have been reported.

Central to many theories of the assembly of TMV or the larger TMV protein aggregates is the concept of the two-turn helix, or "lockwasher" (Caspar, 1963). Various models for the assembly have included the lockwasher as a transitory intermediate, and it has often been presumed to be present in equilibrium with disks (Durham et al., 1971; Schuster et al., 1979; Shire et al., 1979a). Because the lockwasher would have almost the same hydrodynamic properties as the disk, it has not been possible to unequivocally establish its presence in solution, but electron microscopy has shown stacks of disks passing through a lockwasher structure during formation of helices (Durham et al., 1971; Klug and Durham, 1971).

1.3.4. Long Helices

At low pH (see Figure 1), TMV protein forms long helices (Schramm, 1947b), which have a morphology very similar to that of the intact virus (Franklin, 1956a; Durham and Finch, 1972). Mandelkow et al. (1976) found that there are in fact two forms of protein helix, one having the same symmetry as the intact virus ($16\frac{1}{3}$ subunits per turn) and one having $17\frac{1}{3}$ subunits per turn. The technique for preparing the samples used in these studies (Gregory and Holmes, 1965) is such that it has not been possible to correlate the conditions of formation with the type of helix formed, but the conditions for the two types are clearly very similar. The intersubunit bonding in the two forms is almost identical, and the molecular conformations are extremely similar to each other and to the virus (Section 2.2; and Mandelkow et al., 1981).

1.3.5. Transitions between Forms

The transitions between the protein forms described above have been considered for a wide variety of conditions. Caspar (1963) proposed that the most stable small aggregates would be a trimer, a cyclic (6 around 1) heptamer, the disk, and the three-turn helix; and Durham and Klug (1971, 1972) presented a thermodynamic model for the assembly of disks from the A protein. Butler et al. (1972) titrated the protein between pH 4 and 10, and developed a mathematical treatment of the assembly of the protein helix from the A protein to explain the titration curves.

From circular dichroism experiments, Vogel (1973) and Vogel and Jaenicke (1974) showed that the molecular conformation of the protein changed when the solution conditions changed, although the change in aggregation state may take many hours or even days. The disk formed by lowering the pH to neutrality has a different conformation from that formed by increasing the ionic strength at pH 8 (Vogel and Jaenicke, 1976; Shalaby and Lauffer, 1977).

Schuster and colleagues (Schuster et al., 1979; Shire et al., 1979a) have characterized an "overshoot" phenomenon that occurs when protein helices are formed from the A protein by raising the temperature at pH 6.5. Long helices are formed at first, but these slowly depolymerize to short helices. If the heating is very slow, three-turn helices are formed directly. They conclude from their results that under these conditions, the protein helix is formed via a slow nucleation process

$$A \rightleftharpoons D \rightleftharpoons H_2$$

and conversion to a three-turn helix

$$H_2 + A \rightleftharpoons H_3$$

followed by rapid elongation

$$H_3 + A \rightleftharpoons H$$

where A, D, and H are the A protein, disk, and long helix, while H_n is an *n*-turn helix. A detailed account of the mechanism is given by Scheele and Schuster (1974).

1.3.6. Proton Transfer during Polymerization

It has been known for many years that the transition A protein → disk → helix requires proton binding (Fraenkel-Conrat and Ramachandran, 1959; Ansevin et al., 1964; Scheele and Lauffer, 1967; Butler et al., 1972). Two protons are bound to each subunit. At low ionic strength, one proton is required for the transition A → D, and one for D → H. However, TMV has no amino acids which would normally titrate near pH 7. Fraenkel-Conrat and Ramachandran (1959) suggested that these protons were bound to anomalously titrating carboxyl groups, and Caspar (1963) suggested that the anomalous pK values are caused by pairs of carboxyl groups in the conformation shown in Figure 2. Such a structure is not energetically favorable unless the groups are forced into close proximity by the protein structure. Caspar's hypothesis is supported by titration studies (Butler and Durham, 1972) and by the molecular structures of the disk (Bloomer et al., 1978), the virus (Stubbs et al., 1977), and the protein helix (Mandelkow et al., 1981). The anomalous carboxyl groups are now generally known as "Caspar carboxyls." Their identity and their role in protein polymerization and RNA binding will be discussed further.

1.4. Assembly

1.4.1. Self-Assembly

TMV has been studied for many years as the classic self-assembling system. The "blueprint" for assembly of the constituent parts (in this case, the protein and the RNA) is built into those parts: no external machinery, such as a ribosome, or temporary structure, such as the scaffolding protein of bacteriophage P22 (King and Casjens, 1974) is required. This was shown by the reconstitution experiments of Fraenkel-Conrat and Williams (1955). Later, Fraenkel-Conrat and Singer (1959, 1964) showed that the rate of self-assembly is very dependent on solution conditions, and that assembly is most rapid near neutral pH and at room temperature. It is now realized that this is because protein disks are required for assembly

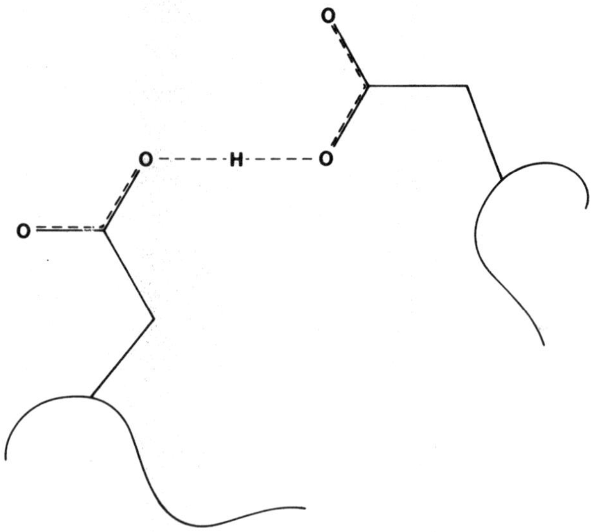

Figure 2. A pair of carboxyl groups forced into proximity so that one proton is more tightly bound and one less so, as proposed by Caspar (1963).

(Butler and Klug, 1971; see Section 2), and are available under those conditions (Figure 1).

1.4.2. Nucleation

The Protein Disk. The assembly of TMV has distinct nucleation and elongation phases. The disk form of TMV protein is essential to the nucleation phase. This was first shown by Butler and Klug (1971), who mixed TMV RNA with protein in different aggregation states. Ohno et al. (1972) chemically modified TMV protein so that it could not form disks, and showed that the modified protein could not initiate assembly, although it could add to existing protein–RNA complexes.

The early stages of TMV assembly are shown in Figure 3. The disk binds to a specific RNA sequence (see next subsection), and subsequently switches to the helical ("lockwasher") form (Butler et al., 1976). Although the disk is assumed to be in equilibrium with the lockwasher (Shire et al., 1979a), it is presumably the disk that binds, since the RNA binding site is exposed in the crystallized disk (Champness et al., 1976), but protected in the helix (Stubbs et al., 1977). The initial binding is sta-

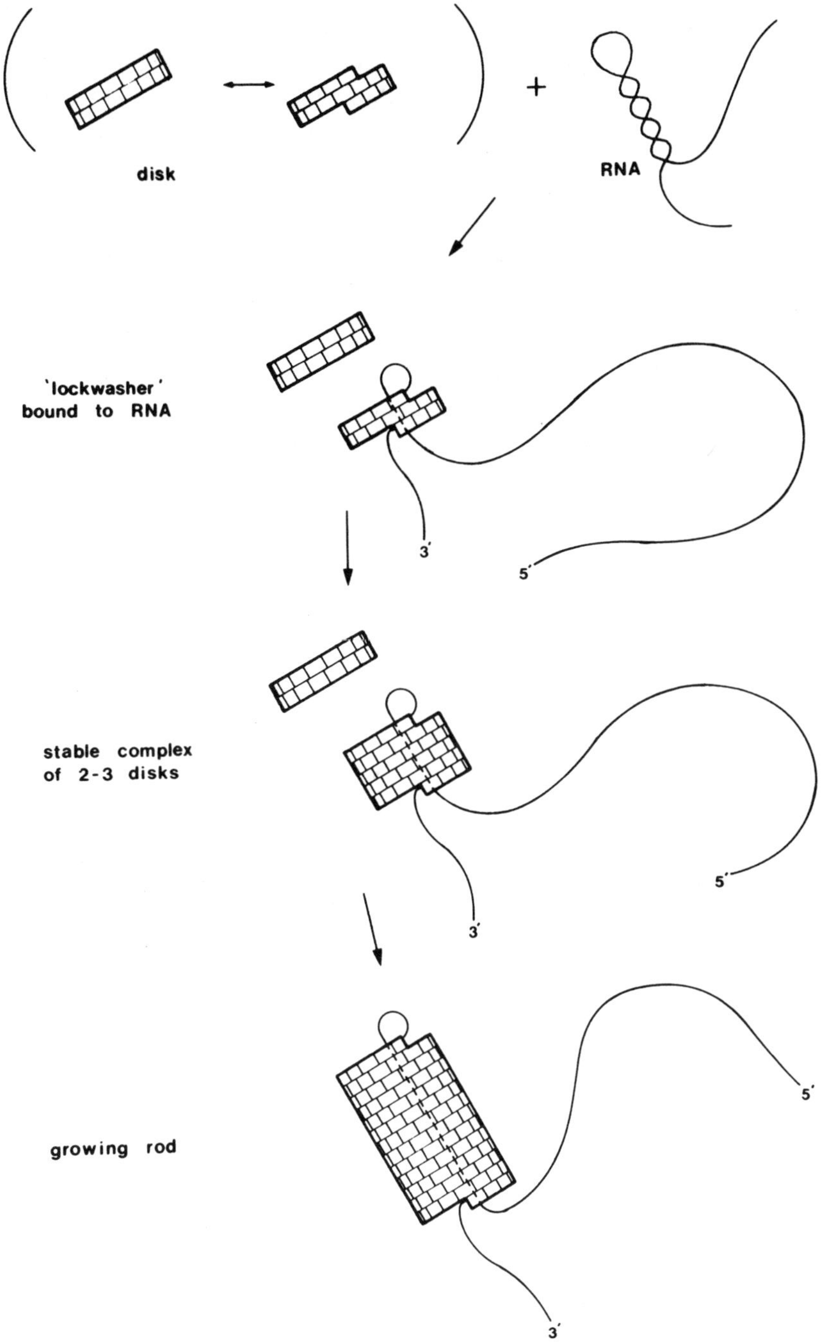

Figure 3. Early stages in the assembly of TMV.

bilized by rapid addition of a few more disks, so that the complete nucleation complex is two to three disks long. This was suggested by Zimmern and Butler (1977), who observed highly cooperative addition of protein to RNA. It is supported by the work of Zimmern (1976), who found that the average RNA fragment protected during nucleation contained about 250 nucleotides (about 2.5 disks), and by Shire et al. (1979b), who found a nucleation site of 2.6 disks by kinetic studies (although these studies were under nonphysiological conditions, at pH 6.5 and 6.5°C).

The RNA Nucleation Sequence. TMV protein assembles efficiently only with its own or closely related RNA. Within the tobamovirus group, reconstitution between RNA and protein of different strains is often very efficient (Atabekov et al., 1970; Okada et al., 1970), although the yield is somewhat lower using the Holmes ribgrass and *vulgare* strains (Holoubeck, 1962). However, whereas many spherical viruses can package a wide variety of viral and nonviral RNA (reviewed by Dodds and Hamilton, 1976) and even such nonbiological polyanions as polyvinyl sulfate (Frist, 1968), TMV protein reconstitutes only in poor yield with unrelated viral RNA (see, for example, Fritsch et al., 1973a; Atabekov et al., 1970). A very small fraction of nonviral RNA has been observed assembling with TMV protein *in vivo* (Siegel, 1971). Assembly is possible with poly-(A), poly-(I), and poly-(U), but not with poly-(C) (Fraenkel-Conrat and Singer, 1964; Lanina et al., 1976). Bearing in mind that TMV has a stoichiometry of three nucleotides per protein subunit, Steckert and Schuster (1982) have determined the binding constants for 25 trinucleoside diphosphates and TMV protein, and found relatively strong binding of AAG, with GAA, GAG, AUG, and CAG also binding strongly. They conclude that the nucleotide binding site on each protein subunit occupied by the base at the 3′ end of the trimer (the third base) is sensitive to base size, favoring purines, whereas the one at the 5′ end is sensitive both to base size and substituent, favoring purines, and NH_2 substituents away from the ribose ring.

Early studies of the binding of TMV-RNA to protein were made difficult by the unexpected "capping" of the 5′ end with an inverted 7-methylguanosine residue, so that the terminal sequence is $m^7G^{5'}ppp^{5'}G$. . . (Zimmern, 1975; Keith and Fraenkel-Conrat, 1975). This is now known to be a feature common to many eukaryotic mRNAs and eukaryotic viral RNAs (see, for example, Matthews, 1981), but it caused a number of early experiments to be misinterpreted. It was therefore initially thought that assembly begins at the 5′ end and proceeds linearly. However, a remarkable feature of the nucleation of TMV assembly is that protein binding is actually initiated at a site about one-sixth of the way along the molecule (Zimmern and Wilson, 1976). The sequence of the site was determined by Zimmern (1977), including sequence information from Jonard et al. (1977). Zimmern proposed the RNA secondary structure shown in Figure 4. The most characteristic feature is the hairpin loop which includes the sequence AAGAAG and has G in six consecutive equivalent (in the trinucleotide binding) positions. G is also in the same position in many nearby triplets. The high A content and low C content of the initiation site are also in accord with the results of Steckert and Schuster (1982) and the polynucleotide binding studies of Fraenkel-Conrat and Singer (1964) and Lanina et al. (1976). Goelet et al. (1982) have now determined the entire nucleotide sequence of TMV RNA, and they find no sequence repeats outside the nucleation region that reflect the protein–RNA binding.

Unlike the *vulgare* strain, the cowpea strain of TMV [already shown by Beachy et al. (1976) to have anomalous stripping behavior (see Section 1.2)] assembles starting near one end (Fukuda et al., 1980). This strain

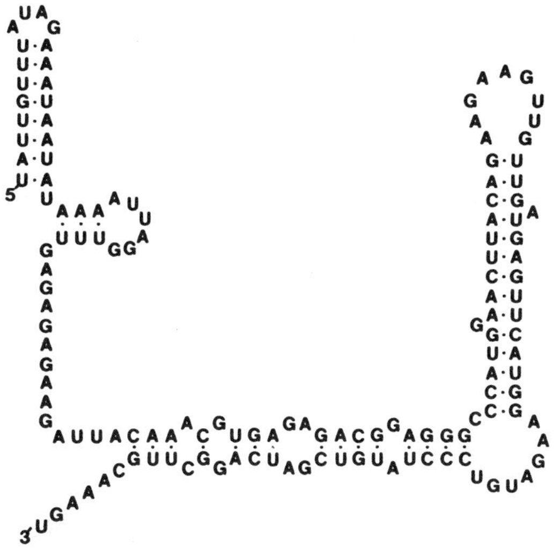

Figure 4. Sequence of the region of TMV RNA which initiates assembly, folded in the secondary structure proposed by Zimmern (1977). Note the AAGAAG sequence in the hairpin loop.

appears to nucleate at a site near the 3′ end of the RNA, inside the coat protein gene. This is also true of both the cucumber and watermelon strains of cucumber green mottle mosaic virus (CGMMV-C and CGMMV-W), tobamoviruses that have been studied by Fukuda et al. (1981). Guilley et al. (1975a,b) found a nucleotide sequence in this region of the *vulgare* strain RNA that binds specifically to protein disks, forming a complex similar to the nucleation complex. They called this sequence SERF (specifically encapsidated RNA fragment). It has a pronounced tendency for purines to appear every three residues, and it can be folded into a stable hairpin structure similar to that of the true assembly nucleation site (Hirth and Richards, 1981). The nucleation sequence of SERF codes for just that part of the protein that binds the RNA (residues 53–129; Stubbs et al., 1977), which allows the intriguing speculation that there might have been an ancestral "protovirus," perhaps containing only a fraction of the present protein gene sufficient to bind the RNA and to act as its own nucleating sequence to initiate assembly (Hendry and Durham, quoted by Hirth and Richards, 1981). However, a less speculative consideration of the evolution of TMV coat protein (McLachlan et al., 1980) will be discussed in Section 2.1.

In intact TMV RNA, the SERF fragment does not act as an alternative nucleation sequence. The RNA specificity of TMV protein is clearly a function of RNA tertiary structure as well as sequence. This is further demonstrated by the fact that removal of a few nucleotides from the 5′ end of TMV RNA, far away from the nucleation site, prevents assembly, whereas further removal of 1000–5000 nucleotides restores this capacity (Wilson et al., 1978). It is also clear [especially in the light of the complete sequence from Goelet et al. (1982)] that the specificity is primarily in the nucleation step rather than in elongation, in accordance with the prediction of Caspar (1963).

Nucleation and the Early Stages of Assembly. In summary, as shown in Figure 3, TMV assembly is initiated by the approach of the specific binding loop of RNA shown in Figure 4 to the disk (or lockwasher). The inner part of this structure is disordered, and open like a pair of jaws (Champness et al., 1976), allowing access to the RNA binding site. The nucleotides bind specifically and cooperatively, forming a stable nucleus. This is further stabilized by rapidly binding approximately two more disks. The RNA is now bound to a very short helix, with the 3′ tail coming from

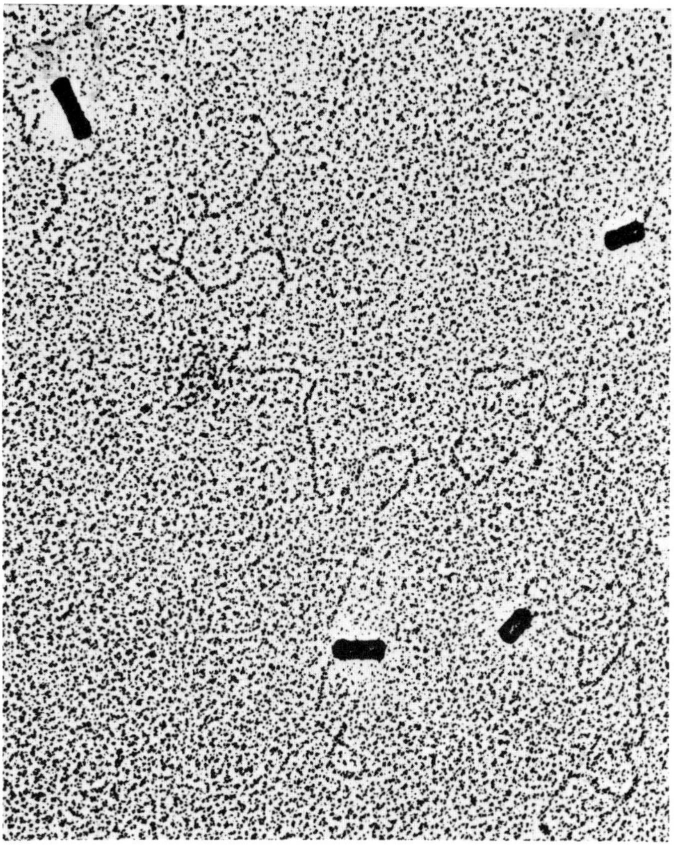

Figure 5. Electron micrograph of partially assembled TMV rods in the presence of dimethyl sulfoxide to separate the RNA tails. Two tails from one end of the growing rod are clearly visible. I thank Dr. G. Lebeurier for providing this picture.

one end of the helix, and the 5′ tail coming from the other end, but looped back down the central hole (Butler et al., 1976; Lebeurier et al., 1977). This was graphically demonstrated by Lebeurier et al. (1977), who used dimethyl sulfoxide to separate the RNA tails of short partially assembled rods, shown in Figure 5. The assembly process continues by "pulling up" the RNA through the hole in the center, and cooperatively binding 51 nucleotides to each turn of the helix as newly added protein is incorporated.

1.4.3. Elongation

Although the need for disks in nucleation has long been widely accepted, there has been continuing controversy about whether elongation from the nucleus proceeds by addition of disks or A protein. Some of the points of view have been expressed in reviews (Richards and Williams, 1976; Butler and Durham, 1977, Hirth and Richards, 1981). Butler and Klug (1971, 1972), following the turbidity of a mixture of TMV RNA and protein (as A protein, disks or a mixture), concluded that elongation occurred by addition of disks. However, Richards and Williams (1972, 1973), using electron microscopy, and Okada and Ohno (1972), using infectivity assays, suggested that A protein is responsible for elongation.

Once it became clear that nucleation occurs away from the ends of the RNA, it was obvious that elongation in the two directions should be considered separately. Lebeurier et al. (1977) showed by electron microscopy that the rods grow toward the 5′ end of the RNA much faster than toward the 3′ end. This was confirmed by Fukuda et al. (1978), who suggested as well that 5′ elongation is complete before 3′ elongation takes place. However, Lomonossoff and Butler (1979) characterized the oligonucleotides protected from nucleases during rod growth, and concluded that although 3′ elongation is very slow, it is simultaneous with 5′ elongation. They suggested a possible function for this bidirectional elongation: the 5′ end of the coat protein cistron, which is a short distance from the initiation site in the 3′ direction (Hunter et al., 1976), is protected very early in assembly. If coat protein mRNA were produced by cleavage of intact TMV RNA, this protection would avoid the formation of defective TMV particles.

Protection studies of this type had already yielded a significant result: the lengths of RNA protected by the growing TMV rod are "quantized," that is, the RNA remaining after nuclease treatment of growing rods gives a distinct banding pattern in gel electrophoresis, with spacing between bands of 100 or 50 nucleotides (Butler and Lomonossoff, 1978). This pattern persists to lengths of almost 3000 nucleotides. The sequence of TMV RNA has recently been determined (Goelet et al., 1982), and contains neither significant repeats at 50 or 100 base intervals, nor predicted hairpin loops at these intervals. Because the 17 subunits of a half-disk would bind 51 nucleotides, this is strong evidence in favor of disk addition, at least in the 5′ direction. The smearing of the bands at later stages could reflect some incorporation of A protein, perhaps in the 3′ direction. Butler and Lomonossoff (1980) suggested on the basis of kinetic studies that A protein was required for elongation in the 3′ direction (see also Lomonossoff and Butler, 1980).

Schuster and colleagues have addressed the question of protein addition directly by measuring the rate of disappearance of A protein and disks from solution as virus rods form. This is possible because equilibration between A protein and disks is much slower than virus growth under the conditions which they used (Butler and Durham, 1977). Schuster's group showed (Shire et al., 1979b) that at pH 6.5 and 6.5°C, A protein is incorporated 50–70 times faster than are disks. Growth is slow enough under these conditions for kinetic studies to be made, and they found biphasic elongation rates which they suggested could reflect different mechanisms of growth in the 3′ and 5′ directions. Under the "standard" conditions used by most groups (pH 7, 20°C, 0.1 *M* ionic strength potassium phosphate buffer), virus rods growing from a mixture of A protein and disk initially at equilibrium (74% disk) incorporated 48% disk and 52% A protein (Shire et al., 1981). Varying the fraction of disks in the solution, they found that rods could grow rapidly from almost pure disks, but that at high A protein:disk ratios there was a minimum requirement for disks: 22% of the protein incorporated was from disks. This represents about 14 disks, considerably more than the ~3 known to be required for nucleation from the work of Zimmern and Butler (1977), but much less than the number of disks which would be required to encapsidate the entire 5′ tail.

It is not known where in the rod these disks are incorporated. Fukuda and Okada (1982) have studied the assembly of RNA from the OM strain of TMV [which differs from the common strain by only two residues in the coat protein (Nozu and Okada, 1968)] with the coat protein of the cucumber green mottle mosaic virus watermelon strain (Nozu et al., 1971). It is difficult to compare this system with the more commonly used TMV *vulgare* system, but they have concluded that elongation toward the 5′ end is by A protein addition, and toward the 3′ end by disk addition. This would fit the requirement for 22% disks noted by Shire et al. (1981). However, as mentioned above, Butler and Lomonossoff (1980) found that elongation in the 3′ direction was by A protein addition. It is not necessary to assume that elongation in a given direction can proceed using only one mechanism. The disks could act as an RNA-melting protein, forming an extended "nucleation" complex or bridging RNA sequences with high melting energy, or they may simply bind regions of RNA with low protein affinity. These mechanisms are not mutually exclusive. The rest of the RNA could bind disks, A protein, or both, depending on what is available in the solution.

2. STRUCTURE

2.1. The Intact Virus

2.1.1. Early Work

Although X-ray solution scattering studies of TMV were described by Wyckoff and Corey in 1936, oriented gels of the virus were first studied by Bawden et al. (1936) and described in detail by Bernal and Fankuchen (1941). They obtained detailed diffraction patterns similar to the one shown in Figure 1, Chapter 4, and recognized that the pattern of layer lines indicates a pattern of repeating units. Watson (1954) realized that the subunits were arranged with helical symmetry, and Franklin and Holmes (1958) showed that there are 49 subunits in three turns of the helix. Franklin (1956a) and Caspar (1956) independently applied the isomorphous replacement method to the equator to calculate a radial density distribution of the virus, and Franklin showed that the RNA is at a radius of 40 Å.

2.1.2. Diffraction from Oriented Gels

Diffraction patterns (see Figure 1, Chapter 4) are obtained from gels, about 15–25% by weight virus, in capillary tubes, prepared by the method of Gregory and Holmes (1965). The long TMV particles are aligned parallel to the capillary axis, but they are randomly oriented about their long axes. Because of the random orientation, the diffraction pattern is cylindrically averaged, and the pattern represents almost the whole of the available data. [For a more detailed discussion, see Holmes and Blow (1980) and Holmes, Chapter 4.)] The problem is, therefore, to determine a three-dimensional structure from a two-dimensional data set. At low resolution, the unaveraged diffraction pattern is cylindrically symmetric because of the high symmetry of the virus, so cylindrical averaging does not cause any loss of information. In this case, the conventional protein crystallographic technique of isomorphous replacement can be used to solve the phase problem and determine the structure. This was done by Barrett et al. (1971), who calculated a map at 10 Å resolution. In that map, density attributable to the RNA was visible, and with hindsight, the map is a good low resolution representation of TMV. However, at higher resolution, cylindrical averaging does cause information to be lost, and extending resolution further requires a method of overcoming this problem. In 1975, Stubbs and Diamond described a procedure for using heavy-atom derivatives to supply the missing information. Their method is analogous to crystallographic isomorphous replacement, but requires large numbers of heavy atom derivatives. The data from the derivatives compensate for the data lost by cylindrical averaging. Using this method, Holmes et al. (1975) calculated an electron density map of intact TMV at 6.7 Å resolution. In 1977, Stubbs et al. described a map at 4 Å resolution, calculated using six heavy atom derivatives. This is the minimum number of derivatives required at this resolution.

2.1.3 Interpretation of the 4Å Map

Thanks to the large amount of α-helix in the protein structure, an abundance of chemical information, such as known heavy atom sites (Barrett et al., 1971; Mandelkow and Holmes, 1974) and the accessibility of the C terminus to carboxypeptidase (Harris and Knight, 1952), it was possible to trace much of the protein chain in the 4 Å map. At a radius greater than 65 Å, this was not possible, because the map is unclear in this region, but it has since been possible to use an improved map and the known structure of the protein disk (Bloomer et al., 1978) to interpret all the electron density (Stubbs, Namba, and Makowski, unpublished). Figure 6 is a schematic drawing of the protein subunit with the RNA, based on the most recent interpretation.

The structure is folded around a core of four approximately parallel α-helices, in a hemerythrin-type fold (antiparallel α-domain: up-and-down helix bundle; Richardson, 1981). These have been called (Champness et al., 1976) the left and right radial helices (LR and RR) and the left and right slewed helices (LS and RS). Considering the virus axis as vertical, LS and RS lie on top of LR and RR, at angles of 10–20°, the same angles

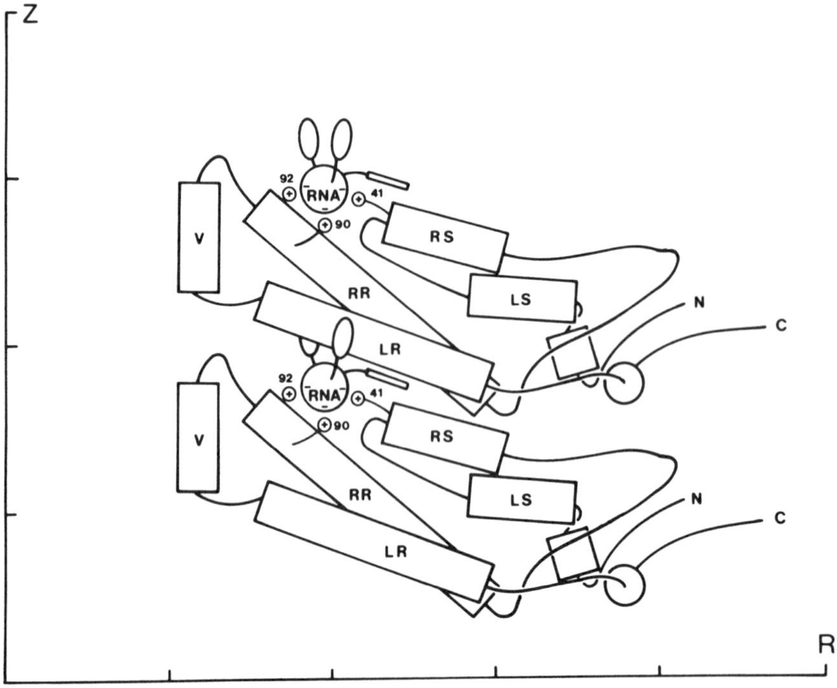

Figure 6. Schematic drawing of two subunits of TMV. The *Z* axis is the virus axis. Fiducial marks are at 20 Å intervals. α-Helices are indicated by rectangles, except for the one near the C terminus, which is seen approximately end-on in this view and is therefore shown as a circle. Interactions of Arg 41, 90, and 92 with the RNA are shown.

as found in coiled-coils of α-helices (Bloomer et al., 1978). LR and RR are connected at the inside wall of the protein (the hole down the center of the virus) by a short length of chain called the V helix (Stubbs et al., 1977). The map is difficult to interpret in this area, but V appears to contain one to three turns of rather irregular α-helix. At a radius of 40 Å, the RNA is bound to the protein, interacting in particular with LR and RR. This will be discussed in detail in Section 3.

The inner end of the protein subunit is formed by the loop of residues 90–115 as shown in Figure 7. Between adjacent subunits in this region is a pocket lined with hydrophobic residues, containing six carboxylic acids. These are Glu 95, Glu 97, Glu 106, Asp 109, Asp 115, and Asp 116. They are partly neutralized by Arg 112, and perhaps also by Arg 113. This concentration of charges has been called the "carboxyl cage" (Stubbs et al., 1977). The four residues 95, 97, 106, and 109 are probably involved in the binding of lead and uranyl fluoride in two of the heavy atom derivatives, and this metal binding site could have physiological significance (Stubbs et al., 1977), perhaps binding calcium or magnesium (Durham and Hendry, 1977; Durham et al., 1977a). The other two residues are about 10 Å away, nearer the RNA. Concentrations of carboxyl groups are extremely common in proteins, especially at cation binding sites, and the arrangement in TMV is reminiscent of the double metal binding sites of thermolysin (Matthews et al., 1972) and concanavalin A (Hardman and Ainsworth, 1972; Edelman et al., 1972), although the sites in TMV are farther apart.

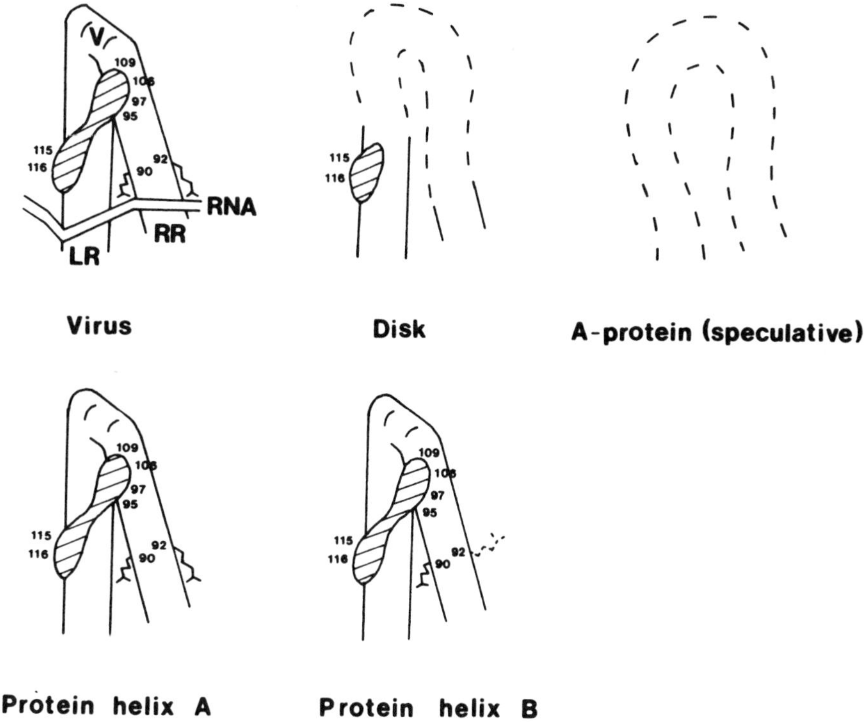

Figure 7. Schematic representation of the central loop (residues 85–120) in TMV aggregates. Numbers indicate approximate residue locations. Solid lines: ordered structure. Broken lines: disordered structure. Shading: carboxyl cage. From Stubbs et al. (1979).

Stubbs et al. (1977) identified the carboxyl cage as the site of the anomalous pK values (Caspar, 1963; Section 1.3.6) and discussed its role in virus assembly. This structure serves as a switch between different states of the protein (Caspar, 1976). In the disk, the mutual repulsion of the negative charges in the carboxyl cage is too great to allow the inner loop of protein to fold (Section 2.3.2), but protonation of the carboxyl groups at low pH (in the protein helix) or the binding energy of the RNA (in the virus) is sufficient to fold the chain and force the formation of the Caspar carboxyl pairs. The role of calcium, magnesium, or other metal ions in this process, if any, is not known, but it is of interest that these ions protect the virus from alkaline degradation (Brakke and Van Pelt, 1969; Powell, 1975; Section 1.2).

At higher radii, the virus map can be interpreted with the aid of the disk interpretation of Bloomer et al. (1978). A "cap" of β-sheet connects the ends of the four core α-helices at about 65–70 Å radius, and outside this are two short α-helices, one near the N terminus and one near the C terminus. Most of the aromatic residues are in this outer region, and together with several proline residues, form what has been called the "hydrophobic girdle," extending from one subunit to another and thus encircling the virus (Bloomer et al., 1978).

The overall protein structure contains a twice repeated pattern of two core α-helices and two strands of the β-sheet "cap." This pattern has led McLachlan et al. (1980) to propose that the TMV protein has evolved by gene duplication of a primitive dimeric protein, with the inner flexible loop developing later.

2.2. Helical Protein Forms

X-ray diffraction studies of oriented gels of the long protein helices that form at low pH were first made by Franklin (1955, 1956a). She concluded that they have a structure very similar to that of the intact virus, other than the absence of RNA. However, Mandelkow et al. (1976) showed that there are two forms of protein helix: Franklin's form, isomorphous to the virus with $16\frac{1}{3}$ subunits per turn, which they called A, and a second form, B, which has $17\frac{1}{3}$ subunits per turn. The bonding between subunits is approximately conserved despite this difference. In the B form the subunits are 5 Å further from the central axis to accommodate the extra subunit.

Mandelkow et al. (1981) calculated electron density maps of these forms at 4 Å resolution, using difference Fourier methods and the virus structure of Stubbs et al. (1977). They found only very small changes from the virus structure. The RNA is replaced by one or more anions, and there are several movements of side chains (discussed in Section 2.4), but they found no evidence of any substantial difference in main chain structure among the virus, protein helix A, and protein helix B.

2.3. The Disk

2.3.1. Crystallographic Studies

The 34-subunit disk (Section 1.3.3) crystallizes in a form suitable for X-ray diffraction studies (Finch et al., 1966). Early studies (Leberman et al., 1974; Finch et al., 1974; Gilbert and Klug, 1974) determined the symmetry and packing of the disk, and Champness et al. (1976) determined the structure of the protein at 5 Å resolution. Because there is one whole disk, MW 600,000, in the asymmetric unit, data collection posed a substantial problem. The technical difficulties were somewhat compensated for by the phasing power of the 17-fold noncrystallographic symmetry (Jack, 1973; Bricogne, 1974, 1976), which was used in conjunction with phasing information from two isomorphous heavy atom derivatives. Bloomer et al. (1978) extended the resolution to 2.8 Å. They used one heavy atom derivative which diffracted to 2.95 Å, then improved the single isomorphous replacement phases and extended them to 2.8 Å using noncrystallographic symmetry.

2.3. Structure

Most of the disk structure, including the core of four α-helices and the hydrophobic girdle, is broadly similar to that of the virus. Some of the differences will be discussed in Section 2.4.2, but the most remarkable difference is that the loop of residues 89–113, which forms the vertical rod of density V and the inner ends of LR and RR in the virus (including most of the carboxyl cage), is disordered in the disk and therefore absent from the map. Jardetzky et al. (1978) have shown by nuclear magnetic resonance studies that this loop is dynamically disordered in the disk in solution, but ordered in the virus.

Bloomer et al. (1978) were able to describe the intersubunit contacts in the disk in some detail. Between the layers, there is an extended salt bridge system involving three residues from the upper subunit, six from the lower, and two water molecules. These contacts are quite different from those in the virus, as will be discussed in Section 2.4. By contrast, the contacts within the layers are essentially the same in the disk and the virus. These consist chiefly of four alternating patches (running radially) of hydrophobic and polar residues, the outermost patch being part of the hydrophobic girdle.

2.3.3. *Oligonucleotide Binding Studies*

A number of oligonucleotides can be diffused into the disk crystals. The binding of the trinucleotide AAG (since observed by Steckert and Schuster (1982) to bind strongly to the protein helix; see Section 1.4.2), and a mixture of hexanucleotides have been studied crystallographically by Graham and Butler (1979). More recently, work has begun on the binding of the hexanucleotide AAGAAG to the disk (Bloomer et al., 1981). In all of these cases, high concentrations of oligonucleotides disrupt the crystals, presumably because the disk → helix switch is activated. However, at low concentrations the crystals remain intact and changes are seen in the electron density maps. Graham and Butler (1979) reported a peak in a 5 Å resolution difference Fourier map below the RR helix and next to LR, which they tentatively identified as the nucleotide binding site. This site appears to correspond roughly to one or two of the base binding sites in the virus (bases 1 and 3: see Stubbs and Stauffacher, 1981), as is illustrated in Figure 8. However, Bloomer et al. (1981) pointed out that this peak only appears after extensive phase refinement, whereas another peak, observed by Graham (1977), disappears during phase refinement. This alternative peak is below the LR helix, 38 Å from the disk center, which is in the region where two of the ribose units and one of the phosphates bind (Stubbs et al., 1977; Stubbs and Stauffacher, 1981). It is evident that a full structural interpretation of the binding of nucleotides to the disk will not be possible until a satisfactory map of a bound oligonucleotide (probably AAGAAG) has been obtained at high resolution.

Graham and Butler (1979) observed other, clearer effects of the binding of the oligonucleotides. There are some movements of helices relative to each other, which will be discussed further in Section 2.4. In addition,

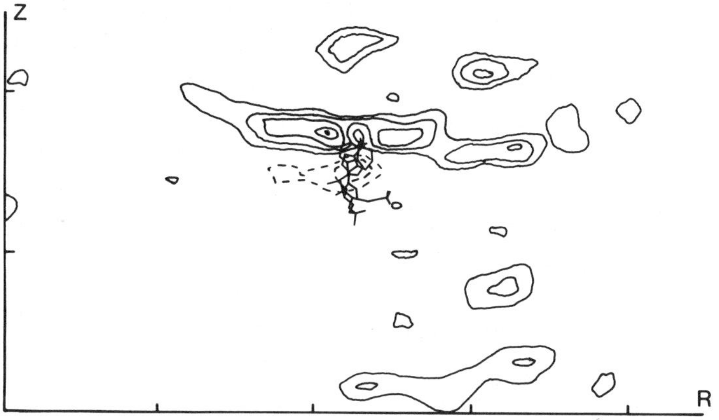

Figure 8. A section of the TMV disk hexanucleotide difference Fourier map (from Graham and Butler, 1979). Only positive contours are shown, so that the large, rodlike feature is the LR helix. The broken contours represent the peak tentatively identified as bound nucleotide, which was found on the section through the adjacent RR helix. The superimposed molecular structure represents adjacent nucleotides 3 and 1 of the RNA (Stubbs and Stauffacher, 1981). Fiducial marks are at 20 Å intervals.

the LR helix extends about 10 Å further to low radius than it does in the unreacted disk. There is also a somewhat smaller extension of the RR helix, shortening the other end of the disordered loop. All these changes indicate that the nucleotide binding causes part of the disordered peptide loop in the disk to take up the ordered virus conformation.

2.4. Comparisons of the Structures

2.4.1. Helical Structures

The various aggregates of TMV protein (including the virus) have been compared in molecular detail by Stubbs et al. (1979). Their conclusions are summarized in Figure 7. There is only one definite difference between the molecular conformations of the A and B forms of protein helix, a peak of electron density near the inner end of RR which is present in A (the $16\frac{1}{3}$ subunits per turn form) and absent in B ($17\frac{1}{3}$). This peak had been tentatively attributed by Stubbs et al. (1977) to the guanidinium group of Arg 92, forming an ion pair with an RNA phosphate. If this is the case, then the difference between the A and B forms is that this arginine retains its conformation in the A form, but is disordered in the B form. Because Arg 92 is in the side-to-side intersubunit boundary, it is not surprising that a change in its conformation could induce the observed change in helical symmetry. Mandelkow et al. (1981) also found that there could be differences in the conformation of the side chain of Arg 41, another possible phosphate binding group, but this result was not as clear as the Arg 92 difference.

Even at a molecular level, the virus and the isomorphous A form of the protein are extremely similar. The RNA in the virus is replaced by at least one bound anion in the protein helix, binding in the high radius phosphate site (see Section 3). Mandelkow et al. (1981) also found some differences at low radius, in the vicinity of the carboxyl cage. These could correspond to the fact that the anomalously titrating groups are ionized in the virus under the conditions studied (pH 8), but protonated in the protein helix at pH 5. Nuclear magnetic resonance experiments (Jardetzky et al., 1978, and Jardetzky, quoted by Mandelkow et al., 1981) indicate that the flexible loop of the disk, which includes these pairs, is slightly more mobile in the protein helix than in the virus, but the difference is too small to be detected by X-ray diffraction at 4 Å resolution.

2.4.2. *Helical Structures and the Disk*

The quaternary structure of the disk is very different from that of the virus (Fig. 9). Whereas the subunits in the virus are one-third of a subunit to the left of their lower neighbors, those of the disk are about one-fifth of a subunit to the right. In the helical core of the subunit, which makes most of the vertical contacts, these are the two possible ways of close packing such a set of α-helices. The only other gross difference is the loop of residues 89–113, referred to in Section 2.3.2, which is ordered in the virus and disordered in the disk.

Graham and Butler (1978) suggested from comparisons of the disk iodinated at Tyr 139 and the Ni2068 mutant of the virus with methylmercury nitrate bound to Cys 139 (Holmes et al., 1975) that the disk ↔ helix transition requires only a rigid body transformation. Bloomer et al. (1981) came to a similar conclusion after comparing 15 corresponding features in the electron density maps and two heavy atom markers, but found that the markers on residue 139 did not fit the transformation that best de-

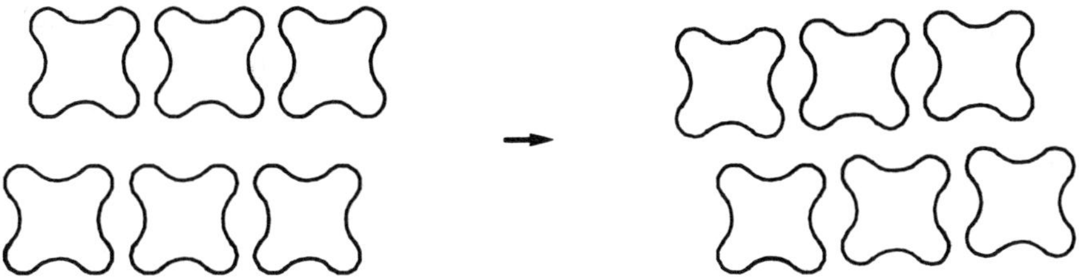

Figure 9. The quaternary structure of the disk and the virus helix. The lateral subunit interactions are the same, but the axial contacts are dramatically altered.

scribed the other features. Our most recent maps suggest that this is because Cys 139 in Ni2068 takes up a different side chain conformation from that of Tyr 139 in *vulgare*, so that the mercury marker may well be as much as 7 Å from the position of the tyrosine ring.

Although the rigid body transformation makes a good approximate correspondence between the two structures, there are many indications of small (1–4 Å) internal differences. Graham and Butler (1978, 1979) observed that either raising the ammonium sulfate concentration of the mother liquor surrounding the crystal or binding of oligonucleotides caused relative movements of the core α-helices. They found a small movement of LR, LS, and RR on binding of nucleotides. We find by comparing the disk structure (Bloomer et al., 1978) with the virus model (a revised model very similar to that of Stubbs et al., 1977) that these three helices are also in a slightly different position relative to RS in the virus. Thus, it appears that RNA binding to TMV protein is accompanied by a small change in the relative positions of the core α-helices.

Side chain movements are more difficult to find, because the virus structure is not yet at sufficiently high resolution to see more than a few aromatic and guanidinium arginine groups. Obviously, in view of the change in quaternary structure (Fig. 9), there must be considerable changes in the salt bridge system between the disk layers. Stubbs et al. (1979) and Bloomer et al. (1981) have suggested that an alternative salt bridge system could be formed in the virus, retaining part of the inner end of the network found by Bloomer et al. (1978), but making several changes at the outer end.

Spectroscopic studies have been valuable in considering side chain interactions in the different forms of TMV protein. Vogel and Jaenicke (1976) showed by circular dichroism studies that Trp 17 and Trp 52 interact in the monomer, A protein, and disk, but not in the helix. This interaction is consistent with the disk structure of Bloomer et al. (1978), and the change is consistent with the observation above that RS, which includes Trp 52, moves relative to the other helices during the transition disk ↔ helix. The NMR studies of Jardetzky et al. (1978), which showed that the disordered loop in the disk crystals is also disordered in solution, have already been mentioned. Other studies by CD and NMR have been made which examined differences among mutants of TMV (see Vogel et al., 1980, and references therein).

2.4.3 Comparisons with the A Protein

As no direct structural studies of the A protein have been made, comparisons must use data from chemical and spectroscopic experiments. Vogel (1973) pointed out that circular dichroism spectra indicate that the gross conformations of the A protein and the disk are the same, although there are some differences in aromatic side chain conformation. Vogel and Jaenicke (1976) identified these differences as coming from Tyr 70 and Tyr 72, and considering ultraviolet spectra as well, attributed them to an increase in side chain mobility as the disk is formed. These tyrosines are near Tyr 139, which is known from chemical experiments to be involved in disk formation (Ohno et al., 1972).

Titration studies have given information about the conformation of the anomalous carboxyl groups (Caspar, 1963) located in the carboxyl cage (see Section 2.1.3). Durham et al. (1977a) suggested that while there are two or three anomalous pKs in the virus and protein helices (Caspar, 1963) and one in the disk (Vogel and Jaenicke, 1974), there are none in the A protein. Stubbs et al. (1979) interpreted this as an indication of a progressive unfolding of the flexible loop. In the virus, all six carboxyl groups in the loop are ordered and ionized anomalously, giving rise to three anomalous pKs. In the disk, only the pair 115–116 remains ordered. They suggested that the LR helix may unwind further in the A protein, so that this pair is also disordered. This possibility is included in Figure 7.

3. THE NUCLEIC ACID BINDING SITE

3.1. The Nucleic Acid Conformation

3.1.1. Determination by Fiber Diffraction Methods

Although the RNA was visible at low resolution in the maps of Barrett et al. (1971) and Holmes et al. (1975), the first attempt at model building based on an electron density map was made by Stubbs et al. (1977), using their 4 Å map, which was unusually clear for a 4 Å nucleic acid map. The difference Fourier maps of the A form of the protein helix (Mandelkow et al., 1981) suggested a small change in the RNA model, and Stubbs and Stauffacher (1980, 1981) rebuilt this model by computer methods. They used a map which was essentially the same as the 1977 map, but improved by the "box function" method of Makowski (1981). Their final model fitted the electron density well (Fig. 10), with no excessively close interatomic contacts, and had the required helical symmetry. Good stereochemistry was maintained using the method of Dodson et al. (1976). Bases were constrained to be planar, and bond lengths and angles were restrained to lie within the ranges observed in small molecules (Arnott, 1970; Saenger, 1973).

The RNA lies between the protein layers in the virus, in contact with the left radial helix of the subunit above it and the right radial helix of the subunit below it (Fig. 6 and 11). All three sugars in the basic repeating unit have the 3′-endo conformation, with σ angles (see Stubbs and Stauffacher, 1981, or Arnott, 1970, for nomenclature) of 83°, 88°, and 81°, all well within the range usually found for 3′-endo sugars (Arnott, 1970). The bases in this model are all adenines, which is a reasonable approximation at 4 Å resolution. Two of the bases are in the anti conformation, but purines in the third position (base 3 in Fig. 11) are in the much more unusual syn conformation (Fig. 10b). The conformation of the pyrimidines in this position is not clear, but they could also be syn. Anti bases are overwhelmingly favored by simple energetic considerations (Haschemeyer and Rich, 1967), and syn bases have been observed only rarely in small structures (Arnott, 1970; Sundaralingam, 1973), but a syn base was observed in the structure of ApA:proflavine by Neidle et al. (1978), who suggested that this structure could serve as a model for single-stranded

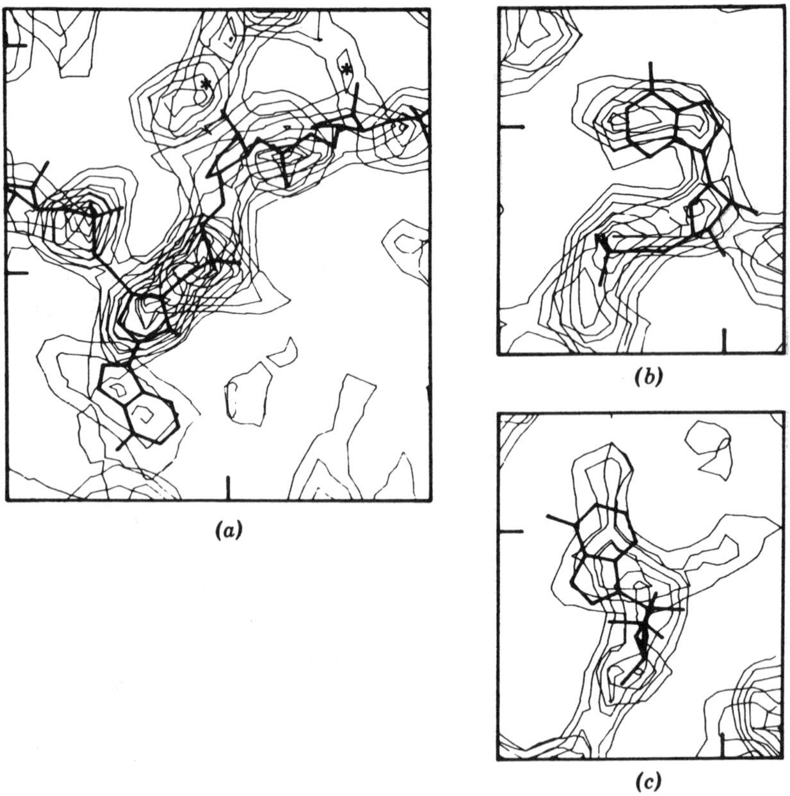

Figure 10. The RNA model (heavy lines) with the corresponding electron density (light lines). Fiducial marks are at 10 Å intervals. (*a*) Viewed down the *Z* axis. Bases 1 and 3 protrude out of the plane toward the reader, so they are omitted. (*b*) The syn base of nucleotide 3, with its sugar and 3′-phosphate. (*c*) Nucleotide 1. (*b*) and (*c*) are viewed approximately from right to left of (*a*). From Stubbs and Stauffacher (1981).

regions of RNA bound to drugs or proteins. Syn bases in small molecules have been observed in solution using nuclear magnetic resonance (Tran-Dinh et al., 1972; Guéron et al., 1973; Neumann et al., 1979) and circular dichroism (Guschlbauer and Courtois, 1968). They had not been observed in larger structures at all until recently, but examination of the data of Haschemeyer and Rich (1967) shows that they could exist with only small distortions of the sugar ring. Syn purines have now been found by crystallographic methods in Z-DNA (Wang et al., 1979; Crawford et al., 1980;

Drew et al., 1980; Arnott et al., 1980), and by NMR in tRNA (Roy et al., 1982).

3.1.2. Confirmation by Other Methods

Some aspects of the structure described by Stubbs and Stauffacher (1981) are confirmed by other methods. The orientation of the phosphate groups relative to the particle axis has been confirmed within experimental error by solid-state NMR (Cross, Opella, Stubbs, and Caspar, unpublished results), in which the three separate resonances from the phosphate groups have been observed. The orientation of two of the three bases is consistent with the observations of Franklin and Commoner (1955) and Franklin

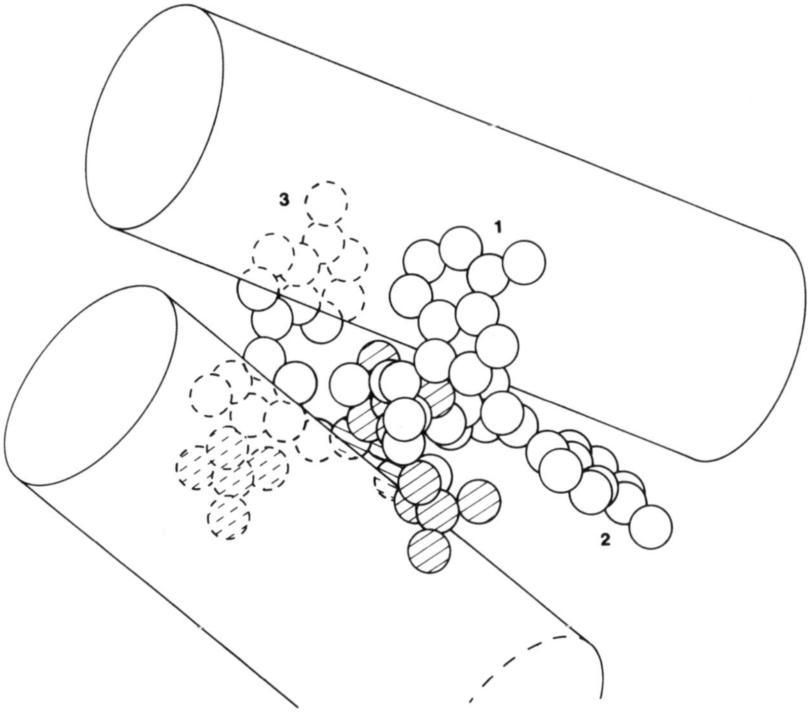

Figure 11. The RNA of TMV between the left radial α-helix of the protein (top) and the right radial helix of the subunit below (bottom). Numbers identify the three nucleotides referred to in the text. The 5′ end is toward the reader. Phosphate groups are shaded.

(1955), who compared the birefringence of TMV and TMV protein, and concluded that the planes of the bases were more parallel to the direction of the particle axis than perpendicular to it. This was in turn consistent with the results of UV dichroism studies by Perutz et al. (1950) and Seeds and Wilkins (1950). The absence of base stacking and, in part, the deviation of sugar conformation from the standard A-RNA structure (Arnott et al., 1976) have been confirmed by Raman spectroscopy (Fish et al., 1981).

3.2. Protein–Nucleic Acid Interactions

Descriptions of the RNA binding site are hampered by the relatively low resolution of the 4 Å virus map, so arguments have often had to depend on the invariance of key amino acid residues and the disk map (Bloomer et al., 1978) as well as on the virus map (Stubbs et al., 1977). Figure 12 illustrates the structure of TMV with invariant amino acids shown. There are two main clusters of such residues: one at 75 Å radius, which includes

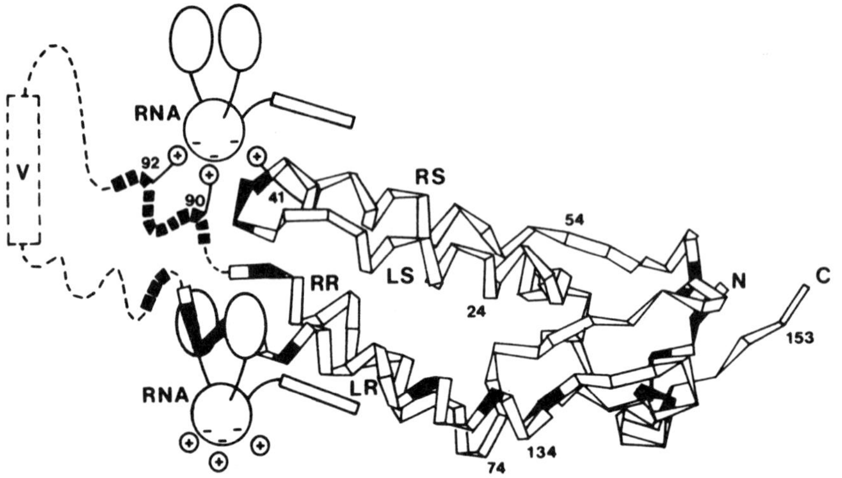

Figure 12. The TMV subunit with invariant amino acids shown in black. Invariance is based on data from the references quoted in Table 1 and by Durham and Butler (1975). Note the clusters of invariant residues around the RNA binding site and near the outer surface of the virus. [The molecular structure shown is from Holmes (1982), and is based on a rigid body transformation of the disk coordinates, so it is only an approximate representation of the detailed virus structure. This does not affect the point of the illustration.]

part of the hydrophobic girdle, and a much larger one at 40 Å radius, which includes most of the RNA binding site. The binding site includes three parts of the peptide chain: residues 88–94 on the right radial helix, residues 113–120 (and possibly further) on the left radial helix, and the connecting loop between the slewed helices, 35–41.

The RNA phosphates form ion pairs with arginines (Caspar, 1963). Arginines are potentially better binding sites for phosphate groups than are lysines or histidines, because of their greater hydrogen bonding capacity (Cotton et al., 1973). Stubbs et al. (1977) identified two of these as Arg 90 and Arg 92, but were unable to determine whether the third was Arg 41 or Arg 113. [The possibility of more than one arginine binding a phosphate, as for example in *Staphylococcus* nuclease (Cotton et al., 1973), cannot be completely excluded.]. Arg 41 points away from the RNA binding site in the disk (Bloomer et al., 1978), but this part of the protein is on the edge of the disordered region in the disk, and at the end of the RS helix, which probably moves slightly during the disk ↔ helix transition (Section 2.4.2), so a conformational change would be possible. Arg 113 is disordered in the disk, and difficult to locate in the virus map, so an improved map of the virus will be needed to completely resolve this matter.

Durham et al. (1977b) have suggested on the basis of titration experiments that one of the phosphates might interact with a carboxyl group to form a carboxyl phosphate pair similar to the Caspar carboxyl carboxylate pairs. However, preliminary NMR results (Cross and Stubbs, unpublished) do not show any change in ^{31}P NMR spectrum of TMV between pH 5.5 and 8.0.

The "saddle" or "claw" shape of the RNA triplet wraps around the LR helix of the subunit above it. Stubbs and Stauffacher (1981) suggested that the ribose hydroxyl groups of nucleotides 1 and 3 could be hydrogen bonded to protein side chains. Holmes (1979) had suggested, on the basis of model building, that this could be the function of the invariant carboxylic acid pair Asp 115-Asp 116, located on the LR helix, although the electron density map indicates that the ribose hydrogen bonds are with atoms from the RR helix below the RNA, rather than LR above. Once again, the matter cannot be resolved without an improved map. Whether ribose 2 is hydrogen bonded to protein is not clear. Holmes (1979) suggested that if it is, it might be with the "hairpin" loop connecting LS and RS below the RNA, perhaps Asn 33.

The three RNA bases lie almost flat against the surface of the LR helix, in contact with the methyl or methylene groups of the protein side chains. This hydrophobic surface can accommodate any base sequence, as required. Holmes (1979) suggested that Ser 123 might form a hydrogen bond with the base in position 2.

3.3. Control of Assembly

The carboxyl cage, located at the inner end of the protein subunit, just inside the RNA binding site (Fig. 8), plays a vital part in the assembly of TMV and the binding of the RNA. However, the details of its structure remain uncertain. Only two of the six groups involved (115 and 116) are invariant, although it is of interest to note (Table 1) that when residue 112 is positively charged (arginine or lysine), there are three or more acidic groups in this region, and when 112 is uncharged, only two are observed. The only definitely known exception to this rule, the mutant PM5, which has Cys instead of Arg 112, is defective, being unable to assemble into virus rods (Hariharasubramanian and Siegel, 1969). The invariant pair 115/116 is near the RNA and may interact with it. To what degree this pair also functions as part of the disk ↔ helix electrostatic switch (or perhaps as a switch between A protein and disk) is not clear. The remaining four carboxylic acids probably form one or two metal binding sites (Section 2.1.3). It seems possible (Table 1) that these sites sometimes utilize acids introduced in other positions, and that the detailed structure of the cage varies considerably among the strains. The U2 strain of TMV retains the anomalous titration behavior of *vulgare* (Butler and Durham, 1972), but it has been shown at low resolution to have a considerably altered structure in the vicinity of the carboxyl cage (Holmes and Franklin, 1958; and an unpublished 7 Å difference Fourier map by Stubbs, Warren, and Morris). An analogous situation exists in two strains of southern bean mosaic virus. The cowpea strain has a carboxyl cage binding Ca^{2+} which includes Glu 194 from three neighboring subunits (Hermodson et al., 1982). Many of the amino acids in this cage are surprisingly different in the bean strain (Mang et al., 1982), perhaps with other groups assuming their function (Rossmann et al., 1983a). Thus, it appears that despite their importance in viral assembly (see also Section 4.2.2), the detailed structure of carboxyl cages is not strongly conserved, in sharp contrast to functional sites in most other proteins.

Table 1. Carboxylic Acids (Including Uncertain Possibilities) in the Inner Loop of Various TMV Strains[a]

Residue Number in *vulgare*	*vulgare*	Dahlemense	U2	HRG	CGMMV-C	CGMMV-W	Cow Pea	PM5
95	Glu	Glu	Glu	Gln	Glu	Glu	Glu	Glu
97	Glu	Glu	Asn	Glu	Glx	Val	Glu	Glu
98	Asn	Asn	Asn	Glu	Asx	Asp	Asn	Asn
99	Gln	Gln	Gln	Gln	Pro	Pro	Ser	Gln
101	Asn	Ser	Ala	Arg	Glu	Asn	Glu	Asn
106	Glu	Glu	Glu	Gln	Glu	Glu	Glu	Glu
109	Asp	Asp	Val	Asn	Asn	Asn	Asn	Asp
111	Thr	Thr	Ala	Thr	Thr	Val	Val	Thr
114	Val	Val	Val	Val	Asx	Thr	Thr	Val
115	Asp	Asp	Asp	Asp	Asx	Asp	Asp	Asp
116	Asp	Asp	Asp	Asp	Asx	Asp	Asp	Asp
112	Arg	Arg	Gln	Gln	Lys	Lys	Arg	Cys

[a] Sequences are from Wittmann-Liebold and Wittmann (1967) (*vulgare*, Dahlemense, U2 and Holmes ribgrass); Nozu et al. (1971) (cucumber green mottle mosaic virus, cucumber and watermelon strains); Rees and Short (1975) (cow pea); and Hariharasubramanian and Siegel (1969) (PM5). The U2 sequence has been revised in accordance with the findings of Altschuh et al. (1981).

4. TOBACCO MOSAIC VIRUS AND OTHER STRUCTURES

4.1. General

TMV occupies a position in the taxonomy of known protein structures similar to that of the platypus in the animal kingdom. It is a virus, and it has an assembly control site (carboxyl cage) similar to that of other viruses, although the site bears more resemblance to the carboxyl cages of concanavalin A and thermolysin. However, its protein fold is similar to that of the hemerythrins and some of the cytochromes, whereas the nucleic acid binding site binds bases and sugars in a way that has some features in common with the dehydrogenases, and phosphates like those in the complex of staphylococcal nuclease and thymidine 3′, 5′-bisphosphate. The conformation of one of the bound nucleotides is almost identical to that of a nucleotide bound to ribonuclease T_1.

The four antiparallel helix fold occurs in a wide variety of apparently unrelated proteins (Richardson, 1981). This fold is a very stable structure (Hol et al., 1981; Sheridan et al., 1982), and may play an important part in stabilizing the quaternary structure of multisubunit proteins including TMV (Sheridan et al., 1982). The positive ends of the dipoles of one or two of the α-helices in TMV may stabilize the negatively charged nucleic acid phosphates (Hol et al., 1978), but if so, this is presumably a feature of the individual α-helices, and not of the four helix fold, because none of the other proteins containing this fold is known to bind nucleotides to the α-helical domain.

Although TMV protein is not known to resemble any other nucleotide or nucleic acid binding protein topologically, there are several groups of proteins with functional and chemical similarities to TMV, which are discussed next.

4.2. Nucleotide Binding Proteins

The classic protein structure for binding individual nucleotides (as opposed to nucleic acids) is the Rossmann fold, originally found by comparing lactate dehydrogenase with other structures which bind nucleotides (Rao and Rossmann, 1973). This fold (a particular arrangement of three β-strands and two α-helices or something similar to it) is found in most nucleotide binding proteins (Schulz and Schirmer, 1979). It has also been found in tyrosyl-tRNA synthetase and suggested as a possible binding site for the 3′-terminal adenosine of the tRNA (Bhat et al., 1982).

Studies of these proteins with bound nucleotides (e.g., Eklund et al., 1981; Grau et al., 1981) have found that the aromatic base lies in a pocket lined by aliphatic hydrophobic side chains. This is similar to what is seen in TMV, although the tertiary structure providing the side chains is totally different. A strongly conserved residue in the Rossmann fold is an aspartic acid at the end of a β-strand which forms a hydrogen bond with the ribose 2′-hydroxyl. Holmes (1979) has suggested that Asp 115 and Asp 116 in TMV (on the LR helix) form such hydrogen bonds with the ribose 2′-hydroxyl in bases 1 and 3 (see Section 3.2). Phosphate groups bound to Rossmann folds interact with arginines or lysines, as they do in many proteins, including TMV. However, both phosphate and ribose groups make several hydrogen bonds with main chain amide groups in the nucleotide binding proteins, a type of interaction not possible in TMV because of the largely α-helical nature of the binding site.

4.3. Nucleic Acid Binding Proteins

Proteins that bind nucleic acids bind either single-stranded or double-stranded nucleic acids. The structures of three double-stranded nucleic acid binding proteins, the catabolite gene activator protein (McKay and Steitz, 1981; McKay et al., 1982), the *cro* repressor (Anderson et al., 1981), and the operator binding domain of the λ repressor (Pabo and Lewis, 1982) have been determined, and found to have similar structures to each other at the proposed DNA binding sites (Steitz et al., 1982). Sequence comparisons and model building suggest that the similarity extends to many other such proteins (Anderson et al., 1982; Matthews et al., 1982; Sauer et al., 1982). However, the nucleic acid binding site bears neither structural nor chemical resemblance to that of TMV, because these proteins are required to recognize a specific secondary structure, the double helix, and to recognize specific base sequences, which they appear to do by forming hydrogen bonds with the bases (Anderson et al., 1981; Ohlendorf et al., 1982).

The gene 5 product of bacteriophage fd has more in common functionally with TMV, being a single-stranded nucleic acid binding protein. However, this protein binds DNA in a long groove formed by β-strands, apparently interacting with bases by stacking them with aromatic side chains (McPherson et al., 1979; McPherson et al., 1980). This is not at all like the RNA binding site in TMV, which may be because there is a sense in which gene 5 protein is a double-stranded nucleic acid binding protein (McPherson et al., 1980).

By contrast, the nucleases are true single-stranded nucleic acid binding proteins, and several of these show structural similarity to TMV. The structure of staphylococcal nuclease with thymidine 3′,5′-bisphosphate and calcium bound has been determined and refined at 1.5 Å resolution by Cotton et al. (1979). This structure has many chemical similarities to TMV. The 5′-phosphate of the inhibitor is bound by the guanidinium ions of two arginine residues, and also acts as a Ca^{2+} ligand. The other ligands of the calcium ion include two aspartic acids and, via a water molecule, a glutamic acid. The latter, Glu 43, is very close to Asp 19, and could form a carboxyl pair (Caspar, 1963). Although the structure is topologically unrelated to TMV and the nucleotide binding serves a totally different purpose, the proximity of carboxylic acids, arginines, a calcium ion, and a phosphate group has much in common with the region of TMV near Asp 115/Asp 116 (Section 2.1.3). More detailed parallels cannot be drawn given the present resolution of the TMV map.

Ribonucleases also use carboxylic acids in their active sites, in a way that provides an interesting parallel with the viruses and other carboxyl cage structures. Ribonuclease A and ribonuclease T_1, which are structurally quite different, both contain a carboxylic acid–histidine pair in the active site. Asp 121 is hydrogen-bonded to His 119 in ribonuclease A (Wlodawer et al., 1982), whereas Glu 58 is hydrogen bonded to His 40 in ribonuclease T_1 (Heinemann and Saenger, 1982). The bridge proton in such a pair has a raised pK (Iida and Ooi, 1969; Heinemann and Saenger, 1982), analogous to the raised pK of the bridge proton in the Caspar carboxyl pairs of TMV. A charge cluster including a histidine is also seen in southern bean mosaic virus (Rossmann et al., 1983a).

An interesting feature of the ribonuclease T_1 structure is the conformation of the bound 2′-guanylic acid (Fig. 13). Like the base in position 3 of TMV, it is in the syn conformation and is attached to a 3′-endo sugar. This provides yet another example (see Section 3.1.2) of the syn conformation stabilized by intermolecular interactions, in this case a protein–nucleic acid interaction analogous to TMV.

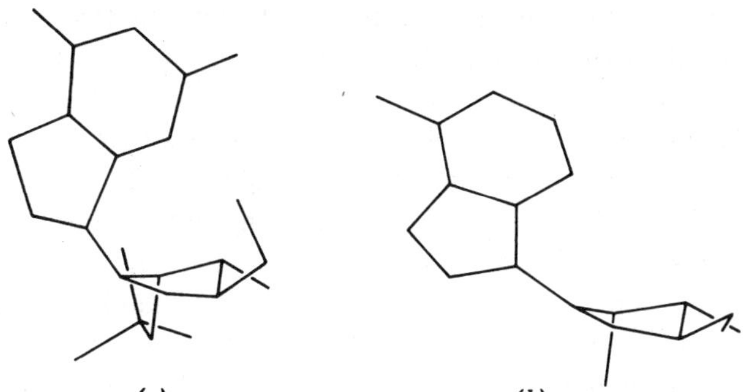

Figure 13. (*a*) Conformation of 2′-guanylic acid bound to ribonuclease T_1 (Heinemann and Saenger, 1982). (*b*) The extremely similar conformation of nucleotide 3 in TMV.

4.4. Viruses

4.4.1. *Nucleic Acid Binding*

TMV is the only virus for which the nucleic acid binding site has been directly visualized. However, there is considerable indirect evidence that the protein–nucleic acid binding in TMV is quite different from that in many other viruses. Molecular models of four other viruses have been put forward on the basis of electron density maps: the filamentous phage Pf1 at 7 Å resolution (Makowski et al., 1980), recently extended to 4 Å resolution (Nave et al., 1981; Bryan et al., 1983), and the three icosahedral plant viruses, tomato bushy stunt virus (TBSV) at 2.9 Å resolution (Harrison et al., 1978; Harrison, 1980), southern bean mosaic virus (SBMV) at 2.8 Å (Abad-Zapatero et al., 1980; Rossmann et al., 1983a), and satellite tobacco necrosis virus (STNV) at 3.0 Å (Liljas et al., 1982).

Pf1 is a filamentous bacteriophage containing a single, circular DNA molecule encapsidated in a helical assembly of protein subunits. Although all the filamentous bacteriophages studied so far have coat proteins made up of α-helices running almost parallel to the particle axis, the different symmetries and stoichiometries of different bacteriophages suggest that the protein–DNA binding is not always the same (Makowski and Caspar, 1981, see also Chapter 6). Furthermore, although the particles have helical symmetry, their molecular structures have little in common with TMV.

The DNA appears to be near the center of the particle (Day et al., 1979; Makowski et al., 1980), and models have been proposed for several bacteriophages in which the bases point out and interact with the protein, so that the phosphates do not bind to the protein at all (Day et al., 1979). More reliable models will require higher resolution studies, but it seems extremely unlikely that the nucleic acid binding sites in these models will particularly resemble that of TMV.

The three icosahedral viruses all have an approximately spherical outer protein shell surrounding a core of RNA. They have many structural similarities with each other, including one domain (the S or shell domain) of very similar tertiary structure, despite the fact that TBSV and SBMV are $T = 3$ viruses of 180 subunits, while STNV is a $T = 1$ virus of 60 subunits (Caspar and Klug, 1962). The RNA in the center of the virion is not seen in the electron density map of any of these viruses. The N terminus of the protein and a number of residues varying from 13 in STNV to perhaps 50 or 60 in TBSV are inside the outer protein shell and are also not seen in the maps. Nuclear magnetic resonance experiments suggest that this is not because of dynamic disorder (Munowitz et al., 1980; McCain et al., 1982, 1983), but because the RNA structure does not conform to the symmetry of the virion. Although the protein–RNA interactions have not been observed directly, models have been proposed on the basis of sequence for alfalfa mosaic virus, brome mosaic virus and STNV (Argos, 1981), and for SBMV (Rossmann et al., 1983b). Argos predicted that the highly basic, disordered N-terminal arm could form an α-helix which would interact with the RNA; in fact, in STNV, where much more of the N-terminal chain is visible in the electron density map than in SBMV or TBSV, three turns of α-helix are seen (residues 14–24). Although this model has some features in common with the RR helix of TMV (basic residues interacting with phosphate groups), there is little detailed correspondence. In the prediction of Rossmann et al. (1983b), basic residues from several different secondary structure elements interact with the phosphates in double-helical RNA. Because this proposal is for double-helical RNA, it is not surprising that it is also very unlike TMV RNA binding.

4.4.2 Carboxyl Groups as Control Sites

Although the viruses mentioned above do not bind RNA in a way comparable to TMV, many of them contain structures analogous to the carboxyl cage of TMV. Many icosahedral viruses swell with increasing pH, and this has been attributed to ionization of carboxyl-carboxylate pairs (Bancroft et al., 1967; Bancroft, 1970) in an analogous manner to that of TMV (Caspar, 1963; Stubbs et al., 1977). Furthermore, it was suggested that these pairs are divalent metal binding sites, again in close analogy to TMV (see Section 2.1.3).

All three icosahedral viruses that have been visualized in molecular detail contain carboxyl cages that bind calcium ions. In TBSV, there are two ion binding sites (probably binding calcium) close together in the intersubunit contacts, using at least four aspartic acid side chains (87, 115, 118, and 120) as ligands (Harrison, 1980). Removal of the metal ions at pH above 7 allows a reversible expansion of the virus (Hogle et al., 1983). The expanded form has been studied by X-ray diffraction to 8 Å resolution by Robinson and Harrison (1982). SBMV has a cluster of carboxylic acids and a lysine (Asp 204, Asp 214, Asp 138, which is equivalent in the sequence to one of the TBSV calcium liganding residues, and Lys 200) at this site, despite the fact that the main chain conformation in this region differs significantly between the two viruses (Rossmann et al., 1983a). However, there is a larger cluster in SBMV, the calcium-binding carboxyl cage of Abdel-Meguid et al. (1981), which includes not only Glu 194 from three subunits, acting as ligands, but Asp 188, Glu 191, Glu 252, Arg 196, and Arg 249 (Fig. 14). The calcium-binding site associated with this cluster is located on the quasi-threefold axis. Like TBSV, SBMV swells when divalent metal ions are removed (Rayment et al., 1979), and under certain conditions, calcium is required for assembly of SBMV (Savithri and Erickson, 1983). In STNV there is a calcium-binding site on the icosahedral

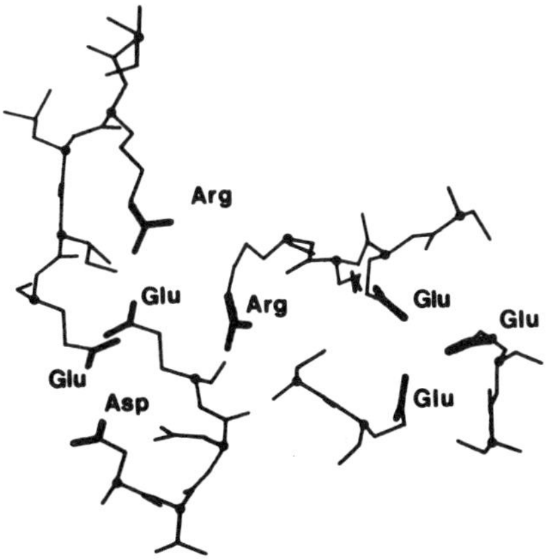

Figure 14. The calcium binding carboxyl cage of southern bean mosaic virus. The three glutamic acid residues on the right are from different subunits, related by the quasi-threefold axis. I thank Dr. Michael Rossmann for providing details of this structure.

threefold axis, with three Asp 55 side chains acting as ligands (Liljas et al., 1982). Rossmann et al. (1983c) have compared the structures of the small plant viruses (see also Chapters 1–3).

In addition to this structural evidence, there is a considerable body of chemical evidence relating carboxyl groups and divalent metal ions to virus assembly. Mg^{2+} is required for the assembly of cowpea chlorotic mottle virus (CCMV), brome mosaic virus, and broad bean mottle virus. Ca^{2+} can be used instead of Mg^{2+} for CCMV. These viruses are believed to use carboxyl–carboxylate pairs binding metal ions to control assembly (Bancroft, 1970). The addition of Mg^{2+} causes cucumber mosaic virus (CMV) to aggregate (Kaper, 1969; Bancroft, 1970), but Durham et al. (1977b) have suggested that this could be a specific degradative process catalyzed by Mg^{2+}, and need not mean that CMV does not have an assembly process like that of the viruses mentioned above. Tobacco rattle virus, a helical virus unrelated to TMV, has very different aggregation behavior from TMV; for example, it forms disks from low molecular weight protein as pH is raised rather than lowered (Fritsch et al., 1973b; Morris and Semancik, 1973). Nonetheless, it does bind divalent cations, and the binding is associated with groups in the virus (but not in the RNA-free protein) titrating with anomalous pK values (Durham and Abou Haidar, 1973).

On the basis of these results, it is tempting to speculate that assembly and disassembly of all simple plant viruses is controlled by carboxyl cages binding divalent metals. [The significance of metal binding in plant virus assembly has been stressed by several authors, including Bancroft (1970) and Durham et al. (1977b).] Carboxyl cages provide a sensitive switch between alternative conformational states of the protein, easily activated at physiological pH because of the raised pKs of the hydrogen bonded pairs of carboxylic acids (Caspar, 1963; Bancroft, 1970). An analogy could be drawn between the helical states of TMV, in which the carboxyl groups are stabilized by protons or RNA binding, and the intact icosahedral viruses, and similarly between the disk form of TMV protein and the swollen forms of the icosahedral viruses. All the viruses whose structures are known in molecular detail contain carboxyl cages with a number of common features. They contain several carboxyl side chains in close proximity, which are often stabilized by arginine or lysine side chains. They bind divalent metal ions (often calcium). The amino acid sequences forming the cage are highly variable, even among related viruses, so that side chains having equivalent function may occur in different parts of the sequence in different virus strains. Carboxyl cages occur at protein–protein interfaces, and the side chains that form the cage come from more than one protein subunit.

5. CONCLUSION

5.1. Summary

The structure of the nucleic acid in TMV is now reasonably clear, as are the types of interactions that it makes with the protein: ionic (and probably hydrogen-bonded) through the phosphates, hydrogen-bonded through the ribose rings, and hydrophobic through the bases (Stubbs and Stauffacher, 1981). It is clear that arginine residues are used to make the phosphate binding sites, but there are ambiguities about which side chains are involved in the other interactions. The protein fold is clear, but the conformations of most of the side chains cannot be determined from the present density map (Stubbs et al., 1977). The structure of the protein in the absence of the RNA is reasonably well understood for the helical forms of TMV protein from 4 Å resolution maps (Mandelkow et al., 1981), but much better understood for the disk form, which has been crystallized and studied at 2.8 Å resolution (Bloomer et al., 1978). To understand the process of assembly adequately, these forms must be compared, and although this has been done at low resolution (e.g., Holmes, 1979; Stubbs et al., 1979; Bloomer et al., 1981), it cannot be done in detail until a higher resolution map of the virus is available.

Nonetheless, two points are clear. The first is that TMV binds RNA in a way that is unique among protein–nucleic acid interactions that are understood at present, although the binding has a number of chemical features which are found in other systems. This uniqueness stems from the fact that TMV protein must bind single-stranded RNA (in contrast to the activators and repressors, and perhaps the icosahedral viruses), and must not bind it in a strongly sequence-specific way, although some sequence specificity is probably required to recognize the RNA initiation site. The second is that control of the assembly of TMV resides in the "carboxyl cage" (Stubbs et al., 1977), a group of six carboxylic acids partially neutralized by one or two arginines and probably one or more divalent metal ions, close to and perhaps interacting with the RNA binding site. Such carboxylic acid clusters are found in all four of the viruses which have been visualized in molecular detail, and may be a general mechanism for control of assembly in simple viruses.

5.2. Prospects

The most important requirement for further understanding of the TMV structure is increased resolution of the electron density map. However, it will also be valuable to examine other strains of TMV, particularly those that differ in structure around the carboxyl cage or the RNA binding site. Such studies will substantially improve our understanding of the relationship between the carboxyl cage and the assembly process. Early X-ray diffraction studies of the strains cucumber virus 4 and U2 were made by Franklin (1956b), and these and the cowpea (Nigerian bean) strain were studied by Holmes and Franklin (1958). The U2 strain is particularly interesting because of its altered structure in the vicinity of the carboxyl cage (see Section 3.2), and is particularly suited for study because of its enhanced layer-line splitting (Franklin and Klug, 1955). Other strains which have been or are being studied (Stubbs, Namba, Warren, and Morris, unpublished) include aucuba, cucumber green mottle mosaic virus (cucumber strain), Dahlemense, Holmes rib-grass, Japanese OM, and odontoglossum ring-spot virus.

The resolution limit of 4 Å for the TMV map was imposed by the difficulty of separating overlapping terms in the diffraction pattern arising from different Bessel orders (Franklin and Klug, 1955). Stubbs et al. (1977) used six heavy atom derivatives to separate three terms by the method of Stubbs and Diamond (1975). Even at 4 Å resolution, there is a small fourth term which they had to ignore. This caused the effective resolution near the outside of the particle to fall to 5.5 Å. Since that time, new methods have been developed which enable more terms to be separated. Stubbs and Makowski (1982) used measurements of the fine splitting of the layer lines in native TMV and two derivatives to obtain an electron density map at 6.7 Å resolution, which otherwise would require four derivatives. Other methods based on electron density modification such as the box function method of Makowski (1981) and a solvent flattening procedure (Namba, unpublished) have been shown to improve maps and increase their effective resolution. The diffraction patterns of TMV gels extend beyond 3 Å resolution before layer-line overlapping prevents further analysis, and derivative patterns extend to 3.5 Å. It should therefore soon be possible using these new methods to calculate a 3.5 Å map, and then extend phasing to 3 Å.

Even a small increase in the resolution of the electron density map of TMV will add a great deal to its interpretability. At 4 Å resolution, only the folding of the main chain and the positions of a few large groups such as tryptophan and the guanidinium group of arginine can be seen (North and Phillips, 1969). However, at 3.5 Å resolution the carboxyl groups begin to become visible, and many medium-sized groups, including all the aromatics, can be found. By 3 Å resolution all side chains, except those that are disordered, should be visible (Blundell and Johnson, 1976), and the protein–nucleic acid interactions will be significantly clearer. We may therefore expect to add greatly to our understanding of TMV structure and assembly in the next few years.

Acknowledgments. I would like to thank Drs. Cynthia Stauffacher and Keiichi Namba for their criticism and helpful suggestions during the preparation of this manuscript, and Michael Rossmann and Alfred Redfield for providing manuscripts prior to publication. This work was supported by National Institute of Health grant GM25236.

REFERENCES

Abad-Zapatero, C., Abdel-Meguid, S. S., Johnson, J. E., Leslie, A. G. W., Rayment, I., Rossmann, M. G., Suck, D., and Tsukihara, T. (1980). *Nature (London)* **286,** 33–39.

Abdel-Meguid, S. S., Yamane, T., Fukuyama, K., and Rossmann, M. G. (1981). *Virology* **114,** 81–85.

Altschuh, D., Reinbolt, J., and van Regenmortel, M. H. V. (1981). *J. Gen. Virol.* **52,** 363–366.

Anderer, F. A. (1959). *Z. Naturforsch.* **14b,** 24–28.

Anderson, W. F., Ohlendorf, D. H., Takeda, Y., and Matthews, B. W. (1981). *Nature (London)* **290,** 754–758.

Anderson, W. F., Takeda, Y., Ohlendorf, D. H., and Matthews, B. W. (1982). *J. Mol. Biol.* **159,** 745–751.

Ansevin, A. T., and Lauffer, M. A. (1959). *Nature (London)* **183,** 1601–1602.

Ansevin, A. T., Stevens, C. L., and Lauffer, M. A. (1964). *Biochemistry* **3,** 1512–1518.

Argos, P. (1981). *Virology* **110,** 55–62.

Arnott, S. (1970). *Prog. Biophys. Mol. Biol.* **21,** 265–319.

Arnott, S., Campbell Smith, P. J., and Chandrasekaran, R. (1976). In *Handbook of Biochemistry and Molecular Biology*, G. D. Fasman, Ed., 3rd ed., *Nucleic Acids*, Vol. 2. CRC Press, Cleveland, Ohio, pp. 411–422.

Arnott, S., Chandrasekaran, R., Birdsall, D. L., Leslie, A. G. W., and Ratliff, R. L. (1980). *Nature* (*London*) **283,** 743–745.

Atabekov, J. G., Novikov, V. K., Vishnichenko, V. K., and Kaftanova, A. S. (1970). *Virology* **41,** 519–532.

Banerjee, K., and Lauffer, M. A. (1966). *Biochemistry* **5,** 1957–1964.

Bancroft, J. B. (1970). *Adv. Virus Res.* **16,** 99–134.

Bancroft, J. B., Hills, G. J., and Markham, R. (1967). *Virology* **31,** 354–379.

Barrett, A. N., Barrington Leigh, J., Holmes, K. C., Leberman, R., Mandelkow, E., von Sengbusch, P. and Klug, A. (1971). *Cold Spring Harb. Symp. Quant. Biol.* **36,** 433–448.

Bawden, F. C., and Pirie, N. W. (1937). *Proc. R. Soc. London, Ser. B* **123,** 274–320.

Bawden, F. C., Pirie, N. W., Bernal, J. D., and Fankuchen, I. (1936). *Nature* (*London*) **138,** 1051–1053.

Beachy, R. N., Zaitlin, M., Bruening, G., and Israel, H. W. (1976). *Virology* **73,** 498–507.

Bernal, J. D., and Fankuchen, I. (1941). *J. Gen. Physiol.* **28,** 111–165.

Bhat, T. N., Blow, D. M., Brick, P., and Nyborg, J. (1982). *J. Mol. Biol.* **158,** 699–709.

Bloomer, A. C., Champness, J. N., Bricogne, G., Staden, R., and Klug, A. (1978). *Nature* (*London*) **276,** 362–368.

Bloomer, A. C., Graham, J., Hovmoeller, S., Butler, P. J. G., and Klug, A. (1981). In *Structural Aspects of Recognition and Assembly in Biological Macromolecules*, M. Balaban, J. L. Sussman, W. Traub, and A. Yonath, Eds., Vol. 2. Balaban ISS, Rehovot and Philadelphia, pp. 851–864.

Blowers, L. E., and Wilson, T. M. A. (1982). *J. Gen. Virol.* **61,** 137–141.

Blundell, T. L., and Johnson, L. N. (1976) *Protein Crystallography*. Academic Press, New York.

Brakke, M. K., and Van Pelt, N. (1969). *Virology* **39,** 516–533.

Bricogne, G. (1974). *Acta Crystallog. A* **30,** 395–405.

Bricogne, G. (1976). *Acta Crystallog. A* **32,** 832–847.

Bryan, R. K., Bansal, M., Folkhard, W., Nave, C., and Marvin, D. A. (1983). *Proc. Natl. Acad. Sci. U.S.A.* **80,** 4728–4731.

Butler, P. J. G., and Durham, A. C. H. (1972). *J. Mol. Biol.* **72,** 19–24.

Butler, P. J. G., and Durham, A. C. H. (1977). *Adv. Protein Chem.* **31,** 187–251.

Butler, P. J. G., and Klug, A. (1971). *Nature* (*London*) *New Biol.* **229,** 47–50.

Butler, P. J. G., and Klug, A. (1972). *Proc. Natl. Acad. Sci. U.S.A.* **69,** 2950–2953.

Butler, P. J. G., and Lomonossoff, G. P. (1978). *J. Mol. Biol.* **126,** 877–882.

Butler, P. J. G., and Lomonossoff, G. P. (1980). *Biophys. J.* **32,** 295–312.

Butler, P. J. G., Durham, A. C. H., and Klug, A. (1972). *J. Mol. Biol.* **72,** 1–18.

Butler, P. J. G., Bloomer, A. C., Bricogne, G., Champness, J. N., Graham, J., Guilley, H., Klug, A., and Zimmern, D. (1976). In *Structure-Function Relationships of Proteins*, R. Markham and R. W. Horne, Eds. North Holland, Amsterdam, pp. 101–110.

Caspar, D. L. D. (1956). *Nature (London)* **177,** 928.

Caspar, D. L. D. (1963). *Adv. Protein Chem.* **18,** 37–121.

Caspar, D. L. D. (1976). In *Structure-Function Relationships of Proteins*, R. Markham and R. W. Horne, Eds. North Holland, Amsterdam, pp. 85–99.

Caspar, D. L. D., and Holmes, K. C. (1969). *J. Mol. Biol.* **46,** 99–133.

Caspar, D. L. D., and Klug, A. (1962). *Cold Spring Harb. Symp. Quant. Biol.* **27,** 1–24.

Champness, J. N., Bloomer, A. C., Bricogne, G., Butler, P. J. G., and Klug, A. (1976). *Nature (London)* **259,** 20–24.

Cohen, S. S., and Stanley, W. M. (1942). *J. Biol. Chem.* **144,** 589–598.

Cotton, F. A., Hazen, E. E., Day, V. W., Larsen, S., Norman, J. G., Wong, S. T. K., and Johnson, K. H. (1973). *J. Am. Chem. Soc.* **95,** 2367–2369.

Cotton, F. A., Hazen, E. E. and Legg, M. J. (1979). *Proc. Natl. Acad. Sci. U.S.A.* **76,** 2551–2555.

Crawford, J. C., Kolpak, F. J., Wang, A. H. J., Quigley, G. J., van Boom, J. H., van der Marel, G., and Rich, A. (1980). *Proc. Natl. Acad. Sci. U.S.A.* **77,** 4016–4020.

Day, L. A., Wiseman, R. L., and Marzec, C. J. (1979). *Nucl. Acids Res.* **7,** 1393–1403.

Dodds, J. A., and Hamilton, R. I. (1976). *Adv. Virus Res.* **20,** 33–86.

Dodson, E. J., Isaacs, N. W., and Rollett, J. S. (1976). *Acta Crystallogr. A* **32,** 311–315.

Drew, H. R., Takano, T., Tanaka, S., Itakura, K., and Dickerson, R. E. (1980). *Nature (London)* **286,** 567–573.

Durham, A. C. H. (1972a). *J. Mol. Biol.* **67,** 289–305.

Durham, A. C. H. (1972b). *FEBS Lett.* **25,** 147–152.

Durham, A. C. H., and Abou Haidar, M. (1973). *Virology* **77,** 520–523.

Durham, A. C. H., and Butler, P. J. G. (1975). *Eur. J. Biochem.* **53,** 397–404.

Durham, A. C. H., and Finch, J. T. (1972). *J. Mol. Biol.* **67,** 307–314.

Durham, A. C. H., and Hendry, D. A. (1977). *Virology* **77,** 510–519.

Durham, A. C. H., and Klug, A. (1971). *Nature (London) New Biol.* **229,** 42–46.

Durham, A. C. H., and Klug, A. (1972). *J. Mol. Biol.* **67,** 315–332.

Durham, A. C. H., Finch, J. T., and Klug, A. (1971). *Nature (London) New Biol.* **229,** 37–42.

Durham, A. C. H., Vogel, D., and de Marcillac, G. D. (1977a). *Eur. J. Biochem.* **79,** 151–159.

Durham, A. C. H., Hendry, D. A., and Von Wechmar, M. B. (1977b). *Virology* **77,** 524–533.

Edelman, G. M., Cunningham, B. A., Reeke, G. N., Becker, J. W., Waxdal, M. J., and Wang, J. L. (1972). *Proc. Natl. Acad. Sci. U.S.A.* **69,** 2580–2584.

Eklund, H., Samama, J.-P., Wallén, L., Bränden, C.-I., Åkeson, Å., and Jones, T. A. (1981). *J. Mol. Biol.* **146,** 561–587.

Finch, J. T., Leberman, R., Chang, Y. S., and Klug, A. (1966). *Nature* (*London*) **212,** 349–350.

Finch, J. T., Gilbert, P. F. C., Klug, A., and Leberman, R. (1974). *J. Mol. Biol.* **86,** 183–192.

Fish, S. R., Hartman, K. A., Stubbs, G. J., and Thomas, G. J. (1981). *Biochemistry* **20,** 7449–7457.

Fraenkel-Conrat, H., and Ramachandran, L. K. (1959). *Adv. Protein Chem.* **14,** 175–229.

Fraenkel-Conrat, H., and Singer, B. (1954). *J. Am. Chem. Soc.* **76,** 180–183.

Fraenkel-Conrat, H., and Singer, B. (1959). *Biochim. Biophys. Acta* **33,** 359–370.

Fraenkel-Conrat, H., and Singer, B. (1964). *Virology* **23,** 354–362.

Fraenkel-Conrat, H., and Williams, R. C. (1955). *Proc. Natl. Acad. Sci. U.S.A.* **41,** 690–698.

Franklin, R. E. (1955). *Biochim. Biophys. Acta* **18,** 313–314.

Franklin, R. E. (1956a). *Nature* (*London*) **177,** 928–930.

Franklin, R. E. (1956b). *Biochim. Biophys. Acta* **19,** 203–211.

Franklin, R. E., and Commoner, B. (1955). *Nature* (*London*) **175,** 1076–1077.

Franklin, R. E., and Holmes, K. C. (1958). *Acta Crystallogr.* **11,** 213–220.

Franklin, R. E., and Klug, A. (1955). *Acta Crystallogr.* **8,** 777–780.

Frist, R. H. (1968). *Biophys. J.* **8,** A69.

Fritsch, C., Stussi, C., Witz, J., and Hirth, L. (1973a). *Virology* **56,** 33–45.

Fritsch, C., Witz, J., Abou Haidar, M., and Hirth, L. (1973b). *FEBS Lett.* **29,** 211–214.

Fukuda, M., and Okada, Y. (1982). *Proc. Natl. Acad. Sci. U.S.A.* **79,** 5833–5836.

Fukuda, M., Ohno, T., Okada, Y., Otsuki, Y., and Takebe, I. (1978). *Proc. Natl. Acad. Sci. U.S.A.* **75,** 1727–1730.

Fukuda, M., Okada, Y., Otsuki, Y., and Takebe, I. (1980). *Virology* **101,** 493–502.

Fukuda, M., Meshi, T., Okada, Y., Otsuki, Y., and Takebe, I. (1981). *Proc. Natl. Acad. Sci. U.S.A.* **78,** 4231–4235.

Gilbert, P. F. C., and Klug, A. (1974). *J. Mol. Biol.* **86,** 193–207.

Goelet, P., Lomonossoff, G. P., Butler, P. J. G., Akam, M. E., Gait, M. J., and Karn, J. (1982). *Proc. Natl. Acad. Sci. U.S.A.* **79,** 5818–5822.

Graham, J. (1977). Thesis, Univ. Cambridge.

Graham, J., and Butler, P. J. G. (1978). *Eur. J. Biochem.* **83,** 523–528.

Graham, J., and Butler, P. J. G. (1979). *Eur. J. Biochem.* **93,** 333–337.

Grau, U. M., Trommer, W. E., and Rossmann, M. G. (1981). *J. Mol. Biol.* **151,** 289–307.

Gregory, J., and Holmes, K. C. (1965). *J. Mol. Biol.* **13,** 796–801.

Guéron, M., Chachaty, C., and Tran-Dinh, S. (1973). *Ann. N.Y. Acad. Sci.* **222,** 307–323.

Guilley, H., Jonard, G., Richards, K. E., and Hirth, L. (1975a). *Eur. J. Biochem.* **54,** 135–144.

Guilley, H., Jonard, G., Richards, K. E., and Hirth, L. (1975b). *Eur. J. Biochem.* **54,** 145–153.

Guschlbauer, W., and Courtois, Y. (1968). *FEBS Lett.* **1,** 183–186.

Hardman, K. D., and Ainsworth, C. F. (1972). *Biochemistry* **11,** 4910–4919.

Hariharasubramanian, V., and Siegel, A. (1969). *Virology* **37,** 203–208.

Harrington, W. F., and Schachman, H. K. (1956). *Arch. Biochem. Biophys.* **65,** 278–295.

Harris, J. I., and Knight, C. A. (1952). *Nature (London)* **170,** 613–614.

Harrison, S. C. (1980). *Biophys. J.* **32,** 139–153.

Harrison, S. C., Olson, A. J., Schutt, C. E., Winkler, F. K., and Bricogne, G. (1978). *Nature (London)* **276,** 368–373.

Hart, R. G. (1955). *Proc. Natl. Acad. Sci. U.S.A.* **41,** 261–264.

Haschemeyer, A. E. V., and Rich, A. (1967). *J. Mol. Biol.* **27,** 369–384.

Heinemann, U., and Saenger, W. (1982). *Nature (London)* **299,** 27–31.

Hermodson, M. A., Abad-Zapatero, C., Abdel-Meguid, S. S., Pundak, S., Rossmann, M. G., and Tremaine, J. H. (1982). *Virology* **119,** 133–149.

Hirth, L., and Richards, K. E. (1981). *Adv. Virus Res.* **26,** 145–199.

Hogle, J. M., Kirchhausen, T., and Harrison, S. C. (1983). *J. Mol. Biol.*, **171,** 95–100.

Hol, W. G. J., van Duijnen, P. T., and Berendsen, H. J. C. (1978). *Nature (London)* **273,** 443–446.

Hol, W. G. J., Halie, L. M., and Sander, C. (1981). *Nature (London)* **294,** 532–536.

Holmes, K. C. (1979). *J. Supramol. Struct.* **12,** 305–320.

Holmes, K. C. (1982) In *Structural Molecular Biology*, D. B. Davies, W. Saenger, and S. S. Danyluk, Eds. Plenum, New York, pp. 475–505.

Holmes, K. C., and Blow, D. M. (1980). *The Use of X-Ray Diffraction in the Study of Protein and Nucleic Acid Structure*. Krieger, New York.

Holmes, K. C., and Franklin, R. (1958). *Virology* **6,** 328–336.

Holmes, K. C., Stubbs, G. J., Mandelkow, E., and Gallwitz, U. (1975). *Nature (London)* **254,** 192–196.

Holoubeck, V. (1962). *Virology* **18,** 401–404.

Hunter, T. R., Hunt, T., Knowland, J., and Zimmern, D. (1976). *Nature (London)* **260,** 759–764.

Iida, S., and Ooi, T. (1969). *Biochemistry* **8,** 3897–3901.

Iwanowski, D. (1892). *Bull. Acad. Imp. Sci. St. Petersbourg, (N. S.)* **3,** 65–70.

Jack, A. (1973). *Acta Crystallogr. A* **29,** 545–554.

Jardetzky, O., Akasaka, K., Vogel, D., Morris, S., and Holmes, K. C. (1978). *Nature (London)* **273,** 564–566.

Jonard, G., Richards, K. E., Guilley, H., and Hirth, L. (1977). *Cell* **11,** 483–493.

Kaper, J. M. (1969). *Virology* **37,** 134–139.

Keith, J., and Fraenkel-Conrat, H. (1975). *FEBS Lett.* **57,** 31–33.

King, J., and Casjens, S. (1974). *Nature (London)* **251,** 112–119.

Klug, A., and Durham, A. C. H. (1971). *Cold Spring Harbor Symp. Quant. Biol.* **36,** 449–460.

Lanina, T. P., Terganova, G. V., Ledneva, R. K., and Bogdanov, A. I. (1976). *FEBS Lett.* **67,** 160–170.

Leberman, R., Finch, J. T., Gilbert, P. F. C., Witz, J., and Klug, A. (1974). *J. Mol. Biol.* **86,** 179–182.

Lebeurier, G., Nicolaieff, A., and Richards, K. E. (1977). *Proc. Natl. Acad. Sci. U.S.A.* **74,** 149–153.

Liljas, L., Unge, T., Jones, T. A., Fridborg, K., Lövgren, S., Skoglund, S., and Strandberg, B. (1982). *J. Mol. Biol.* **159,** 93–108.

Lomonossoff, G. P., and Butler, P. J. C. (1979). *Eur. J. Biochem.* **93,** 157–164.

Lomonossoff, G. P., and Butler, P. J. G. (1980). *FEBS Lett.* **113,** 271–274.

Lonchampt, M., Lebeurier, G., and Hirth, L. (1972). *FEBS Lett.* **22,** 297–300.

Loring, H. S. (1939). *J. Biol. Chem.* **130,** 251–258.

McCain, D. C., Virudachalam, R., Markley, J. L., Abdel-Meguid, S. S., and Rossmann, M. G. (1982). *Virology* **117,** 501–503.

McCain, D. C., Virudachalam, R., Santini, R. E., Abdel-Meguid, S. S., and Markley, J. L. (1982). *Biochemistry,* **21,** 5390–5397.

McKay, D. B., and Steitz, T. A. (1981). *Nature (London)* **290,** 744–749.

McKay, D. B., Weber, I. T., and Steitz, T. A. (1982). *J. Biol. Chem.* **257,** 9518–9524.

McLachlan, A. D., Bloomer, A. C., and Butler, P. J. G. (1980). *J. Mol. Biol.* **136,** 203–224.

McPherson, A., Jurnak, F. A., Wang, A. H. J., Molineux, I., and Rich, A. (1979). *J. Mol. Biol.* **134,** 379–400.

McPherson, A., Jurnak, F., Wang, A., Kolpak, F., Rich, A., Molineux, I., and Fitzgerald, P. (1980). *Biophys. J.* **32,** 155–173.

Makowski, L. (1981). *J. Appl. Crystallogr.* **14,** 160–168.

Makowski, L., and Caspar, D. L. D. (1981). *J. Mol. Biol.* **145,** 611–617.

Makowski, L., Caspar, D. L. D., and Marvin, D. A. (1980). *J. Mol. Biol.* **140,** 149–181.

Mandelkow, E., and Holmes, K. C. (1974). *J. Mol. Biol.* **87,** 265–273.

Mandelkow, E., Holmes, K. C. and Gallwitz, U. (1976). *J. Mol. Biol.* **102,** 265–285.

Mandelkow, E., Stubbs, G. and Warren S. (1981). *J. Mol. Biol.* **152,** 375–386.

Mang, K. Q., Ghosh, A., and Kaesberg, P. (1982). *Virology* **116,** 264–274.

Matthews, B. W., Colman, P. M., Jansonius, J. N., Titani, K., Walsh, K. A., and Neurath, H. (1972). *Nature (London) New Biol.* **238,** 41–43.

Matthews, B. W., Ohlendorf, D. H., Anderson, W. F., and Takeda, Y. (1982). *Proc. Natl. Acad. Sci. U.S.A.* **79,** 1428–1432.

Matthews, R. E. F. (1981). *Plant Virology*, 2nd ed. Academic Press, New York.

May, D. S., and Knight, C. A. (1965). *Virology* **25,** 502–507.

Morris, T. J., and Semancik, J. S. (1973). *Virology* **53,** 215–224.

Munowitz, M. G., Dobson, C. M., Griffin, R. G., and Harrison, S. C. (1980). *J. Mol. Biol.* **141,** 327–333.

Nave, C., Brown, R. S., Fowler, A. G., Ladner, J. E., Marvin, D. A., Provencher, S. W., Tsugita, A., Armstrong, J., and Perham, R. N. (1981). *J. Mol. Biol.* **149,** 675–707.

Neidle, S., Taylor, G. L., Sanderson, M. R., Berman, H. M., and Shieh, H. S. (1978). *Nucl. Acids Res.* **5,** 4417–4422.

Neumann, J.-M., Guschlbauer, W., and Tran-Dinh, S. (1979). *Eur. J. Biochem.* **100,** 141–148.

North, A. C. T., and Phillips, D. C. (1969). *Prog. Biophys. Mol. Biol.* **19,** 1–132.

Nozu, Y., and Okada, Y. (1968). *J. Mol. Biol.* **35,** 643–646.

Nozu, Y., Tochihara, H., Komuro, Y., and Okada, Y. (1971). *Virology* **45,** 577–585.

Ohlendorf, D. H., Anderson, W. F., Fisher, R. G., Takeda, Y., and Matthews, B. W. (1982). *Nature (London)* **298,** 718–723.

Ohno, T., and Okada, Y. (1977). *Virology* **76,** 429–432.

Ohno, T., Yamaura, R., Kuriyama, K., Inoue, H., and Okada, Y. (1972). *Virology* **50,** 76–83.

Okada, Y., and Ohno, T. (1972). *Mol. Gen. Genet.* **114,** 205–213.

Okada, Y., Ohashi, Y., Ohno, T., and Nozu, Y. (1970). *Virology* **42,** 243–245.

Pabo, C. O., and Lewis, M. (1982). *Nature (London).* **298,** 443–447.

Perham, R. N. (1969). *J. Mol. Biol.* **45,** 439–441.

Perham, R. N., and Wilson, T. M. A. (1976). *FEBS Lett.* **62,** 11–15.

Perham, R. N., and Wilson, T. M. A. (1978). *Virology* **84,** 293–302.

Perutz, M. F., Jope, M., and Barer, R. (1950). *Discuss. Faraday Soc.* **9,** 423–427.

Powell, C. A. (1975). *Virology* **64,** 75–85.

Rao, S. T., and Rossmann, M. G. (1973). *J. Mol. Biol.* **76,** 241–256.

Rayment, I., Johnson, J. E., and Rossmann, M. G. (1979). *J. Biol. Chem.* **254,** 5243–5245.

Rees, M. W., and Short, M. N. (1975). *Biochim. Biophys. Acta* **393,** 15–23.

Richards, K. E., and Williams, R. C. (1972). *Proc. Natl. Acad. Sci. U.S.A.* **69,** 1121–1124.

Richards, K. E., and Williams, R. C. (1973). *Biochemistry* **12,** 4574–4581.

Richards, K. E., and Williams, R. C. (1976). In *Comprehensive Virology*, H. Fraenkel-Conrat and R. R. Wagner, Eds., Vol. 6. Plenum, New York, pp. 1–37.

Richardson, J. (1981). *Adv. Protein Chem.* **34,** 167–339.

Robinson, I. K., and Harrison, S. C. (1982). *Nature (London)* **297,** 563–568.

Rossmann, M. G., Abad-Zapatero, C., Hermodson, M. A., and Erickson, J. W. (1983a). *J. Mol. Biol.*, **166,** 37–73.

Rossmann, M. G., Chandrasekaran, R., Abad-Zapatero, C., Erickson, J. W., and Arnott, S. (1983b). *J. Mol. Biol.* **166,** 73–80.

Rossmann, M. G., Abad-Zapatero, C., Murthy, M. R. N., Liljas, L., Jones, T. A., and Strandberg, B. (1983c). *J. Mol. Biol.* **165,** 711–736.

Roy, S., Papastavros, M. Z., and Redfield, A. G. (1982). *Nucl. Acids Res.* **10,** 8341–8349.

Saenger, W. (1973). *Angew. Chem.* (Engl. ed.) **12,** 591–601.

Sauer, R. T., Yocum, R. R., Doolittle, R. F., Lewis, M., and Pabo, C. O. (1982). *Nature (London)* **298,** 447–451.

Savithri, H. S., and Erickson, J. W. (1983). *Virology*, **126,** 328–335.

Scheele, R. B., and Lauffer, M. A. (1967). *Biochemistry* **6,** 3076–3081.

Scheele, R. B., and Schuster, T. M. (1974). *Biopolymers* **13,** 275–288.

Schramm, G. (1943). *Naturwissenschaften* **31,** 94–96.

Schramm, G. (1947a). *Z. Naturforsch.* **2b,** 112–121.

Schramm, G. (1947b). *Z. Naturforsch.* **2b,** 249–257.

Schramm, G., and Zillig, W. (1955). *Z. Naturforsch.* **10b,** 493–499.

Schramm, G., Schumacher, G., and Zillig, W. (1955). *Z. Naturforsch.* **10b,** 481–492.

Schulz, G. E., and Schirmer, R. H. (1979). *Principles of Protein Structure*. Springer-Verlag, New York.

Schuster, T. M., Scheele, R. B., and Khairallah, L. H. (1979). *J. Mol. Biol.* **127,** 461–485.

Seeds, W. E., and Wilkins, M. H. F. (1950). *Discuss. Faraday Soc.* **9,** 419–423.

Shalaby, R. A. F., and Lauffer, M. A. (1977). *J. Mol. Biol.* **116,** 709–725.

Sheridan, R. P., Levy, R. M., and Salemme, R. (1982). *Proc. Natl. Acad. Sci. U.S.A.* **79,** 4545–4549.

Shire, S. J., Steckert, J. J., and Schuster, T. M. (1979a). *J. Mol. Biol.* **127,** 487–506.

Shire, S. J., Steckert, J. J., Adams, M. L., and Schuster, T. M (1979b). *Proc. Natl. Acad. Sci. U.S.A.* **76,** 2745–2749.

Shire, S. J., Steckert, J. J., and Schuster, T. M. (1981). *Proc. Natl. Acad. Sci. U.S.A.* **78,** 256–260.

Siegel, A. (1971). *Virology* **46,** 50–59.

Sreenivasnya, M., and Pirie, N. W. (1938). *Biochem. J.* **32,** 1707–1710.

Steckert, J. J., and Schuster, T. M. (1982). *Nature (London)* **299,** 32–36.

Steitz, T. A., Ohlendorf, D. H., McKay, D. B., Anderson, W. F., and Matthews, B. W. (1982). *Proc. Natl. Acad. Sci. U.S.A.* **79,** 3097–3100.

Stubbs, G. J., and Diamond, R. (1975). *Acta Crystallogr. A* **31,** 709–718.

Stubbs, G. J., and Makowski, L. (1982). *Acta Crystallogr. A* **38,** 417–425.

Stubbs, G. J., and Stauffacher, C. V. (1980). *Biophys. J.* **32,** 244–246.

Stubbs, G. J., and Stauffacher, C. V. (1981). *J. Mol. Biol.* **152,** 387–396.

Stubbs, G. J., Warren, S. G., and Holmes, K. C. (1977). *Nature (London)* **267,** 216–221.

Stubbs, G. J., Warren, S. G., and Mandelkow, E. (1979). *J. Supramol. Struct.* **12,** 177–183.

Sundaralingam, M. (1973). *Jerusalem Symp. Quant. Chem. Biochem.* **5,** 417–455.

Symington, J. (1969). *Virology* **38,** 317–323.

Tran-Dinh, S., Guschlbauer, W., and Guéron, M. (1972). *J. Am. Chem. Soc.* **94,** 7903–7911.

Vogel, D. (1973). *Biochem. Biophys. Res. Commun.* **52,** 335–341.

Vogel, D. (1982). *Biochim. Biophys. Acta* **706,** 65–79.

Vogel, D., and Jaenicke, R. (1974). *Eur. J. Biochem.* **41,** 607–615.

Vogel, D., and Jaenicke, R. (1976). *Eur. J. Biochem.* **61,** 423–431.

Vogel, D., Durham, A. C. H., and de Marcillac, G. D. (1977). *Eur. J. Biochem.* **79,** 161–171.

Vogel, D., de Marcillac, G. D., Hirth, L., Gregori, E., and Jaenicke, R. (1979). *Z. Naturforsch.* **34c,** 782–792.

Vogel, D., de Marcillac, G. D., Hirth, L., and Akasaka, K. (1980). *Z. Naturforsch.* **35c,** 482–494.

Wang, A. H. J., Quigley, G. J., Kolpak, F. J., Crawford, J. L., van Boom, J. M., van der Marel, G., and Rich, A. (1979). *Nature (London)* **282,** 680–686.

Watson, J. D. (1954). *Biochim. Biophys. Acta* **13,** 10–19.

Wilson, T. M. A., Perham, R. N., Finch, J. T., and Butler, P. J. G. (1976). *FEBS Lett.* **64,** 285–289.

Wilson, T. M. A., Perham, R. N., and Butler, P. J. G. (1978). *Virology* **89,** 475–483.

Wilson, T. M. A., Lomonossoff, G. P., and Glover, J. F. (1981). *J. Gen. Virol.* **53,** 225–234.

Wittmann-Liebold, B., and Wittmann, H. G. (1967). *Mol. Gen. Genet.* **100,** 358–363.

Wlodawer, A., Bott, R., and Sjolin, L. (1982). *J. Biol. Chem.* **257,** 1325–1332.

Wyckoff, R. W. G., and Corey, R. B. (1936). *J. Biol. Chem.* **116,** 51–55.

Zimmern, D. (1975). *Nucl. Acids Res.* **2,** 1189–1201.

Zimmern, D. (1976). *Philos. Trans. R. Soc. London, Ser. B* **276,** 189–204.

Zimmern, D. (1977). *Cell* **11,** 463–482,

Zimmern, D., and Butler, P. J. G. (1977). *Cell* **11,** 455–462.

Zimmern, D., and Wilson, T. M. A. (1976). *FEBS Lett.* **71,** 294–298.

6

Structural Diversity in Filamentous Bacteriophages

LEE MAKOWSKI
College of Physicians and Surgeons
Columbia University
New York, New York

CONTENTS

1. INTRODUCTION

The filamentous, single-stranded DNA phages constitute a family of similar viruses (Marvin and Kohn, 1969; Denhardt et al., 1978). The best known of these, fd, f1, and M13 are very similar indeed, with nucleotide sequences that differ at only a few percent of their bases (van Weczenbeek et al., 1980). However, at least a half dozen distinctly different strains of

* When discussing phage structure in general, I will refer to fd although similar comments will almost always apply equally to M13 and f1. Exceptions will be noted explicitly.

filamentous phage have been characterized in detail. These phages are all flexible cylinders about 60 Å in diameter made up largely of several thousand copies of a small, relatively hydrophobic structural protein coating the circular, single-stranded DNA. They vary substantially in host range, virion length, protein to DNA ratio, nucleotide sequence, coat protein amino acid sequence, and in a variety of physicochemical properties.

Structurally, the best characterized filamentous phage is Pf1. However, the genetics, replication, morphogenesis and life cycle of fd (M13, f1)* have been studied far more extensively than those of Pf1. In this chapter, the structure of Pf1 as determined by X-ray diffraction studies is described in detail and interpreted in terms of the results of other biochemical and biophysical studies of the phage. The temperature dependent structural transition of Pf1 is discussed in terms of what is known about Pf1 structure. Comparison of biochemical and biophysical studies of Pf1 with similar studies of other phages is used as a basis for discussing the structure of other strains of filamentous phage.

2. STRUCTURAL ORGANIZATION OF THE VIRIONS

Two distinct structural classes of filamentous phage have been identified based on their X-ray diffraction patterns (Marvin et al., 1974a,b), which distinguish them on the basis of helical symmetry. In Table 1 the viruses discussed in this chapter are classified according to the helical symmetry

Table 1. Structure of Classes of Filamentous Phages

Class I	Class II
fd, f1, M13, ZJ-2[a]	Pf1
If1	Xf
IKe	Pf3[b]

[a] fd, f1, M13, and ZJ-2 are nearly identical phages with very similar genomes and no more than one difference among the amino acid sequences of their coat proteins.
[b] Pf3 lacks several features common to Pf1 and Xf such as an integral nucleotide/protein ratio. It has been suggested (L. Day, personal communication) that Pf3 may represent a new Class III of filamentous phage.

of their coat proteins. Class I and Class II viruses differ not only in helical symmetry but in a variety of other physical and chemical properties. All Class I virions are quite similar. The amino acid sequences of their coat proteins exhibit substantial homologies as can be seen in the listing of coat protein sequences in Table 2. In spite of the extensive homologies,

Table 2. Amino Acid Sequences of Coat Proteins[a]

Class I			Class II		
fd	If1	IKe	Pf1	Xf	Pf3
		Ala			
		Glu		Ser	
	Ala	Phe	Gly	Gly	Met
Ala	Asp	Asn	Val	Val	Gln
Glu	Asp	Ala	Ile	Gly	Ser
Gly	Ala	Ala	Asp	Asp	Val
Asp	Thr	Thr	Thr	Gly	Ile
Asp	Ser	Asn	Ser	Val	Thr
Pro	Gln	Tyr	Ala	Asp	Asp
Ala	Ala	Ala	Val	Val	Val
Lys	Lys	Thr	Gln	Val	Thr
Ala	Ala	Glu	Ser	Ser	Gly
Ala	Ala	Ala	Ala	Ala	Gln
Phe	Phe	Met	Ile	Ile	Leu
Asp	Asp	Asp	Thr	Glu	Thr
Ser	Ser	Ser	Asp	Gly	Ala
Leu	Leu	Leu	Gly	Ala	Val
Gln	Thr	Lys	Gln	Ala	Gln
Ala	Ala	Thr	Gly	Gly	Ala
Ser	Gln	Gln	Asp	Pro	Asp
Ala	Ala	Ala	Met	Ile	Ile
Thr	Thr	Ile	Lys	Ala	Thr
Glu	Glu	Asp	Ala	Ala	Thr
Tyr	Met	Leu	Ile	Ile	Ile
Ile	Ser	Ile	Gly	Gly	Gly

Table 2. (*Continued*)

Class I			Class II		
fd	If1	IKe	Pf1	Xf	Pf3
Gly	Gly	Ser	Gly	Gly	Gly
Tyr	Tyr	Gln	Tyr	Ala	Ala
Ala	Ala	Thr	Ile	Val	Ile
Trp	Trp	Trp	Val	Leu	Ile
Ala	Ala	Phe	Gly	Thr	Val
Met	Leu	Val	Ala	Val	Leu
Val	Val	Val	Leu	Met	Ala
Val	Val	Thr	Val	Val	Ala
Val	Leu	Thr	Ile	Gly	Val
Ile	Val	Val	Leu	Ile	Val
Val	Val	Val	Ala	Lys	Leu
Gly	Gly	Val	Val	Val	Gly
Ala	Ala	Ala	Ala	Tyr	Ile
Thr	Thr	Gly	Gly	Lys	Arg
Ile	Val	Leu	Leu	Trp	Trp
Gly	Gly	Val	Ile	Val	Ile
Ile	Ile	Ile	Tyr	Arg	Lys
Lys	Lys	Lys	Ser	Arg	Ala
Leu	Leu	Leu	Met	Ala	Gln
Phe	Phe	Phe	Leu	Met	Phe
Lys	Lys	Lys	Arg		Phe
Lys	Lys	Lys	Lys		
Phe	Phe	Phe	Ala		
Thr	Val	Val			
Ser	Ser	Ser			
Lys	Arg	Arg			
Ala	Ala	Ala			
Ser	Ser	Ser			

[a] Sequences taken from: fd: Nakashima and Konigsberg, 1974; Ike, If1: Nakashima et al., 1981; Pf1: Nakashima et al., 1975; Xf: Frangione et al., 1978; Pf3: D. Putterman and L. Day, personal communication.

there are also some differences suggestive of structural variations. The members of Class II are much less similar. This is clear not only in the amino acid sequences listed in Table 2, but also in the physicochemical properties listed in Table 3.

In the virions, the coat proteins form a cylindrical shell around the circular single-stranded DNA. The coat protein is a product of gene 8 and is sometimes referred to as B protein. The DNA extends from one end of the virion to the other where it must loop back on itself to form a covalent circle. For most of the length of the phage, the DNA exists as a pair of non base-paired, antiparallel strands. The length of the virion is determined by the length of the DNA. If a length of exogenous DNA is inserted into the genome, a longer phage particle results (e.g., Herrmann et al., 1980). Very short "miniphage" are observed when extensive regions of the genome are deleted (Horiuchi et al., 1978). In fd, the DNA is oriented with a portion of the intergenic region capable of forming a stable hairpin located at one end of the virus (Ikoku and Hearst, 1981; Webster et al., 1981). It seems likely that the hairpin formed by these nucleotides plays a role in phage morphogenesis (Schaller, 1978; Ikoku and Hearst, 1981; Webster et al., 1981).

Table 3. Composition and Physical Properties[a]

	fd	Pf1	Xf	Pf3
Length (Å)[b]	8,900	19,400	9,800	7,200
Mass/unit length (daltons/Å)[b,c]	1,830	1,860	1,830	1,860
Number of coat proteins	2,710	7,620	3,670	2,630
Number of nucleotides	6,408	7,390	7,420	5,960
Nucleotides/coat protein	2.3	1.0	2.0	2.4
Axial translations (Å)				
Of coat protein[b]	3.1	2.85–3.06[d]	2.6	2.75
Of nucleotides (each strand)[a]	2.7	5.7–6.1[d]	2.6	2.3

[a] Based on Table 1 from Thomas et al. (1983). For references and estimates of uncertainties see Thomas et al. (1983).

[b] Virion length, mass per unit length and the axial translations per nucleotide and per coat protein are dependent on the virion environment. In particular, temperature, relative humidity, or salt concentration may affect these parameters.

[c] See, for example, Wiseman et al. (1976) and Newman et al. (1982).

[d] Figures used are for high temperature and low temperature forms of Pf1 (Nave et al., 1979). The low temperature form exhibits the larger axial translations.

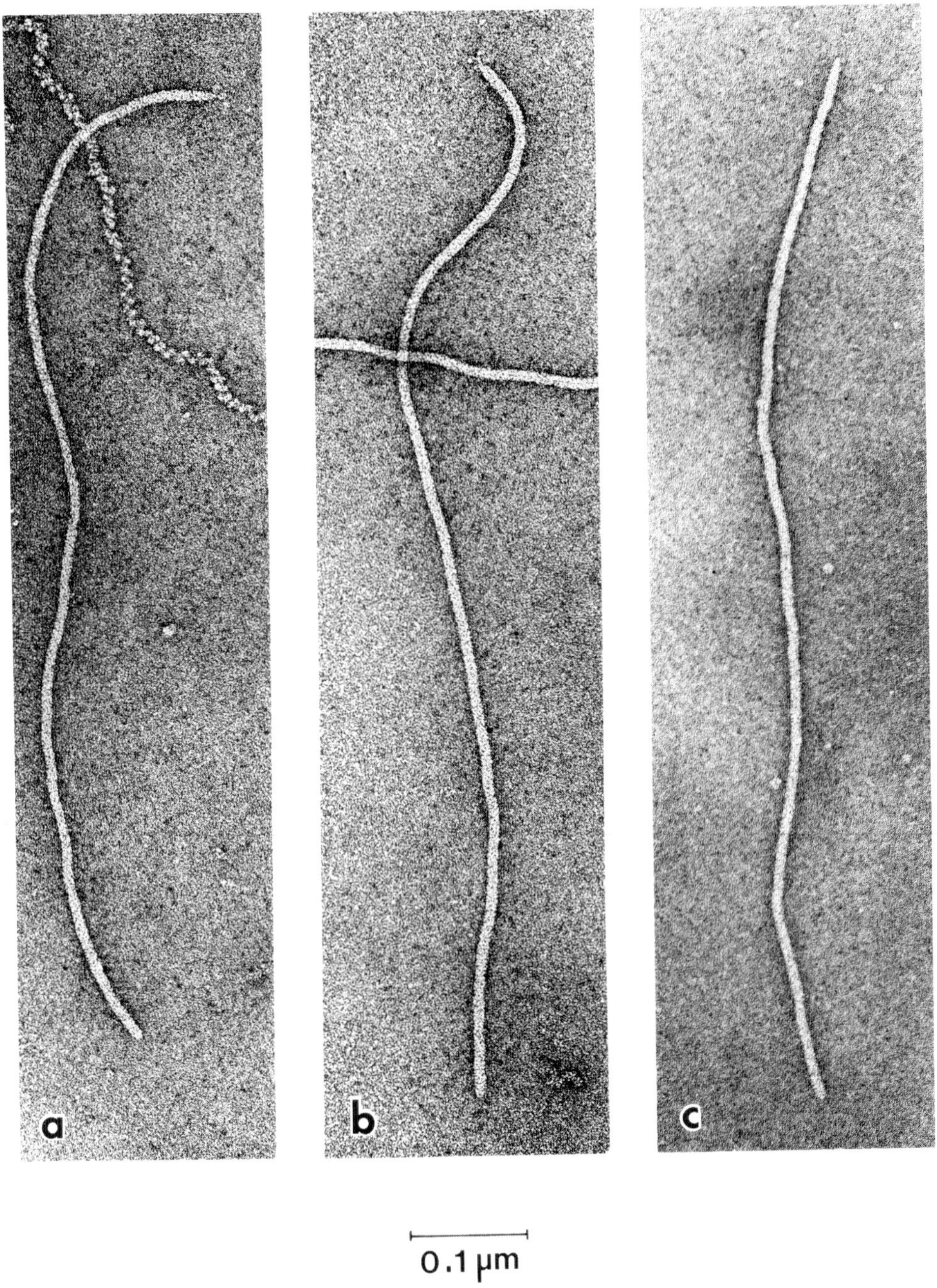

Figure 2. Electron micrographs of fd virions. The micrographs are oriented with the adsorption complex composed of A and D proteins at the top. (*a*) and (*b*) Electron micrographs of intact fd virions showing faint knobs at the top. (*c*) Virion treated for 60 min at 37°C with subtilisin at 200 μg/ml (virions at 9 mg/ml). This treatment removes the knobs. A segment of the twisted complex of fd DNA and fd gene 5 protein (Gray et al., 1981) can be seen in (*a*). Electron micrographs courtesy of Dr. C. W. Gray; reprinted with permission (Gray, Brown, and Marvin, 1981).

In addition to the DNA and the large number of coat proteins, each phage has a small number of minor proteins at each end. In fd there are five copies each of the products of gene 3 (A protein), gene 6 (D protein), and genes 7 and 9 (C protein) (Lin et al., 1980; Grant et al., 1981; Simons et al., 1981). The C protein is located near the nucleotide sequence implicated as a recognition site for the initiation of morphogenesis (Webster et al., 1981). The A and D proteins are located at the other end as diagrammed in Figure 1.

The A and D proteins make up an adsorption complex which, in electron micrographs of negatively stained phage, has the appearance of a 40–50 Å diameter knob on the end of a stem approximately 70 Å long (Gray et al., 1981). These structures are visible at the top of the micrographs in Figure 2*a* and *b*. Subtilisin treatment removes the knobs as demonstrated by the micrograph in Figure 2*c*. The D protein is not large enough to account for the knobs, which must be made up principally of A protein. This complex is involved in the attachment of phage to the pilus of host bacteria, the first step in infection (Caro and Schnös, 1966). It has been suggested (Gray et al., 1981) that the knob may function to provide a large surface area to support tight noncovalent binding to the receptor and the stem may provide flexibility to permit the extremely long virion to sway without breaking contact with the receptor.

Within the body of the virions, no clear substructure has yet been discerned by electron microscopy. As seen in the micrographs in Figure 2, away from the ends of the virus, the phage appear as relatively uniform cylinders. Under some conditions, a thin region of stain accumulation appears in the center of the particles (Bradley, 1964; Marvin, 1966). How-

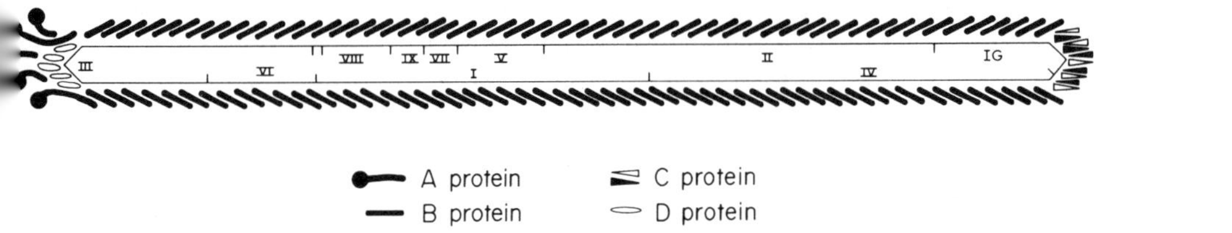

Figure 1. Schematic representation of the positions of structural proteins and genes in bacteriophage fd (f1, M13). The diagram shows only the relative positions of molecules and is not meant to imply any specific interactions between proteins or between protein and DNA. Reproduced with the kind permission of Dr. Robert Webster (Webster et al., 1981).

ever, in order to obtain higher resolution images of virion substructure, it is necessary to use X-ray diffraction from oriented fibers or gels.

3. FIBER DIFFRACTION

In fiber diffraction, each virion acts as a one-dimensional crystal (see also Holmes, Chapter 4). Diffracted intensities from the virions fall on layer planes oriented perpendicular to the virion axis. Because the virions in a fiber are disordered with regard to rotation about their long axes, the intensity on the observed layer lines is cylindrically averaged. Analysis of this cylindrically averaged intensity requires special methods (Stubbs and Diamond, 1975; Makowski, 1978, 1980; Makowski et al., 1980; Nave et al., 1981; Stubbs and Makowski, 1982; Bryan et al., 1983), which are still under development. These methods have been used to construct electron density maps of the high temperature (Makowski et al., 1980) and low temperature (Nave et al., 1981) forms of Pf1 to 7 Å resolution, a preliminary map of the low temperature form of Pf1 to 4 Å resolution (Bryan et al., 1983), and 4 Å maps of the structure of tobacco mosaic virus (TMV) to 4 Å resolution (Stubbs et al., 1977; Stubbs, Namba, and Makowski, unpublished). TMV is discussed in Chapters 4 and 5.

Figure 3 contains three diffraction patterns from oriented fibers of filamentous phages. Figure 3*a* is a diffraction pattern from M13. The layer lines in the diffraction pattern are spaced at intervals of approximately $\frac{1}{32}$ $Å^{-1}$. This indicates that the axial repeat of the helical array of coat proteins in M13 is 32 Å. The diffraction pattern in Figure 3*b* is from Pf1. The layer lines in this diffraction pattern are spaced approximately every $\frac{1}{75}$ $Å^{-1}$ indicating that the axial repeat of Pf1 is 75 Å under these conditions. Figure 3*c* is another diffraction pattern from Pf1 similar to that in Figure 3*b*. However, in the pattern in Figure 3*c* there are distinct reflections falling half way between the layer lines corresponding to those seen in Figure 3*b*. This means that under some conditions the axial repeat of Pf1 is doubled to 150 Å. It is not immediately apparent from these patterns how many structural subunits are present in each axial repeat and with what sort of helical symmetry the subunits are arranged. Determination of these parameters, crucial for structural study, requires measurement of chemical composition, mass per unit length, and often, analysis of diffraction from heavy atom derivatives or processing of electron micrographs.

Analysis of a set of fiber diffraction patterns results in an electron density map, which represents an average image of the structural subunits in the virus particle. If each subunit has the same structure, then an image of the entire virus particle can be built up from many identical images of the subunits. However, if a portion of the subunit structure varies from subunit to subunit, the variable portion of the structure will not be visible in the resulting averaged image. For instance, when the viral DNA is associated differently with different coat proteins, then the resulting electron density map may not show details of the DNA structure.

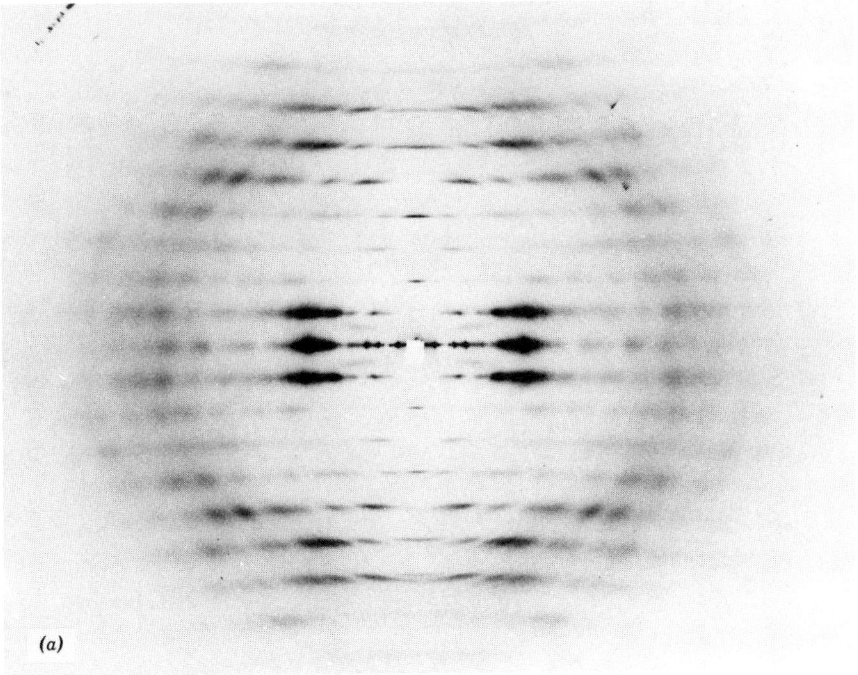

Figure 3. X-ray diffraction patterns from fibers of filamentous phages oriented in a magnetic field and partially dried to 98% relative humidity. (*a*) Diffraction pattern from M13. On page 212: (*b*) Diffraction pattern from Pf1 exhibiting a 75 Å axial repeat. (*c*) Diffraction pattern from Pf1 exhibiting a 150 Å axial repeat.

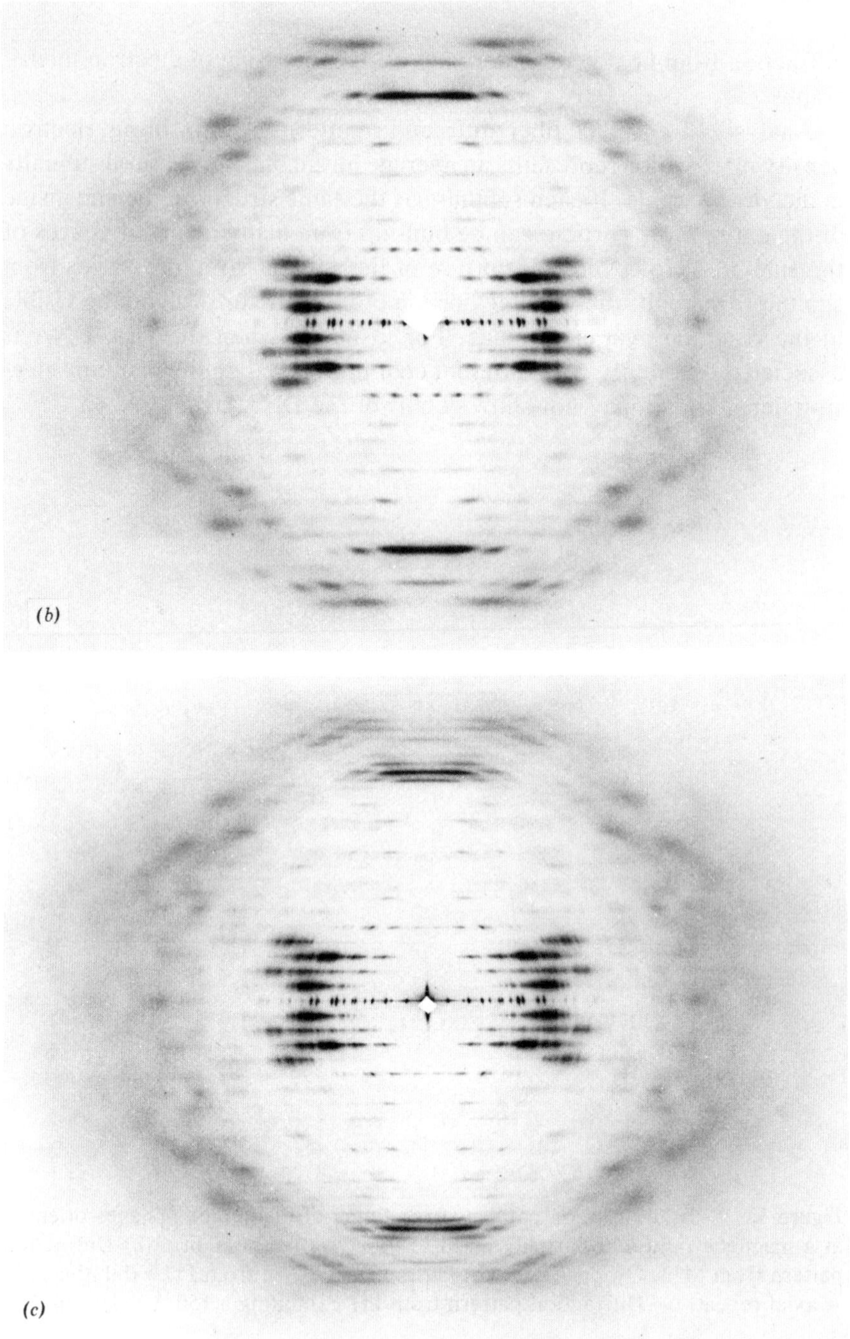

Figure 3. Continued

4. HELICAL SYMMETRY

A helical assembly can be defined by the structure of a single asymmetric unit and the symmetry operations that relate the positions of asymmetric units to one another. An asymmetric unit is the smallest portion of the structure which, by use of the symmetry operations, can be used to generate the entire structure. It may consist of a single peptide or it may include several chemically identical proteins if these proteins are structurally distinct or in structurally distinct environments. Generally, in filamentous phages, the asymmetric unit consists of a single coat protein subunit. The asymmetric unit may also include DNA if the DNA is arranged with the same helical symmetry as the coat protein. In most filamentous phages, the DNA appears to have a different helical symmetry from the coat protein. One of the most challenging problems in the study of filamentous phage structure is to determine the molecular details of the pattern of interactions between the DNA and the coat proteins.

The symmetry operations used to define a helical symmetry include the basic screw operation (a rotation coupled to a translation along the helix axis) and the rotational point group symmetry (Klug et al., 1958). The point group symmetry must be either the cyclic n-fold (C_n) or the dihedral n-fold (D_n) operation. The simplest example is a one-fold rotation C_1, corresponding to a 1-start helix. A twofold axis perpendicular to the helix axis as in double-helical DNA structure is designated D_1. No special distinction need be drawn between the point groups with $n = 1$ or $n > 1$. A screw operation consists of a unit translation by a distance h parallel to the axis coupled with a rotation by an angle ϕ about this axis. This rotation can equally well be represented by the ratio $\sigma = 2\pi/n\phi$ where n is the rotational order of the point group. When $n = 1$, σ is equal to the number of units per turn. The hand of rotation for σ can be either to the right or to the left. The helical symmetry of a structure can be represented by the symbols C_n or D_n and S_σ where S_σ designates the σ-fold screw symmetry (Makowski et al., 1980). For example, in this notation the symmetry of an α-helix is $C_1S_{3.6}$ with $h = 1.5$ Å. The observed helical symmetries of filamentous phages are listed in Table 4.

The helical symmetry of a structure can be visualized by construction of a helix net or surface lattice (Klug et al., 1958). A helix net is constructed by placing on a cylinder of given radius a set of lattice points related to one another by the symmetry operations of the structure of

Table 4. Symmetries Observed[a]

fd, M13	$C_5S_{2.0}$
If1	$C_5S_{1.9}$
IKe	$C_5S_{1.9}$
Pf1 ($T > 8$°C)	$C_1S_{5.40}$
Pf1 ($T < 8$°C)	$C_1S_{5.46}$

[a] Exact symmetries are sensitive to temperature and salt conditions and the binding of heavy atoms. Symmetries taken from data presented in Marvin et al. (1974a,b), Makowski and Caspar (1978), and Nave et al. (1979, 1981).

interest. Each lattice point corresponds to one asymmetric unit or subunit. This cylinder is then cut along a line parallel to the helix axis and flattened to provide a two-dimensional view of the lattice points of the helix as drawn in Figure 4. In Figure 5, drawings of helix nets for four different filamentous phage structures are shown.

As determined from diffraction patterns similar to the one from M13 in Figure 3*a*, the symmetry of the fd protein coat is $C_5S_{2.0}$, consisting of a fivefold rotation axis and a twofold screw axis (Makowski and Caspar, 1978, 1981; Banner et al., 1981). Five subunits are arranged around the

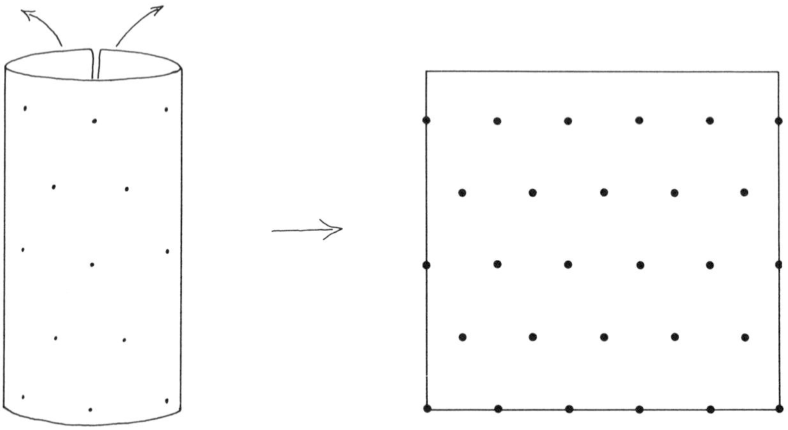

Figure 4. The construction of a helical net or surface lattice. The lattice points of the helical array are drawn on a cylinder, which is then cut down a line parallel to its axis and unfolded to lay flat for viewing. In this chapter, the helical nets are oriented to correspond to the view from the outside of the helix.

helix axis at the same axial position. Starting at a particular position in one subunit, an identical position on a different subunit can be reached either by a rotation of 72° or by a rotation of 36° followed by an axial translation of about 16 Å. The axial repeat distance is approximately 32 Å and corresponds to the spacing of layer lines in the diffraction pattern from M13 in Figure 3*a*. Although related by a fivefold rotation axis, the subunits sharing the same axial position should not be thought of as pentamers because it is quite possible that they have no direct interactions with one another. The helix net for the fd protein coat is shown in Figure 5*a*.

Each subunit in the fd structure consists of a single coat protein. The DNA making up the two antiparallel strands in the phage cannot have the same symmetry as the coat protein because it cannot have a fivefold

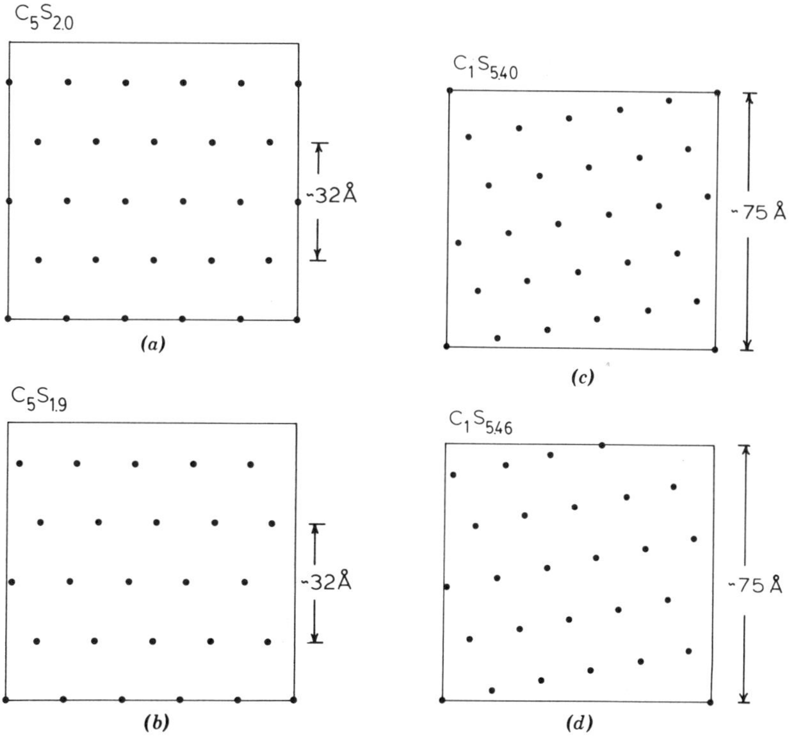

Figure 5. Four helical nets showing four symmetries taken on by the filamentous phages: (*a*) $C_5S_{2.0}$. (*b*) $C_5S_{1.9}$. (*c*) $C_1S_{5.40}$. (*d*) $C_1S_{5.46}$.

rotation axis. The nonintegral ratio of nucleotides to coat proteins is further confirmation that the DNA and protein are arranged with different symmetries. On overexposed diffraction patterns from fd it is possible to see pairs of weak intensity maxima corresponding to layer lines spaced with a 26.7 Å periodicity (Banner et al., 1981). These have been interpreted as being due to the DNA. The subunits arranged in the $C_5S_{2.0}$ helical lattice must interact with the DNA in a large number of different ways because there does not appear to be a simple relationship between the symmetry of the coat protein and that of the DNA.

The fivefold rotation axis is characteristic of all Class I phages but the screw axis varies slightly from phage to phage. Both the screw axis and the axial repeat distance may change slightly with changes in temperature, relative humidity, or ionic strength. Figure 5*b* is a helix net with symmetry $C_5S_{1.9}$ as observed for Ike and If1 (Marvin et al., 1974b). In this case, the screw operation is a rotation of 34.2° (or equivalently, 37.8° in the opposite direction) coupled with an approximately 16 Å axial translation. The absolute hand of this helix net has not yet been determined.

Class II viruses are distinguished by their lack of a rotation axis (except the C_1 identity). Pf1 (at room temperature), Xf, and Pf3 all appear to have symmetry $C_1S_{5.40}$ (Marvin et al., 1974a; Makowski and Caspar, 1978; Peterson et al., 1982). A preliminary 4 Å resolution electron density map of Pf1 indicates that the primary helix of Pf1 is right-handed. The axial repeat is about 75 Å and there are 27 subunits in five turns (5.40 units per turn) in the axial repeat. The axial repeat corresponds to the spacing of layer lines seen in the diffraction pattern from Pf1 in Figure 3*b*. A helix net corresponding to symmetry $C_1S_{5.40}$ is shown in Figure 5*c*. Although lacking a fivefold rotation axis, comparison of this helix net with those in Figures 5*a* and 5*b* shows that the helical lattices of Class I and Class II viruses are quite similar. This suggests that the packing of subunits in the two classes of phage are similar.

Mass per unit length and chemical composition measurements (Table 3) indicate that the subunits of Pf1 consist of one protein and one nucleotide. It is conceivable that the DNA could take on a conformation that would mimic the symmetry of the coat proteins in Pf1. However, because half of the nucleotides are in one strand and the other half are in the antiparallel strand, they cannot actually have the same symmetry. Strictly, each asymmetric unit in Pf1 must consist of at least two coat proteins and two nucleotides (one from each strand). An arrangement of this sort cannot have an axial repeat of less than 150 Å. A few weak layer lines are often observed half way between the 75 Å layer lines. One of these can be seen on the meridian of the diffraction pattern in Figure 3*b*. The absence of other half-order layer lines might indicate that the two strands look approximately the same in spite of the fact that they are running in opposite directions. Or it might indicate that the DNA has symmetry not simply related to that of the coat protein. However, no layer lines, which might indicate a different symmetry for the DNA, have been observed, and the integer ratios of nucleotides to proteins in both Pf1 and Xf suggest an ordered interaction between DNA and protein.

Occasionally very strong half-order layer lines are observed as in the diffraction pattern of Pf1 in Figure 3*c*. These layer lines are too strong to be due to diffraction from DNA alone. They indicate that the structure of the protein coat has been perturbed so that the coat proteins have taken on at least two different conformations. This perturbation may be due to protein–DNA interactions or to interactions between virions in the fiber.

The protein coat of Xf has the same $C_1S_{5.40}$ helical symmetry as Pf1 but with two nucleotides per coat protein. It is possible that, in this case, the DNA has exactly the same symmetry as the coat protein. Each asymmetric unit would then consist of one coat protein plus one nucleotide from each of the two DNA strands. Pf3, unlike the other Class II viruses, has a nonintegral nucleotide to protein ratio. This suggests that, as in the case of Class I phages, the DNA and protein in Pf3 may have completely different helical symmetries.

Depending on temperature, heavy atom binding, and possibly other environmental factors, Pf1 takes on a variety of helical symmetries varying from $C_1S_{5.40}$ to $C_1S_{5.46}$ (Wachtel et al., 1976; Nave et al., 1979; Nave et al., 1981). The two principal symmetries observed for native virus are $C_1S_{5.40}$ at temperatures above about 10°C and $C_1S_{5.46}$ at temperatures below about 6°C (Nave et al., 1979). At about 8°C, a mixture of the two forms appears to exist. The helical net for the symmetry $C_1S_{5.46}$ is shown in Figure 5*d*. In addition to the temperature transition resulting in a slight shift in the angular rotation between subunits, it also changes the axial translation between subunits from 3.06 Å below the phase transition temperature to 2.85 Å above that temperature (Nave et al., 1979).

The helix parameters of all the filamentous phages appear to be dependent to some extent on phage environment. This is not indicative of weak interactions between subunits because the phage are stable and remain infective over a wide range of temperature, pH, and ionic strength. It may, however, suggest that either the subunits or the bonds between subunits are capable of some distortion. This is also suggested by the apparent flexibility of the phages which is indicated by the curvature of the phage axis in electron micrographs such as those in Figure 2. Although the virions are subjected to substantial surface forces when placed on the electron microscope grid, the observed curvature is in sharp contrast to the straight rods observed in electron micrographs of, for instance, TMV, and suggestive of some flexibility in solution. The local environment of subunits in curved portions of the phage may resemble that of phages with slightly altered helical symmetry and axial translation between subunits, so that the observed variation in helix parameters is consistent with the apparent flexibility of the phage particles.

5. STRUCTURE OF Pf1 AS DETERMINED BY X-RAY DIFFRACTION

Electron density maps of Pf1 have been calculated to 7 Å resolution for both the high temperature (Makowski et al., 1980) and low temperature (Nave et al., 1981) forms and a preliminary electron density map of the low temperature form at 4 Å resolution has been calculated (Bryan et al., 1983). The methods used for the analysis of the fiber diffraction data were different in all three studies, but the resulting electron density maps appear consistent with one another and most of the differences can be interpreted in terms of structural differences between the two forms.

X-ray diffraction (Marvin, 1966) and a wide variety of spectroscopic probes (Marvin and Kohn, 1969; Thomas et al., 1983) have indicated that the coat protein of Pf1 is largely α-helical. The strongest reflections in the diffraction patterns in Figure 3 are at about 10 Å spacing near the equator and at about 5 Å spacing near the meridian. Because α-helices pack about 10 Å apart in proteins and have a pitch of about 5.4 Å, this strong diffraction is diagnostic of a high proportion of α-helical conformation in the virus protein. The orientation of this diffraction indicates that the α-helices must be approximately parallel to the virion axis.

The sharp reflections on the equator to the right and left of the center of the diffraction patterns in Figure 3 are due to the two-dimensional hexagonal packing of the virions in the fiber. Analysis of these reflections can be used to construct a cylindrically averaged image of the virus particle and this is shown for the high temperature form of Pf1 in Figure 6. The very high electron density at low radius corresponds to the DNA. The lower electron density from 10 to 30 Å radius corresponds to the protein coat. An unusally low electron density region at about 20 Å radius is indicative of a large number of hydrophobic residues at this radius. This is consistent with the large proportion of hydrophobic residues in the coat protein. Because α-helices are about 10 Å in diameter, this cylindrical projection suggests that the viral DNA is encased in a double layer of α-helices stabilized largely by the interactions of hydrophobic residues between the two layers.

Analysis of data to 7 Å spacing has been used to calculate three-dimensional electron density maps of Pf1 in the high temperature and low temperature forms (Makowski et al., 1980; Nave et al., 1981). A section through each of these three-dimensional maps is shown in Figure 7. The contouring of the two maps, taken from different publications, is rather different. Only the highest peaks have been contoured in the low temperature form, and for consistency, the negative contours have been removed from the map of the high temperature form. The axial spacing between subunits is about 2.7 Å in the high temperature form and 3.1 Å in the low temperature form. Full three-dimensional maps of Pf1 can be constructed from these sections by operating on them with the symmetry operators, in this case a simple screw operation. In the high temperature

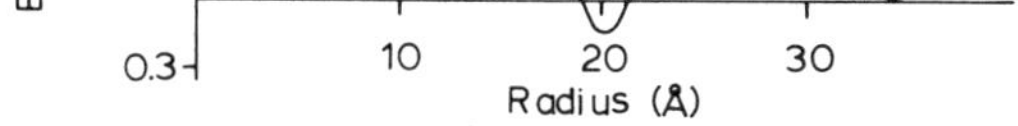

Figure 6. The cylindrically symmetric part of the equatorial projection of the Pf1 virus particle. The high electron density at the peak corresponds to the viral DNA. The electron density peaks at radii of 15 Å and 25 Å correspond to two layers of α-helical segments. The very low electron density at a radius of 20 Å suggests that a large number of nonpolar residues project into this region from the two layers of α-helices (Makowski et al., 1980).

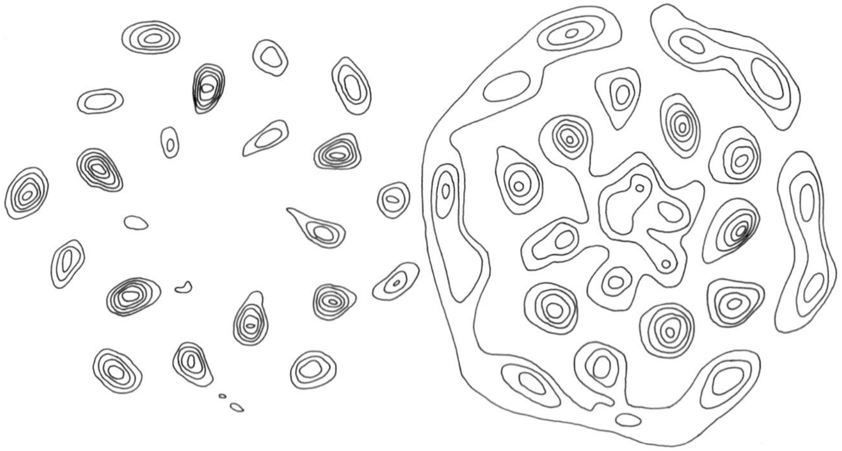

(a) (b)

Figure 7. Single sections through electron density maps of the (*a*) low temperature form and (*b*) high temperature form of Pf1 at 7 Å resolution. The two maps, taken from different publications, are contoured somewhat differently. The electron density at low radius, corresponding to the DNA, is only visible in the electron density map of the high temperature form. Peaks of electron density away from the particle axis correspond to sections through α-helical segments of the coat proteins. Redrawn from Makowski et al. (1980) and Nave et al. (1981).

form, this consists of a translation of 2.7 Å coupled to a rotation of 66.7°; in the low temperature form the parameters are 3.1 Å and 64.9°. At 7 Å resolution identical sections rotated relative to one another and stacked are sufficient to provide a complete three-dimensional electron density map. Figure 8 includes drawings of stacked sections constructed from the drawings in Figure 7. The axial separation between sections in Figure 8 is thus slightly different for the high temperature and low temperature forms.

The electron density maps of the two forms are very similar. Electron density less than 10 Å from the center of the particle is due largely to the DNA. This density is visible only in the map of the high temperature form, possibly because of the choice of contours or the method used for phasing the low temperature form. Density in this region can be observed in the preliminary 4 Å map of the low temperature form (Bryan et al., 1983). No detailed interpretation of the density in this region in terms of DNA structure has been possible. This may be due to the fact that any image of the DNA in these maps must correspond to the average of two antiparallel strands. At 4 Å resolution some interpretable features of the DNA should be visible. A description of the RNA–protein interaction in TMV has been possible based on an electron density map calculated from fiber diffraction data to 4 Å resolution (Stubbs et al., 1977; Stubbs and Stauffacher, 1981; see also Chapters 4 and 5).

Electron density further than about 10 Å from the center of the particle is due largely to the protein coat. Each electron density peak farther than 10 Å from the particle axis corresponds to a section through an α-helix. Because the axial separation between adjacent subunits along the 1-start helix actually changes by at least 0.2 Å during the temperature transition, the corresponding peaks in the two electron density maps may correspond to portions of the peptide chains separated by as much as 4 Å along the α-helices. In spite of this, the positions of these peaks in the two maps are very similar. As measured from the maps in Figure 7, their positions differ by an average of only 2.3 Å.

The two layers of α-helical segments can be seen surrounding the DNA core in both Figures 7 and 8. In Figure 8 the slewing of the rods of electron density around to the right can be seen. The inner layer of α-helices are

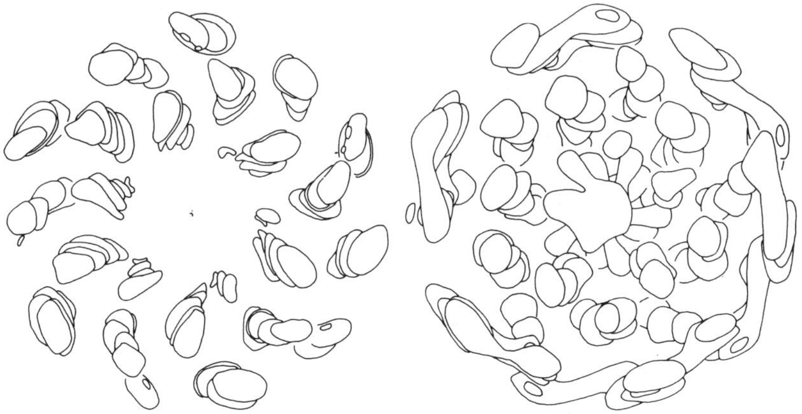

(a) *(b)*

Figure 8. Stacked sections of the electron density maps of the (*a*) low temperature form and (*b*) high temperature form of Pf1 at 7 Å resolution. The axial separation between sections in (a) is about 3.1 Å, and about 2.8 Å in (b), corresponding to the different axial rise per subunit in the two forms of the virion. Constructed from figures taken from Makowski et al. (1980) and Nave et al. (1981).

tilted on average by about 6° relative to the particle axis. The outer layer is tilted by about 26° to the particle axis in the high temperature form and by about 22° in the low temperature form. Drawings of the arrangement of the rods of electron density in the high temperature form of Pf1 are shown in Figure 9. In these drawings the coat protein has been broken up into two rods each drawn as solid cylinders about 9 Å in diameter and 30 Å long. As will be discussed below, the outer rod of electron density may not be entirely α-helical. The center drawing corresponds to about 1% of the total length of the Pf1 virion. The drawings to the left and right correspond to the outer and inner layers of α-helical segments.

Each coat protein consists of one rod of electron density from the inner layer and one from the outer layer. The carboxy-terminal half of the protein occupies one rod in the inner layer with the basic residues at the carboxyl terminus located at the lower radius end of the rod of electron

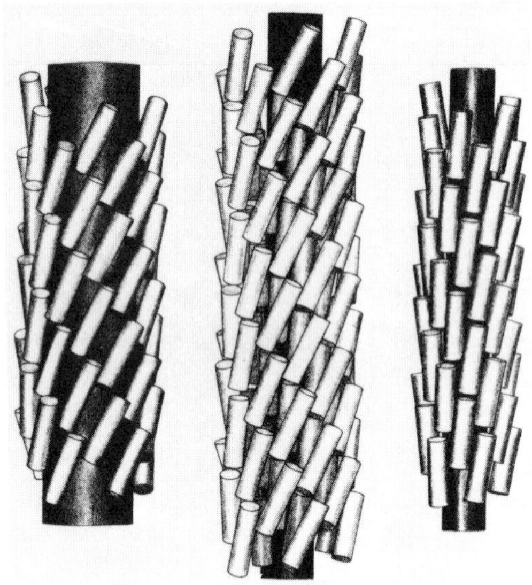

Figure 9. Diagrams showing the arrangement of rods of electron density in the Pf1 virus particle. In the center is a representation of about 2.5 axial repeats ~185 Å of the virion. A complete drawing of the virion would be over 100 times this length. The central core, occupied by the DNA, is represented by a 20 Å diameter cylinder. The two layers of α-helical segments are drawn separately to the right and left for clarity (Makowksi et al., 1980).

density. The amino-terminal half of the coat protein occupies one rod of electron density in the outer layer. The connectivity of the inner helix with the outer is not unambiguously defined in the 7 Å maps. This is because the top of one rod of electron density in the outer layer is close to the bottom of another rod of density. In both maps shown in Figure 8, the two rods appear to be connected end-to-end suggesting that the coat protein is arranged in an extended configuration. However, a hairpin configuration cannot be ruled out from these maps. The preliminary 4 Å density map of the low temperature form (Bryan et al., 1983) appears to confirm the extended configuration of the coat protein. If this is the case, then the entire coat protein is arranged as an elongated α-helical peptide with a bend near its center. The two halves of the protein are connected by a curved region of electron density which cannot be interpreted in terms of the peptide conformation in the preliminary 4 Å resolution map (Bryan et al., 1983).

The protein coat of Pf1 is thus a 20 Å thick cylinder of intricately interlocked α-helical segments. The helix net of Pf1 was initially chosen to be left-handed (Marvin et al., 1974a; Makowski et al., 1980) because the α-helical segments in the inner layer would then interlock with those of the outer layer for a length of about 25 Å in a knobs-into-holes sense as first predicted by Crick (1953) for α-helical coiled-coils based on the surface lattice of an α-helix. However, the hand of the α-helices as seen in the 4 Å map (Bryan et al., 1983), indicate that the Pf1 surface lattice is right-handed. This means that the α-helices cannot interact in a knobs-into-holes sense. Using the notation of Chothia et al. (1977), the helices from the inner and outer layers interlock with a crossing angle of $-20.5°$ for a distance of at least 22 Å. A survey of 25 observed α-helical crossing angles in six globular proteins (Chothia et al., 1977) suggests that this is a relatively unusual but not unprecedented α-helical crossing angle. The sizes of amino acid residues on the interacting faces of the two α-helices must be important in determining the crossing angle (Richmond and Richards, 1978; Lim et al., 1978; Etimov, 1979) because in the hydrophobic interior of the protein coat the residues will be close packed (Richards, 1974).

Two shorter helix–helix interactions occur in the inner layer along the 5-start and 6-start helix directions. In both cases, the interacting helices are nearly parallel. This is a common pattern for the interaction of adjacent

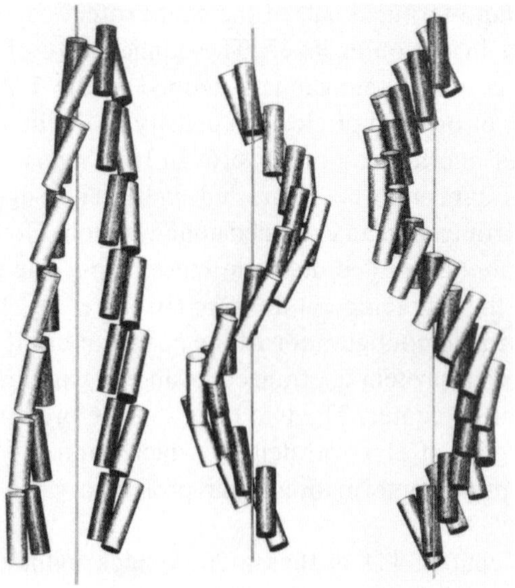

Figure 10. Diagrams representing the contact relationships between structure units along the 11-start (left), 6-start (center), and 5-start (right) helical directions. The rods of electron density in the inner layer are shaded slightly darker than the outer layer of rods (Makowski et al., 1980).

helices (Chothia et al., 1977). Figure 10 is a drawing of subsets of the rods of electron density in Pf1 to provide an easy visualization of the interactions between α-helical segments.

6. INTERPRETATION OF THE ELECTRON DENSITY MAPS

In this section, results from some of the chemical and biophysical studies conducted on Pf1 are used to discuss the possible positions of specific amino acid residues in the electron density maps of Pf1.

All physical and chemical evidence suggests that at least 90% of the Pf1 coat protein is in an α-helical conformation. Analysis of the amino acid sequence based on empirical rules relating the probability of occurrence of various secondary structures to the amino acid sequence (Chou and Fasman, 1974) also indicates that most of the protein is likely to be in an α-helical conformation. Two short portions of the sequence have a low probability of being α-helical. These sections are from Gly 1 to Ser 6 and from Ile 22 to Tyr 25 (Nakashima et al., 1975).

Analysis of the volumes available for DNA, the inner layer of α-helices, and the outer layer of α-helices using a Voronoi construction (Makowski et al., 1980) indicates that the DNA occupies most of the volume of the inner peaks connected together by low electron density in the map in Figure 7 *b*. In this region, the two largest radii peaks connected to the central peak by narrow density features are probably the location of the carboxy terminal amino acids of the coat protein. The inner layer of α-helical segments is defined as occupying those electron density peaks having little or no exposure to the external virus surface in the density map in Figure 7*b*. These peaks have a volume sufficient to contain residues 26–46. The volume analysis places residues 24–26 in the region where the rod of electron density is most highly curved. Because these residues (Gly-Gly-Tyr) have a relatively low probability of being in an α-helical conformation, it is possible that the regular hydrogen bonding of the α-helix is disrupted in the immediate vicinity of the pair of glycines. This would place residues 25–46 in the inner layer of α-helices and residues 1–22 in the outer layer.

Solid state NMR studies (Cross et al., 1983) have shown that the N–H bonds of the amide groups in intact Pf1 phage fall into two populations: in one group the amide bonds are nearly parallel to the virus axis, and in the other group the amide bonds are tilted by about 20° relative to the axis. The N–H bonds in an α-helix are parallel to the α-helix axis to within about 10° so that these two populations of amide groups correspond to the inner and outer layers of α-helical segments indicated by the electron density map. The NMR studies suggest that the α-helical segment tilted by about 20° relative to the particle axis is about 65% as long as the α-helical segment nearly parallel to the particle axis. If the inner α-helical segment contains 22 residues (residues 25–46), the outer segment would, according to this result, contain about 14 residues. If, as suggested by secondary structure prediction, a few residues at the amino terminus are in a nonhelical conformation, then the NMR results suggest that residues 1–8 may be in a nonhelical conformation and residues 9–22 may be in an α-helical conformation.

Further analysis of the amino acid sequence also suggests that the first few residues at the amino terminus are not in an α-helical conformation.

Figure 11 contains diagrams representing a view directly down α-helical segments with the amino acid sequences of the carboxy- and amino-terminal sections of the Pf1 coat protein. The top diagram of the amino terminal section is oriented so that as many polar amino acid residues point "up" as is possible. The bottom diagram is oriented with polar residues

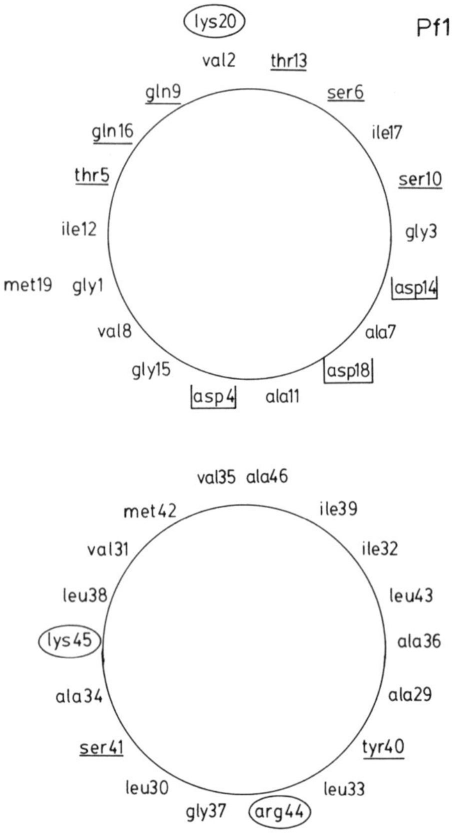

Figure 11. Helical wheel diagram of Pf1 coat protein. The top wheel includes the amino acid residues at the amino terminus of the Pf1 coat protein oriented to place as many polar residues as possible on the top half of the diagram. The top half of the α-helix depicted by this wheel should face the solvent; the bottom half should face into the hydrophobic interior of the protein coat. The bottom wheel includes amino acid residues at the carboxyl terminus of the coat protein. The basic residues are oriented down. The top half of the α-helix depicted by this diagram should face the hydrophobic interior of the protein coat. The bottom half should face the DNA near the center of the virus particle. Uncharged polar residues are underlined; acidic residues boxed; basic residues are circled.

pointing "down." If the Pf1 coat protein is entirely α-helical, Figure 11 should provide a means to make a rough estimate of the radius of a given amino acid residue. The DNA containing center of the particle corresponds to the bottom of the diagram, the outside of the particle to the top, and the hydrophobic interior of the protein coat to the center. For the most part, the residues of Pf1 do divide with the polar or charged groups at the top and bottom of Figure 11 and with the nonpolar residues in the center. On the amino terminal helical wheel, residues 2, 4, 14, 17, and 18 deviate from this pattern. However, Asp 14 and Asp 18 appear, from chemical modification studies, to be inaccessible to solvent (Nakashima and Konigsberg, 1980), and Ile 17 is not far from the polar/nonpolar interface. Thus, only residues 2 and 4 deviate dramatically from positions expected relative to the particle surface. Furthermore, chemical modification studies indicate that Asp 4 is accessible to solvent (Nakashima and Konigsberg, 1980). These results taken together suggest that at least the first four amino acid residues at the amino terminus of the Pf1 coat protein are in a nonhelical conformation. The results of NMR studies suggest that as many as eight residues at the amino terminus may not be in an α-helical conformation. The rods of electron density in the two layers are approximately the same length in the maps in Figure 7. From the available data, it seems likely that a region at the high radius end of the outer rod of electron density is occupied by a few residues in a non-α-helical conformation. This being the case, the amino terminus may extend somewhat beyond the electron density visible in the maps.

Other residues not in their expected orientation in the helical wheel for the outer α-helical segment are Asp 14, Ile 17, and Asp 18, all within four residues or 6 Å along the α-helical axis. It is possible that this region is also non-α-helical. However, chemical modification studies (Nakashima and Konigsberg, 1980), indicate that the carboxyl groups of Asp 14 and Asp 18 are not freely accessible to solvent on the surface of the virus particle. This result suggests that these two residues are buried within the interior of the virus particle. These side chains cannot exist in the hydrophobic interior of the protein coat in isolation. Depending on their ionization state, they must make salt bridges and/or hydrogen bonds to other polar residues, main chain groups not hydrogen bonded in an α-helical conformation, or, as commonly found in virus particles (Stubbs, Chapter 5), a positive ion. Possible candidates for interaction with Asp 14 and Asp 18 may be found in Figure 12. Figure 12 is a drawing of the outer layer of electron density at 28 Å radius showing the relative positions

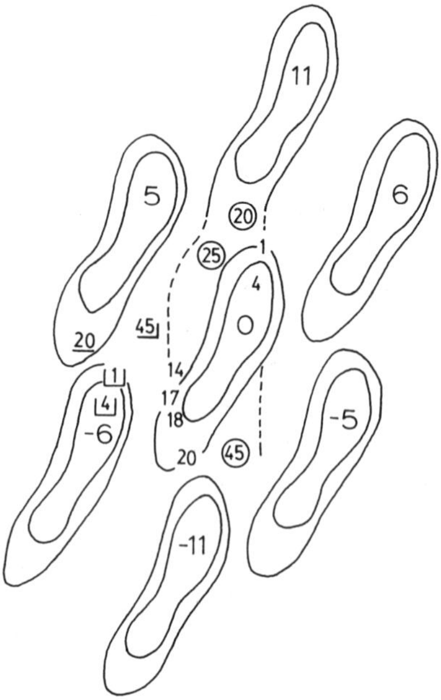

Figure 12. Intermolecular interactions in the Pf1 protein coat. Seven rods of electron density from the outer layer of the protein coat are sketched and labeled according to their positions on the primary, 1-start helix. The subunit labeled 11 is extended in broken lines to show the approximate position of the carboxy terminal portion of this protein. The small numbers mark the approximate positions of amino acid residues from subunit 0. Circled numbers are from subunit 11. The position of lys 20 from residue 5 is underlined; Gly 1 and Asp 4 from subunit −6 are enclosed in boxes; Lys 45 from subunit 16 (not shown) is enclosed in half a box. Potential interactions among polar groups can be identified from this diagram and information about their radial positions from the electron density map in Figure 7*b* and the helical wheels in Figure 11. Asp 14 and Asp 18 appear to be buried on the inside of the outer α-helical rod. Residues that could potentially interact with Asp 14 and Asp 18 from subunit 0 include Lys 45, from subunit 11, the amino terminus from subunit −6, and the carboxyl terminus (Ala 46) from subunit 11 (which would require the presence of a positive ion). Lys 45 from subunit 11 and Lys 20 from subunit 5 are probably too remote to make contacts with either Asp 14 or Asp 18. Lys 20 from subunit 11 may be in a position to interact with Asp 4. At very low radius, Lys 45 could potentially interact with the carboxy terminus of adjacent subunits. There do not appear to be any other potential polar interactions in the virion. The only potential nonpolar interaction between subunits in the outer layer appears to be between Ile 17 in subunit 0 and Ala 2 in subunit −6.

of adjacent subunits. The subunits are labelled according to their positions along the primary, 1-start helix. Subunit 6, for instance, is six positions along the 1-start helix from subunit 0. These proteins are nearest neighbors along 6-start helices. Subunit 0 interacts with those subunits 5, 6, and 11 positions along the primary helix from it. The combination of information from the helical wheel diagrams and the volume calculations from the electron density maps can be used to postulate the approximate positions of amino acid residues in the virus. Volume calculations were used to determine the approximate positions of the residues along each rod of electron density. The helical wheel diagrams suggest the radius of each residue and which side of the α-helices the residues fall on. In addition to this, a variety of biochemical and biophysical studies indicate the environment of some coat protein residues as listed in Table 5. The position of Tyr 25 is indicated by the X-ray diffraction studies of iodinated Pf1 (Nave et al., 1981; Bryan et al., 1983). The position of the amino terminus, labeled with one mercury atom (Nave et al., 1981), is less well defined by the X-ray studies because the mercury atom may be as much as 8 Å from the amino terminus. Based on these sets of data, the approximate positions of selected residues are indicated in Figure 12.

Tyr 25, which from volume calculations (Makowski et al., 1980) ought to be located near the high radius end of the inner α-helical segment, appears to be held rather rigidly in a hydrophobic environment (Table 5). It accepts a hydrogen bond from a nonsolvent donor and is also available for modification to 3,5-diiodotyrosine. The two iodines bind at radii of 26.1 and 28.2 Å (Bryan et al., 1983). The phenolic group appears to extend radially out from the main chain along the bottom of the outer α-helix. This does not put it in a position to interact with either Asp 14 or Asp 18 as can be seen in the diagram on Figure 12.

Lys 20 is predicted to be on the surface of the particle from the helical wheel in Figure 11. Although, it can be completely modified by reaction with SHIMS (methyl 3-mercapto-propionimidate), the product is not susceptible to carboxy methylation by iodoacetate even when the protein is disrupted by sodium dodecyl sulfate (Nave et al., 1981). The structural significance of these observations is not clear. The length of the lysine side chain is probably not sufficient to extend from the protein labeled 5 in Figure 12 into the vicinity of Asp 14 or Asp 18 of protein 0. Lys 20 from subunit 11 may be able to interact with Asp 4 from subunit 0. However, because of the possibility of non-α-helical conformation near the amino terminus, the position of Asp 4 is difficult to estimate.

Table 5. Accessibility and Environment of Residues

Virus	Residue	Environment	Method	Reference
Pf1	Gly 1	Accessible to solvent	Chemical modification	Nakashima and Konigsberg (1980), Nave et al. (1981)
	Asp 4	Accessible to solvent	Chemical modification	Nakashima and Konigsberg (1980)
	Asp 14	Inaccessible to solvent	Chemical modification	Nakashima and Konigsberg (1980)
	Asp 18	Inaccessible to solvent	Chemical modification	Nakashima and Konigsberg (1980)
	Lys 20	Partially accessible	Chemical modification	Nakashima and Konigsberg (1980), Nave et al. (1981)
	Tyr 25	Accessible to solvent	Chemical modification	Nakashima and Konigsberg (1980), Nave et al. (1981)
		Accepts hydrogen bond from nonsolvent donor	Raman	Thomas (1976), Thomas et al. (1983)
		Reversibly deprotonated at pH 11.3	Absorbance	Day and Wiseman (1978)
		Rigidly constrained in nonpolar environment	Fluorescence	Day et al. (1979)
	Tyr 40	Accepts hydrogen bond from nonsolvent donor	Raman	Thomas (1976), Thomas et al. (1983)
		Inaccessible to solvent	Chemical modification	Nakashima and Konigsberg (1980), Nave et al. (1981)
		Hydrogen bonded or near anion	Fluorescence	Day and Wiseman (1978), Day et al. (1979)
		Possibly base-stacked	Fluorescence	Day and Wiseman (1978), Day et al. (1979)

	Arg 44	Inaccessible to solvent	Chemical modification	Nakashima and Konigsberg (1980)
	Lys 45	Inaccessible to solvent	Chemical modification	Nakashima and Konigsberg (1980), Nave et al. (1981)
fd	Phe 11	Rotation hindered, not immobilized	NMR	Gall et al. (1982)
	Tyr 21	Accepts hydrogen bond from nonsolvent donor	Raman	Thomas (1976), Thomas et al. (1983)
		Rotation hindered, not immobilized	NMR	Gall et al. (1982)
	Tyr 24	Accepts hydrogen bond from nonsolvent donor	Raman	Thomas (1976), Thomas et al. (1983)
		Rotation hindered, not immobilized	NMR	Gall et al. (1982)
	Trp 26	Hydrophobic environment	Raman	Thomas et al. (1983)
		Not near DNA	Fluorescence	Day et al. (1982)
		Rigidly immobilized	NMR	Gall et al. (1982)
	Phe 42	Stacked–unstacked by salt, temperature	Raman	Thomas et al. (1983)
		Rotation hindered, not immobilized	NMR	Gall et al. (1982)
	Phe 45	Stacked–unstacked by salt, temperature	Raman	Thomas et al. (1983)
		Rotation hindered, not immobilized	NMR	Gall et al. (1982)

Other candidates for interacting with Asp 14 and Asp 18 are Lys 44 or Arg 45 or the amino terminus. Residues 44 and 45 have been assumed to interact with the DNA. Tyr 40, Ser 41, Arg 44, and Lys 45 all fall on the same side of the inner α-helix as seen in Figure 11. Fluorescence measurements (Day and Wiseman, 1978; Day et al., 1979) indicate that Tyr 40 is involved in DNA interactions. Arg 44 is probably acting to neutralize the charged DNA phosphate groups. Because there is only one nucleotide per coat protein in Pf1, it is possible that only one basic residue, Arg 44, is involved in DNA interactions. Lys 45 may or may not be involved in DNA binding. The long lysine side chain could allow the lysine ε-amino group from protein 11 to extend to a higher radius where it would be in a position to interact with Asp 14 and Asp 18.

Because the residues at the amino terminus may not be in an α-helical conformation, it is difficult to estimate the position of the terminal amino group. It is possible that the amino terminus from protein −6 could interact with Asp 14 or Asp 18 of protein 0.

All the charged residues in Pf1 occur near the ends of the α-helical segments. Because the two α-helical segments end in close proximity it is possible that many of the charged groups of the virus are in close proximity. Concentrations of charged groups appear to be common at the interface between subunits in virus particles and may act as control sites for assembly or disassembly (Stubbs, Chapter 5). This means that the identification of hydrogen bonds or salt bridges between subunits in filamentous phages will be important for understanding control of assembly and disassembly and the structural transitions observed in these viruses.

The position of Ile 17 which is anomalous on the helical wheel diagram of Figure 11 becomes clearer in the cylindrical section of Figure 12. It appears to be located at the protein–protein interface along the 6-start helix. Ile 17 from subunit 0 projects toward subunit −6 in the diagram in Figure 12. This is the only position where the left-hand half of the outer α-helix is in contact with another α-helix in the outer layer. The exposure of Ile 17 to solvent is thus less than is suggested by its position on the helical wheel.

The possible polar interactions between subunits appear to be limited to those between adjacent subunits along 6-start helices and 11-start helices. In the outer layer of α-helices, the only close contacts are along these two directions. In the inner layer, there are also contacts along the

5-start direction but there do not appear to be any polar groups in appropriate positions to act along these directions. Most intersubunit interactions in the inner layer are likely to be between nonpolar residues.

7. DNA STRUCTURE IN Pf1

The electron density in the region of the Pf1 map where the DNA is located is not, in the 4 Å resolution map (Bryan et al., 1983), interpretable in terms of the DNA conformation. That electron density map represents an image of the structural subunits that have helical symmetry $C_1S_{5.46}$. Each of these subunits includes one coat protein and one nucleotide. Because the DNA is composed of two antiparallel strands, the electron density map in the region of the DNA must be the average of at least two different (antiparallel) configurations. Consequently, the absence of interpretable features in the region of the DNA is not surprising.

The axial spacing between nucleotides in the Pf1 DNA is about 5.4 Å in each strand in the high temperature form and over 6 Å in the low temperature form. DNA phosphates cannot be more than about 7.1 Å apart in a fully extended chain. If the strands of DNA follow a 15 Å pitch helix similar to that of the coat protein, the phosphates cannot be further than about 2.5 Å from the helix axis (Day et al., 1979). At this radius, phosphate groups placed in the Pf1 surface lattice would be approximately 3 Å apart. This seems stereochemically and electrostatically unlikely. A more likely possibility is that the two DNA chains interact with the coat proteins in rather different ways. If this were the case, however, half-order layer lines should be observed in diffraction patterns from all Pf1 specimens. In fact, half order layer lines seem to be of highly variable intensity in diffraction patterns from Pf1.

The variable intensity of half-order layer lines in diffraction patterns from Pf1 may be associated with the problems in controlling ionic strength in partially dehydrated fibers of Pf1 and the fact that salt concentration has a substantial effect on Pf1 structure (Thomas et al., 1983). The effect may be simply due to lattice interactions among virus particles, or it may be mediated through the effects of ions on DNA structure (Casadevall and Day, 1982). Absence of half-order layer lines indicates either that the symmetry of the DNA is not simply related to that of the coat protein or that the structures of the two antiparallel strands mimic one another.

Solid-state NMR has been used to determine the orientation of the phosphates in Pf1 DNA (Cross, et al., 1983). These studies have shown that all of the phosphodiester groups have the O–P–O plane of the non-esterified oxygens approximately perpendicular to the virion axis. This is consistent with the axial translation per nucleotide which shows that the DNA chain must be almost fully extended.

Linear dichroism measurements (Torbet and Maret, 1981) indicate that the planes of the DNA bases are tilted to within 30° of being parallel to the virion axis. This observation is consistent with the possibility that the relatively hydrophobic faces of the DNA bases lie flat up against the hydrophobic surfaces of α-helices as observed for the RNA of TMV (Stubbs et al., 1977). The helical wheel diagram in Figure 11 suggests that a large fraction of the inner layer of α-helices facing the DNA is hydrophobic in character. The remainder is polar. The polar and nonpolar residues on this surface are segregated. The polar residues take up about 6 Å of the helix length and the hydrophobic residues (26–39) cover the interior surface for a distance of about 20 Å. Although many of the hydrophobic residues will be at relatively high radius involved in interactions with other subunits, very little of this hydrophobic surface is likely to be in contact with polar groups on the DNA or with DNA-associated solvent. On the surface of fully extended DNA, polar atoms comprise about 54% of the exposed surface with approximately equal contributions from the backbone and bases (Alden and Kim, 1979). This suggests that a large fraction of the nonpolar surfaces of the DNA interact with the protein whereas most of the DNA–DNA interactions are likely to be polar. This argues against substantial base stacking. Tyr 40, located at the very top of the polar portion of the α-helix, may stack with the DNA bases as suggested by fluorescence measurements (Day et al., 1979).

8. THE TEMPERATURE TRANSITION IN Pf1 AT 8°C

X-ray diffraction patterns taken from oriented fibers of Pf1 prepared above 8°C have a distribution of layer lines indicative of helical symmetry $C_1S_{5.40}$, but diffraction patterns from fibers prepared below 8°C indicate that the virus symmetry is $C_1S_{5.46}$ (Wachtel et al., 1976; Nave et al., 1979). On decreasing temperature, the slight decrease in the angular interval between units from 66.7° to 65.9° is accompanied by an increase in the axial displacement between subunits from 2.85 to 3.06 Å. Because there are about 7600 coat proteins in Pf1 this small axial displacement results in a length change of about 1000 Å for the virion as a whole (Nave et al., 1979). Density, circular dichroism and microcalorimetric measurements (Hinz et al., 1980), and Raman spectroscopy (Thomas et al., 1983) all indicate structural changes occur in Pf1 between 5° and 25°C. Measurement of the high field magnetic birefringence of Pf1 in solution (Torbet and Maret, 1981) indicates that the α-helical segments are more nearly parallel to the virion axis in the low temperature form than in the high temperature form.

The structural differences between high and low temperature forms are not apparent in the electron density maps shown in Figures 7 and 8. The structural change is much more obvious in the cylindrical sections shown in Figure 13. These sections have been constructed from the electron density maps in Figure 7 and represent the electron density of the inner layer (Fig. 13*a*) and outer layer (Fig. 13*b*) of α-helices projected onto cylinders of radius 15 and 28 Å, respectively. In Figure 13 the solid lines represent the high temperature form, and the broken lines, the low temperature form. The difference in the surface lattices of the two forms is apparent in the slight displacements of nearest neighbors relative to the central subunits. A small difference in the tilt of the α-helical rods in the outer layer in the two temperature forms can be seen in Figure 13 *b*. The rods of density in the high temperature form are tilted by about 26° relative to the particle axis; those of the low temperature form are tilted by about 22°. There is some movement of adjacent proteins relative to one another due to the change in helical surface lattice. In the outer layer, the major protein–protein contacts are along sixfold and 11-fold helical directions. Along the sixfold direction the axial movement of adjacent subunits is about 1.3 Å. Along the 11-fold direction, the relative axial movement is 2.3 Å. The 25 Å long interaction of each α-helix in the inner layer with one α-helix in the outer layer is between adjacent proteins along the 11-start direction. The displacement of these two α-helices by over 2 Å relative to one another must result in a reorganization of the side chains involved in the helix–helix interactions. Generally, displacements of adjacent helices by more than about 1.5 Å must be accompanied by some repacking of the amino acid side chains at the interface (Chothia et al.,

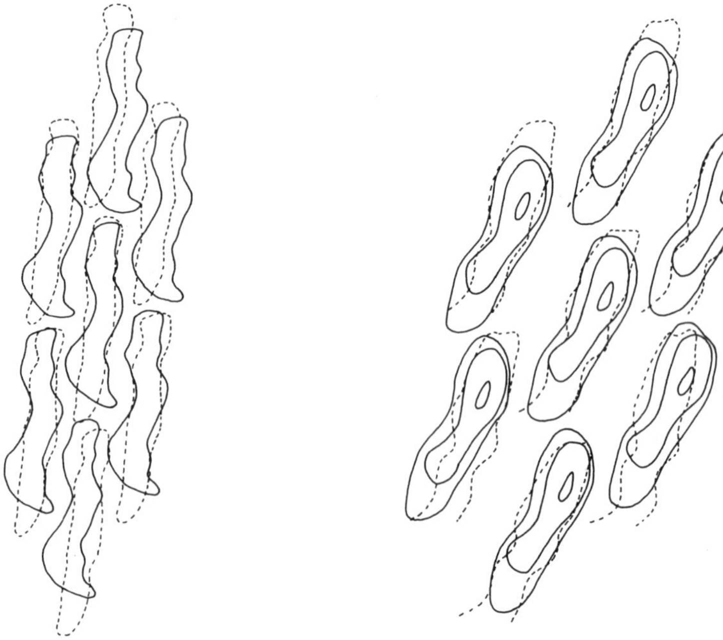

(a) (b)

Figure 13. Projections of the inner and outer layers of rods of electron density onto cylinders cut through the maps of the high temperature form (solid lines) and low temperature (broken lines) of Pf1 (*a*) at 15 Å radius and (*b*) at 28 Å radius. Differences in the surface lattices of the two forms can be seen in the displacements of subunits relative to one another. Figure was constructed from the maps in Figure 7 taken from Makowski et al. (1980) and Nave et al. (1981).

1983). The movement of these two helices relative to one another is large enough to alter the details of protein–protein interactions and the environment of some amino acid side chains.

The α-helices in the inner layer, as drawn in Figure 13*a*, do not measurably alter in average tilt relative to the helix axis. The shape differences near the lower end of the rods of the electron density are exaggerated somewhat because of their low radius. This is also where differences due to different treatment of the data (particularly the equator), different phasing methods, and contouring of the maps are likely to be most significant. The curve in the center of the α-helical segment in the high temperature form is a relatively major feature, also visible in Figure 8. It may represent a curvature of the α-helix backbone or just an asymmetry in the distribution of amino acids around the α-helix. There is also some suggestion of a slight bend in the electron density rods in the outer layer in these projections; the top portion appears to be more parallel to the particle axis than the bottom portion. These features are not apparent in the electron density maps of the low temperature form.

The physical and chemical data available make possible a discussion of the origin of this structural transition. Measurement of the relative apparent isopotential specific volumes (Hinz et al., 1980) of Pf1 and fd (which does not exhibit a comparable structural transition) shows that, relative to fd, Pf1 increases in specific volume by about 65 $Å^3$ per coat protein (about the volume of two water molecules). The total increase in specific volume with increasing temperature is about 130 $Å^3$ across the transition. The increase in specific volume for fd is 1.5 to 3.0 times that observed for most proteins (Hinz et al., 1980). The calorimetrically determined transition enthalpy corresponds to about 3.46 kcal/mol per coat protein. Both of these measurements are consistent with the transition involving a decrease in the contact of hydrophobic side chains with water as temperature increases. The change in enthalpy for moving one molecule of methane from water into benzene is about 2.8 kcal/mol (Frank and Evans, 1945; Kauzmann, 1959). The volume change associated with this transfer is 37.7 $Å^3$, due mainly to the fact that water molecules ordered around a methane molecule occupy less volume than they do in bulk water. Thus, during the increase of temperature across the Pf1 structural transition, the enthalpy change observed corresponds to that of the movement of about 1.2 methane molecules from water into benzene. The observed volume change (relative to fd) corresponds to the movement of 1.7 methane molecules from water into benzene. During the transition it is possible that one or more hydrophobic side chains change position in order to lower the amount of their surface in contact with water. This driving force is responsible for the temperature induced polymerization of many proteins including tubulin and deoxygenated sickle cell hemoglobin. The Pf1 structural transition may be closely akin to a polymerization driven by the hydrophobic effect.

The transition may be somewhat more complex than this simple explanation suggests. The apparent temperature of transition appears to vary depending on the method used to study it. X-ray diffraction measurements on oriented fibers suggest that the transition occurs at 8°C (Nave et al.,

1979; Marvin et al., 1981). The temperature of the midpoint of the transition appears to be 11°C from measurement of partial specific volume, 13°C from microcalorimetry, 15° from circular dichroism measurements (Hinz et al., 1980), and 16°C from Raman spectroscopy (Thomas et al., 1983). These differences may be due to differences in virion concentration and solvent as seen in the high temperature transition of Pf3 (Thomas and Day, 1981). Or it is possible that the physical property probed by each technique changes at slightly different temperatures, and that a spectrum of small structural changes occurring between 8° and 16°C accounts for the measurements. Raman spectra (Thomas et al., 1983) suggest that Pf1 changes structure continuously as a function of temperature throughout the range measured (9°–80°C).

The exact helical symmetry of Pf1 is very sensitive to environment. Binding of heavy atoms to Pf1 alters the helical symmetry (Nave et al., 1981; Marvin et al., 1981). In those cases reported the symmetry switches toward the low temperature form even at room temperature. Reported symmetries range from 5.39 to 5.46 units per turn. This is a much larger range than observed in heavy atom derivatives of TMV (49.019–49.024 units in three turns) (Stubbs and Makowski, 1982). The rigid rod of TMV would be expected to be less sensitive to environment than the flexible rods of the filamentous phages.

Other structural transitions of Pf1 have been observed including transitions induced by ether (Amako and Yasunaka, 1977; Nave et al., 1979) or exposure to a chloroform–water interface (Manning et al., 1981). An additional temperature transition has been observed at high temperatures (>60°C) in some phage (Thomas and Day, 1981). This transition will not be discussed here because, as of yet, no X-ray diffraction studies have been performed at high temperatures.

9. OTHER CLASS II PHAGES

Although X-ray diffraction experiments have been performed on both Xf (Marvin et al., 1974a) and Pf3 (Peterson et al., 1982), there has been no detailed analysis of the diffraction data. The diffraction patterns from Xf look very similar to those of Pf1. Those from Pf3 resemble the diffraction patterns from Pf1 in gross distribution of intensity, but there are also some major differences. Because Pf3 does not readily form highly oriented fibers,

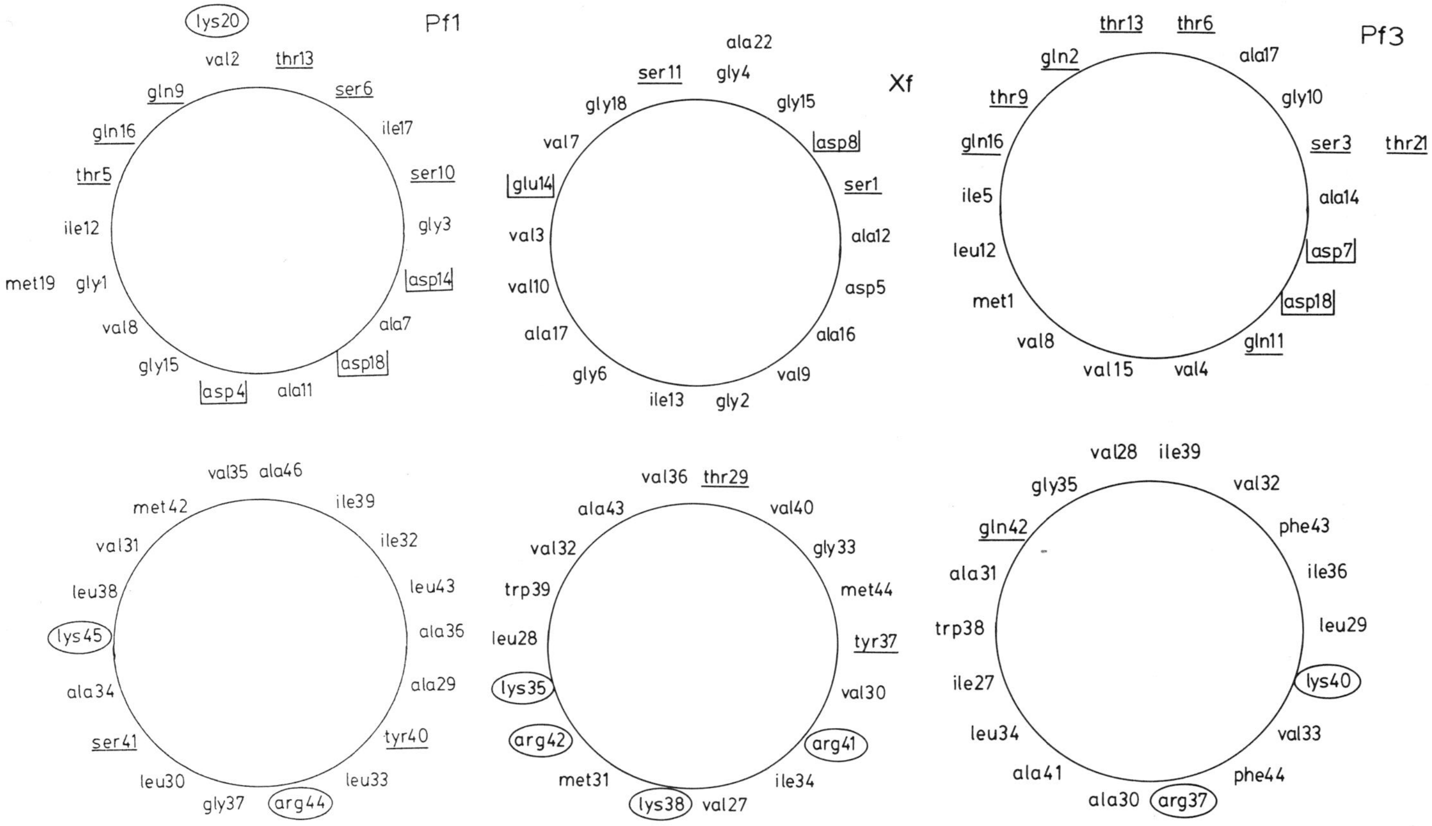

Figure 14. Helical wheels for the Class II viruses Pf1, Xf, and Pf3 arranged as described in Figure 11 so that residues at high radius tend to be near the top of the diagram and residues at low radius tend to be near the bottom.

the intensities in diffraction patterns from Pf3 are arced widely around the center of the pattern. This prevents a detailed comparison of the diffraction patterns with those from Pf1. However, it is clear that the near meridional layer lines spaced at orders of about 15 Å are much stronger in diffraction from Pf3 than from Pf1.

In Figure 14 the helical wheel projections of the amino acid sequences of Xf and Pf3 are arranged in a manner analogous to that of Figure 11 with those residues expected to be on the surface of the virus particle at the top and those expected to be near the center of the virus particle at the bottom. In Xf, both helical wheels exhibit an almost complete segregation of polar and nonpolar residues. Only Val 7, Thr 29, and possibly Asp 5 and Ala 12 deviate from this pattern. Val 7 may be involved in bonding between subunits as postulated for Ile 17 of Pf1. Thr 29 appears positioned such that its hydroxyl group must be in the hydrophobic interior of the protein coat in a manner similar to Asp 14 and Asp 18 in Pf1. These positions could equally well be explained by breaks in the α-helical secondary structure. Results of Raman spectroscopy on Xf (Thomas et al., 1983) indicate that Xf has the lowest proportion of α-helical secondary structure of all the filamentous phages, being perhaps as low as 50%. The similarity of diffraction patterns from Xf with those from Pf1 indicate a substantial proportion of α-helical structure in Xf, and the helical wheel diagrams are consistent with this. The region of the Xf coat protein least likely to be in an α-helical conformation is in the vicinity of Pro 19 and Gly 24-Gly 25. This portion of the peptide is not displayed on the helical wheels. Available evidence is consistent with a relatively long stretch of non-α-helical peptide in Xf connecting two relatively short α-helical segments.

Unlike Pf1, Xf has two nucleotides per coat protein. This is reflected in the larger number of basic residues near the carboxyl terminus of the coat protein. The axial translation per nucleotide is 2.7 Å in Xf, about half that in Pf1. With comparable packing density, the Xf DNA will occupy about twice as much cross sectional area as the Pf1 DNA. This means it could occupy all the volume in Xf out to about 14 Å radius. This could only occur if the α-helical segments in Xf are organized rather differently from those in Pf1.

Pf3 appears in many ways to be grossly different from Pf1 and Xf. It has a nonintegral ratio of nucleotides to proteins (see Table 3). Its behavior and sensitivity to salt and temperature seem anomalous when compared to other filamentous phages (Thomas and Day, 1981; Casadevall and Day, 1982; Thomas et al., 1983). Nevertheless, comparison of the helical wheel diagrams of Pf1 and Pf3 suggest that there may be similar features in the two structures. In the outer α-helical segment, the residues that deviate from complete segregation of polar and nonpolar residues are very similar for the two viruses (Asp 7, Ala 17, and Asp 18 of Pf3 corresponding to Asp 4, Ile 17, and Asp 18 of Pf1). The diagrams also suggest that Glu 11 and Glu 42 may be buried in the hydrophobic interior of the Pf3 protein coat.

The nonintegral ratio of nucleotides to proteins in the Pf3 virion means that, unlike other Class II phages, the nucleotides interact in different ways with different coat proteins. This is similar to the situation in Class I viruses. Furthermore, there are not enough basic residues in the coat proteins to completely neutralize the DNA. This suggests that the presence of cations in association with the DNA may be essential for the stability of the virion. Pf3 appears unique among the filamentous phages in being almost totally insensitive to both silver ions (Casadevall and Day, 1982) and NaCl concentration (Thomas et al., 1983). If cations arc indccd associated with the Pf3 DNA, then this insensitivity to ionic strength changes may be necessary for maintaining virion stability. It is possible that the volume containing the DNA in Pf3 is closed to the penetration or escape of cations (Casadevall and Day, 1982).

10. CLASS I PHAGES

In terms of genetics, life cycle, and biochemistry, more is known about the Class I phages than the Class II phages. Structural studies on Class I virus particles have lagged behind those of Class II for two technical reasons. First, the symmetry of Class I virions was unknown for sometime. Second, there is intrinsically less information in a diffraction pattern from fd or M13 than from Pf1. An examination of the diffraction patterns in Figure 3 shows that, out to 7 Å spacing, Pf1 has about 11 layer lines. M13 has only about five. Only half the number of independent intensity measurements can be made from the diffraction pattern from M13 relative to that from Pf1. Now that the symmetry of M13 (fd, f1) has been resolved (Makowski and Caspar, 1978, 1981; Banner et al., 1981), structural studies are likely to progress.

There are some fundamental differences between the structure of fd and of Pf1. Perhaps most important is the helical symmetry. The fd DNA cannot have the same helical symmetry as the coat protein because of the fivefold rotation axis of the protein lattice. This situation is similar to what has been seen in the icosahedral plant viruses. Different portions of the DNA must interact in different ways with the coat proteins.

In diffraction patterns from fd taken with very long exposures, it is possible to see weak layer lines corresponding to a 26.7 Å periodicity which has been interpreted as due to the pitch of helical DNA in fd (Banner et al., 1981). These reflections indicate a maximum electron density fluctuation at about 8.5 Å radius. The axial translation between nucleotides in each DNA chain is about 2.7 Å suggesting that there are about 9.8 nucleotides per turn. This DNA will not be base-paired and the DNA backbone is not like that of A-DNA or B-DNA (Thomas et al., 1983). However the conformation of the two antiparallel single strands may be related to known conformations of double-stranded DNA because binding of Ag^+ to fd DNA results in a CD spectrum very similar to that of superhelical double-stranded DNA (Casadevell and Day, 1982). Fluorescence measurements have been used to suggest that DNA bases may be stacked in fd (Day, 1969). Solid-state NMR techniques have been used to examine the orientation of phosphodiester groups in fd (Cross et al., 1983). The orientations of the O–P–O planes of the nonesterified oxygens in fd DNA do not seem to be preferentially oriented at any particular angle relative to the virus axis. This need not imply a disordering of the DNA backbone. It does imply that the DNA must have a complex conformation with the phosphodiester groups taking on at least 10 different orientations relative to the virus axis. To be consistent with the NMR results, the orientations of the phosphodiester groups must be distributed rather uniformly over the possible orientation angles. In TMV, the RNA phosphodiester groups take on three distinct orientations (Stubbs et al., 1977; Stubbs and Stauffacher, 1981). The regular interaction of three RNA bases with each coat protein in TMV limits the number of distinct conformations of the nucleotides to three. In fd, if any regular pattern of DNA–protein interactions exists, it must involve a very long axial repeat with a large number of different nucleotide conformations and different kinds of protein–DNA interactions.

Although the difference in helical symmetry between Pf1 and fd is very important for considerations of nucleotide structure, it is much less important in considering protein structure. The helical nets in Figure 5 show that for all observed symmetries of filamentous phages the packing of subunits into the protein coat is very similar. An extended α-helical subunit can be fit into the fd surface lattice as simply as it can be fit into the Pf1 surface lattice. Solid-state NMR measurements (Cross et al., 1983) indicate that the fd coat protein is more likely to contain a single gradually curved α-helical rod than to have two α-helical segments with a bend near the center. This is reminiscent of the models initially proposed for the structures of both fd (Marvin et al., 1974b) and Pf1 (Marvin et al., 1974a; Marvin and Wachtel, 1975, 1976).

The cylindrically averaged structure of f1 (which should be identical to that of fd) has been solved using X-ray diffraction from oriented fibers (Milman, 1968). The cylindrically averaged electron density is drawn in Figure 15 along with that for Pf1 for comparison. The resolution of these two projections is only about 15 Å so it cannot show the same level of detail as that for Pf1 drawn in Figure 6 at 8 Å resolution. Nevertheless, there are clear differences in the radial distribution of electron density in the viruses. In particular, the central peak in the density map of fd extends out to the minimum at a radius of 17 Å, whereas the central peak in the Pf1 map extends only to about 10.5 Å radius (less in this low resolution map). These peaks are due to the DNA in the center of the virions and their relative sizes correspond to the different amounts of DNA per unit length in fd and Pf1. There is enough DNA in Pf1 to fill a cylinder

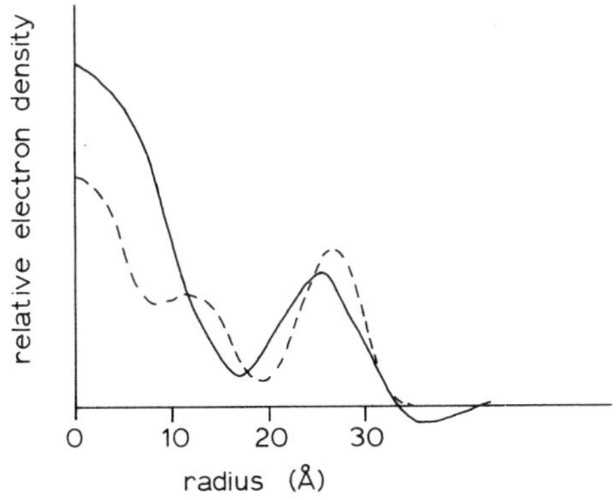

Figure 15. The cylindrically symmetric part of the equatorial projections of f1 (solid line) and Pf1 (broken line) at 14.8 Å resolution. Calculated from data taken from Milman (1968) and Makowski et al., (1980).

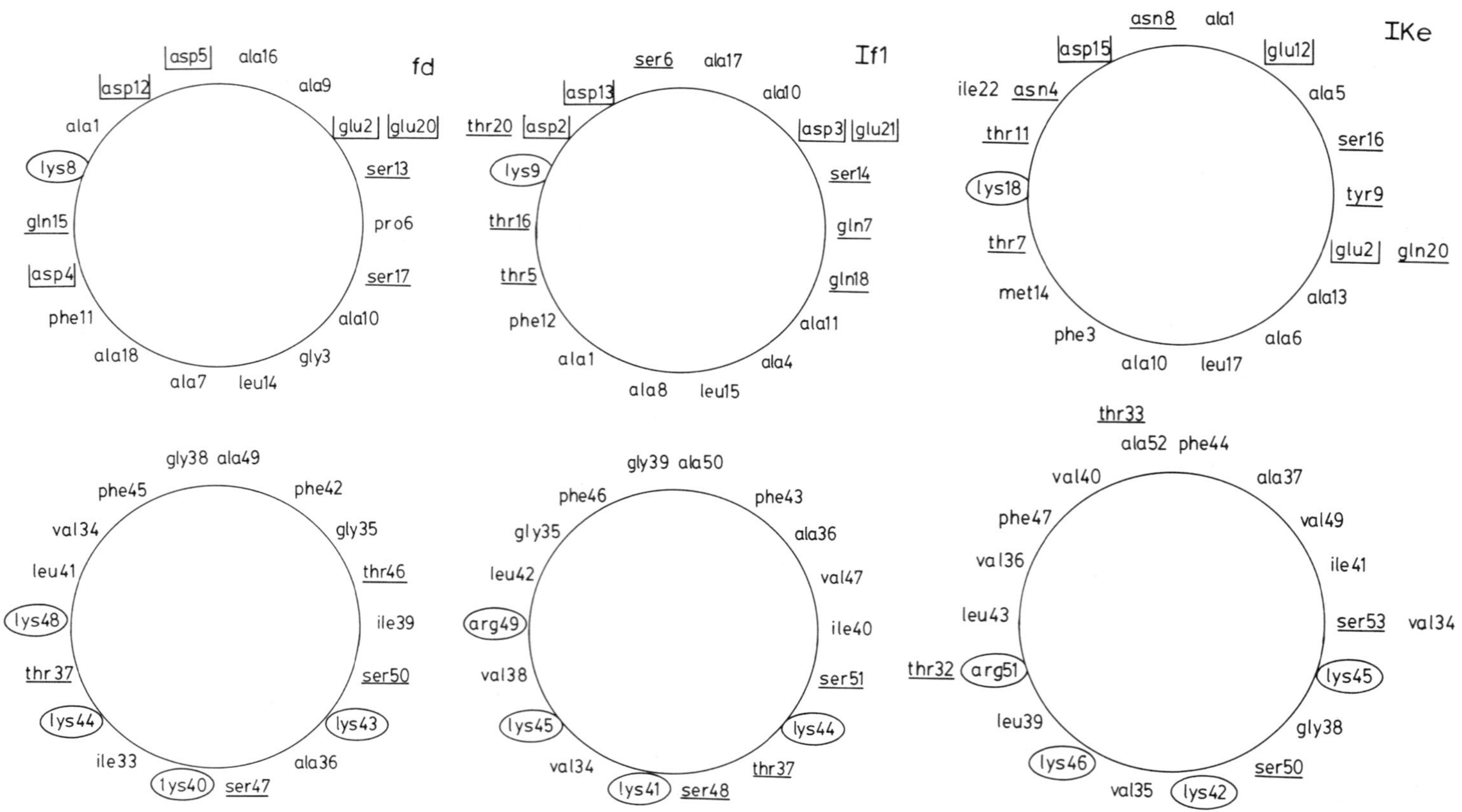

Figure 16. Helical wheels for the Class I viruses fd, If1, and Ike arranged as described in Figure 11 so that residues at high radius tend to be near the top of the diagram and residues at low radius tend to be near the bottom of the diagram.

about 10 Å in radius. In fd, the DNA could fill a cylinder over 14 Å in radius. The 17 Å radius of the central peak of electron density in fd suggests that the DNA in fd may be somewhat less tightly packed than in Pf1. Or, given the relatively high central density of the fd DNA peak, it is possible that some protein extends in to radii less than 17 Å. The axis of the low radius, carboxyl end of the α-helix is likely to be 19–22 Å from the viral axis. The single broad peak of electron density from 17 to 33 Å in radius in Figure 15 must mark the position of the majority of the coat protein. Because the protein region is only about 16 Å thick it cannot be composed of two distinct layers of α-helical segments. The amino- and carboxy-terminal halves of the protein must interlock with one another radially. For instance, in moving around a circle at 25 Å radius, one will encounter alternatively amino acid residues from the carboxyl half of a coat protein extending out from a main chain at lower radius and residues from the amino half of a coat protein extending from a main chain at higher radius.

Drawings of α-helical wheels of fd, Ike, and If1 are shown in Figure 16. For the most part, the polar and nonpolar residues are segregated in these diagrams. In fd, residues that do not follow this pattern appear to be Ala 1, Asp 4, Pro 6, Ala 9, Ala 16, and Ser 17. The relatively large number of residues which are not segregated into polar and nonpolar halves of the α-helical diagram suggest that there may be several breaks in the α-helix. Raman spectra of fd (Thomas et al., 1983) also indicate a lower proportion of α-helical conformation in fd than in Pf1. The lack of an electron density map for fd makes further interpretation difficult. It is clear from the bulk of the available data that there will be some substantial differences between the structure of fd and the structure of Pf1.

11. CONFORMATIONAL SWITCHING IN THE PHAGE LIFE CYCLE

Before virion assembly, the coat protein of fd is inserted into the host cell inner membrane of fd (Smilowitz et al., 1972) with its carboxyl terminus on the cytoplasmic side and its amino terminus on the external, periplasmic side (Wickner, 1975, 1976; Ohkawa and Webster, 1981). Within the host cell, the viral DNA is complexed with gene 5 protein to form a flexible rod-shaped structure (Salstrom and Pratt, 1971). Assembly of the virion involves the dissociation of the DNA from the gene 5 protein and its association with the gene 8 coat protein as the DNA passes through the host cell membrane.

Assembly initiates at the C protein end of the phage (Webster et al., 1981; Lopez and Webster, 1983), where a specific DNA sequence is always found (Webster et al., 1981; Ikoku and Hearst, 1981). This DNA sequence is located at the end of the most potentially stable hairpin structure in the genome. The first steps in assembly must involve the interaction of phage DNA with the C proteins or both C proteins and coat proteins. This event may take place at the membrane. The amino acid sequences of the C proteins (products of genes 7 and 9) suggest that these very small peptides (3000–4000 daltons each) may be initially associated with the host cell membrane. The minimum number of components necessary to initiate assembly would appear to include the DNA, five copies of the C protein (five copies each of the products of genes 7 and 9), and perhaps, five or more copies of the coat protein.

The specificity of the assembly process for phage DNA appears to involve the presence of a particular nucleotide sequence (GCAG) at the end of a stable hairpin structure. Shorter hairpins predicted from the nucleotide sequence of the intergenic space are not found at the end of the virion, but a specific length does not appear to be crucial for initiation of assembly. In chimeric M13 phages which lack a portion of the DNA at the 3′ end of the hairpin the stem is much shorter and viable phage are produced (Schaller, 1978). The position of this hairpin at the C protein end of the phage has not, however, been proved. Attempts to clone DNA fragments with a GCAG sequence at the end of a somewhat shorter stable hairpin structure into M13mp8 or M13mp9 have led to the production of mutant phage with scrambled sequences and deletion of some or all of the inserted stem sequence (Rodi, 1983; Makowski et al., 1983). The presence of two recognition sites may have resulted in the initiation of assembly at two points along the nucleotide chain. The resulting "forked" structure could not lead to the production of viable phage.

The requirement of a hairpin structure to initiate assembly brings into question the configuration of the DNA in the gene 5 complex. Most, if not all, of the phage DNA will be single-stranded in the gene 5 complex. If the disassembly of the gene 5 complex is closely linked to the assembly of phage, then the hairpin must be accessible on the surface of the gene 5 complex in order to interact with C protein and initiate assembly of the phage and the concomitant disassembly of the gene 5 complex.

It should be noted that, although most of the DNA in fd is single-stranded and not base-paired, there may be sufficient room inside the fd protein coat for the double-stranded stem of the recognition hairpin to remain base-paired. The C proteins are too small to completely coat the stem so that much of it must be enclosed in the fd coat protein. The inside diameter of the protein coat is 28–34 Å based on the amount of volume required by the DNA and protein as discussed above. The hydrated diameter of double-stranded DNA in the heads of bacteriophages T2 and T4, is 26–28 Å (Earnshaw and Harrison, 1977; Earnshaw et al., 1975). It is possible that the recognition hairpin loop and other short stretches of DNA are base-paired within the fd virion. By comparison, double-helical structures could not exist in the Pf1 virion even for short stretches.

The assembly of TMV is also initiated by the interaction of a particular RNA sequence, AAG, at the end of the hairpin loop with the TMV coat protein disk (Zimmern, 1977; Zimmern and Butler, 1977; Jonard et al., 1977; Steckert and Schuster, 1982). The binding of RNA to the disk apparently triggers a change in the coat protein conformation which causes the 34 proteins in the two layers of the disk to alter the pattern of intersubunit bonds and switch into slightly more than two turns of a $16\frac{1}{3}$ units-per-turn helical structure.

Similarly, the interaction of the filamentous phage DNA with the C protein and B protein must initiate the complex process of assembling the coat protein around the DNA as it passes through the plasma membrane. In fd, the fivefold rotational symmetry of the protein coat implies that this process is cooperative (Makowski and Caspar, 1981). Assembly must proceed through the addition of five coat proteins at a time as the DNA moves out by ~16 Å steps through the cell membrane. Because the axial length of a subunit is about 60 Å, the protein being assembled from the membrane will not be clear of the surface until it has moved out at least four 16 Å steps. As long as the protein is interacting with membrane lipid, it is unlikely to be in its final conformation in the assembled structure. This means that the switching of the protein from its conformation in the membrane to that in the virus particle must be a multistep process involving at least three discrete intermediate steps.

The assembly of Class II particles has been assumed to be similar to that of Class I particles. Even if, as in Class I, the coat protein is first inserted into the host cell membrane and the virus is assembled as the DNA passes through the membrane, the assembly of Class II particles would have to differ in significant details from Class I assembly (Makowski and Caspar, 1981). If added one at a time, it is likely that each coat protein would have to pass through perhaps two dozen discrete conformational states before reaching the final conformation in the virus particle because addition of a single coat protein would involve only a 2.77 Å translation of the DNA through the membrane. Alternatively, several coat proteins could be added at the same time in a cooperative process which must include a step akin to the disk-to-helix transition in TMV. The difference between the symmetries of the Class I and Class II particles suggests that different assembly processes may have evolved to form these structures with similar protein packing architectures.

The structural transition at 8°C in Pf1 is an instance of conformational switching which can be studied in great detail structurally. Although intermediate steps in the transition cannot be ruled out, most of the data appear to be consistent with it being a two-step process (Hinz et al., 1980). As discussed here, the characteristics of the transition indicate that it may be driven by the hydrophobic effect in a manner similar to the polymerization of macromolecular assemblies such as microtubules. The result is a small change in the pattern of intersubunit interactions. The axial distance change between the closely interacting nearest neighbors along the 11-start direction may change by over 2 Å, indicating that the residues involved in the 25 Å long helix–helix interaction between the α-helices in the inner and outer layers must be rearranged during the transition (Chothia et al., 1983).

Intersubunit interactions in a macromolecular assembly generally involve a combination of hydrophobic and polar interactions (e.g., Bloomer et al., 1978; Harrison et al., 1978). The interactions among nonpolar groups at the subunit interfaces contribute to the stability of the intersubunit bond, but the polar interactions tend to determine the specificity of the bonding (Chothia and Janin, 1975). The existence of two well-defined patterns of intersubunit interactions (high and low temperature forms of Pf1) suggests that the structural transition between them involves a switching in the pattern of polar interactions between subunits. Changes in only hydrophobic interactions might be expected to give rise to a distribution of incrementally different structures as is observed in Pf1 over broader ranges of temperature and salt concentration. The most likely subunit–subunit interface at which a change in polar interactions would occur is between nearest neighbor subunits along the 11-start helices. As summarized in Figure 12, the two candidates for polar interactions between these subunits are (1) Lys 20 with Asp 4 and (2) Lys 45 with Asp 14 and/or Asp 18. A relative movement by the subunits of only 2 Å is probably not enough to break any hydrogen bonds or salt bridges that may exist between these subunits. It could, however, result in a substantial distortion of the bonds. Two stable patterns of intersubunit interactions could exist if the bonding between these subunits had mechanical properties analogous to those of a light switch: two stable states with an unstable intermediate position. The low temperature state of this bond would be intrinsically the more stable state (locally). As temperature rises, the increased importance of the hydrophobic effect will tend to cause hydrophobic residues near the surface to become more buried in the particle interior. At some point, this tendency will put sufficient strain on the polar intersubunit bonds that they will switch conformation to the high temperature form. The details of this process should be made clear by high resolution electron density maps of the two forms.

The pattern of intersubunit bonds are also important in determining the mechanical properties of the phage as a whole. Electron micrographs of the phage particles suggest that they are quite flexible. There are three sets of intersubunit interactions: those along the 5-start helices, the 6-start helices, and the 11-start helices. If polar interactions occurred at all three sets of subunit boundaries then it is likely the phage particles would be rigid. Consider, for instance, a parallelogram constructed of four pieces of wood connected by hinges at the vertices. This structure is mechanically very flexible. Join one pair of opposite vertices with a diagonal rod and a rigid structure is formed. The stable, flexible structure of the filamentous phages is most consistent with polar interactions occuring at two of the three subunit–subunit interfaces. Interactions among nonpolar side chains will contribute less to the mechanical rigidity of the particles. In Pf1, the most extensive interface occurs between nearest neighbors along the 11-start direction, because the outer α-helical segment of one protein interacts with the inner α-helical segment of the next protein along the 11-start helix. There may also be polar interactions between adjacent outer α-helical segments along the 6-start directions. There is no interaction between outer α-helical segments along the 5-start direction and no polar residues likely to form bonds along the 5-start direction among α-helical segments in the inner layer. The lack of polar interactions along the 5-start direction in Pf1 may be crucial for the flexibility of the phage particles. Because of the extreme length of the virions, this flexibility is an important defense against fragmentation by shear. All the filamentous phages appear to be flexible filaments. Presumably, this flexibility is maintained through different patterns of bonding in each strain.

Acknowledgments. The X-ray diffraction patterns in Figure 3 were obtained in this laboratory by Marc Glucksman and Jan van Oostrum. I would like to thank Drs. Tim Cross and Stan Opella for their time and guidance and for the use of their equipment. Many of the ideas presented here have grown out of conversations with Drs. D. L. D. Caspar and Gerald Stubbs. This work was supported by N.I.H. Grant GM 29829, an Alfred P. Sloan Foundation Fellowship, and an Irma T. Hirschl Career Scientist Award.

REFERENCES

Alden, C. J., and Kim, S. H. (1979). *J. Mol. Biol.* **132**, 411–434.

Amako, K., and Yasunaka, K. (1977). *Nature* (*London*) **267**, 862–863.

Banner, D. W., Nave, C. and Marvin, D. A. (1981). *Nature* **289**, 862–863.

Bloomer, A. C., Champness, J. N., Bricogne, G., Staden, R. and Klug, A. (1978). *Nature* **276**, 362–368.

Bradley, D. E. (1964). *J. Gen. Microbiol.* **35**, 471.

Bryan, R. K., Bansal, M., Folkhard, W., Nave, C., and Marvin, D. A. (1983). *Proc. Natl. Acad. Sci. U.S.A.* **80,** 4728–4731.

Caro, L. G., and Schnös, M. (1966). *Proc. Natl. Acad. Sci. U.S.A.* **56,** 126–132.

Casadevall, A., and Day, L. A. (1982). *Nucl. Acids Res.* **10**, 2467–2481.

Chothia, C., and Janin, J. (1975). *Nature* (*London*) **256**, 705–708.

Chothia, C., Levitt, M., and Richardson, D. (1977). *Proc. Natl. Acad. Sci. U.S.A.* **74**, 4130–4134.

Chothia, C., Lesk, A. M., Dodson, G. G., and Hodgkin, D. C. (1983). *Nature* (*London*) **302**, 500–505.

Chou, P. Y., and Fasman, G. D. (1974). *Biochemistry* **13**, 211–222.

Crick, F. H. C. (1953). *Acta Crystallogr.* **6**, 689–697.

Cross, T. A., Tsang, P., and Opella, S. J. (1983). *Biochemistry* **22**, 721–726.

Day, L. A. (1969). *J. Mol. Biol.* **39**, 265–277.

Day, L. A., and Wiseman, R. L. (1978). In *The Single-Stranded DNA Phages*, D. T. Denhardt, D. Dressler, and D. S. Ray, Eds. Cold Spring Harbor Laboratories, Cold Spring Harbor, New York, pp. 605–625.

Day, L. A., Wiseman, R. L., and Marcez, C. J. (1979). *Nucl. Acids Res.* **7**, 1393–1403.

Denhardt, D. T., Dressler, D., and Ray, D. S. (1978). *The Single-Stranded DNA Phages,* Cold Spring Harbor Laboratories, Cold Spring Harbor, New York.

Earnshaw, W. C., and Harrison, S. C. (1977). *Nature (London)* **268**, 598–602.

Earnshaw, W. C., King, J., Harrison, S. C., and Eiserling, F. A. (1975). Cell **14**, 559–568.

Efimov, A. V. (1979). *J. Mol. Biol.* **134**, 23–40.

Frangione, B., Nakashima, Y., Konigsberg, W., and Wiseman, R. L. (1978). *FEBS Lett.* **96**, 381–384.

Frank, H. A., and Evans, M. W. (1945). *J. Chem. Phys.* **13**, 507.

Gall, C. M., Cross, T. A., DiVerdi, J. A., and Opella, S. J. (1982). *Proc.Natl. Acad. Sci. U.S.A.* **79**, 101–105.

Grant, R. A., Lin, T. C., Konigsberg, W., and Webster, R. E. (1981). *J. Biol. Chem.* **256**, 539–546.

Gray, C. W., Brown, R. S., and Marvin, D. A. (1981). *J. Mol. Biol.* **146**, 621–627.

Griffith, J., Manning, M., and Dunn, K. (1981). *Cell* **23**, 747–753.

Harrison, S. C., Olson, A. J., Schutt, C. E., Winkler, F. K., and Bricogne, G. (1978). *Nature (London)* **276**, 368–373.

Herrmann, R., Neugebauer, K., Pirkl, E., Zentgraf, H., and Schaller, M. (1980). *Mol. Gen. Genet.* **177**, 231–242.

Hinz, H. J., Greulich, K. O., Ludwig, H., and Marvin, D. A. (1980). *J. Mol. Biol.* **144**, 281–289.

Horiuchi, K., Ravetch, J. V., and Zinder, N. D. (1978). Cold Spring Harbor Symp. Quant. Biol. **43**, 389–399.

Ikoku, A. S., and Hearst, J. E. (1981). *J. Mol. Biol.* **151**, 245–249.

Jonard, G., Richards, K. E., Guilley, M., and Hirth, L. (1977). *Cell* **11**, 483–493.

Kauzmann, W. (1959). *Adv. Protein Chem.* **14**, 1–63.

Klug, A., Crick, F. H. C., and Wyckoff, H. W. (1958). *Acta Crystallogr.* **11**, 199–213.

Lim, V. I., Mazanov, A. L., and Efimov, A. V. (1978). *Mol. Biol. (U.S.S.R)* **12**, 206–213.

Lin, T. C., Webster, R. E., and Konigsberg, W. (1980). *J. Biol. Chem.* **255**, 10331–10337.

Lopez, J., and R. E. Webster (1983). *Virology* **127**, 177–193.

Makowski, L. (1978). *J. Appl. Crystallogr.* **11**, 273–283.

Makowski, L. (1980). In *Fiber Diffraction Methods,* A. D. French and K. H. Gardner, Eds., *ACS Symp. Ser.* **141**, 139–148.

Makowski, L., and Caspar, D. L. D. (1978). In *The Single-Stranded DNA Phages,* D. T. Denhardt, Dressler, D. and D. S. Ray, Eds. Cold Spring Harbor Laboratories, Cold Spring Harbor, New York, pp. 627–643.

Makowski, L., and Caspar, D. L. D. (1981). *J. Mol. Biol.* **145**, 611–617.

Makowski, L., Caspar, D. L. D., and Marvin, D. A. (1980). *J. Mol. Biol.* **140**, 149–181.

Makowski, D. R., Haas, R. A., Dolan, K. P., and Grunberger, D. (1983). *Nucl. Acids Res.* **11**, 8609–8624.

Manning, M., Chrysogelos, S., and Griffith, J. (1981). *J. Virol.* **40**, 912–919.

Marvin, D. A. (1966). *J. Mol. Biol.* **15**, 8–17.

Marvin, D. A., and Hohn, B. (1969). *Bacteriol. Rev.* **33**, 172–209.

Marvin, D. A., and Wachtel, E. J. (1975). *Nature (London)* **253**, 19–23.

Marvin, D. A., and Wachtel, E. J. (1976). *Philos. Trans. R. Soc. London, Ser. B* **276**, 81–98.

Marvin, D. A., Wiseman, R. L., and Wachtel, E. J., (1974a). *J. Mol. Biol.* **82**, 121–138.

Marvin, D. A., Pigram, W. J., Wiseman, R. L., Wachtel, E. J., and Marvin, F. J. (1974b). *J. Mol. Biol.* **88**, 581–600.

Marvin, D. A., Nave, C., Ladner, J. E., Fowler, A. G., Brown, R. S., and Wachtel, E. J. (1981). In *Structural Aspects of Recognition and Assembly in Biological Macromolecules,* M. Balabain, J. L. Sussman, W. Traub, and A. Yonath, Eds. Balaban ISS, Rehovot.

Milman, G. (1968). Thesis, Harvard Univ. Cambridge, Massachusetts.

Nakashima, Y., and Konigsberg, W. H., (1974). *J. Mol. Biol.* **88**, 598–600.

Nakashima, Y., and Konigsberg, W. H. (1980). *J. Mol. Biol.* **138**, 493–501.

Nakashima, Y., Wiseman, R. L., Konigsberg, W., and Marvin, D. A. (1975). *Nature (London)* **253**, 68–71.

Nakashima, Y., Frangione, B., Wiseman, R. L., and Konigsberg, W. (1981). *J. Biol. Chem.* **256**, 5792–5797.

Nave, C., Fowler, A. G., Malsey, S., Marvin, D. A., Siegrist, H., and Wachtel, E. J. (1979). *Nature (London)* **281**, 232–234.

Nave, C., Brown, R. S., Fowler, A. G., Ladner, J. E., Marvin, D. A., Provemaker, S. W., Tsugita, A., Armstrong, J., and Perham, R. N. (1981). *J. Mol. Biol.* **149**, 675–707.

Newman, J., Day, L. A., Dalack, G. W., and Eden, D. (1982). *Biochemistry* **21**, 3352–3358.

Ohkama, I., and Webster, R. E., (1981). *J. Biol. Chem.* **256**, 9951–9958.

Peterson, C., Dalack, G. W., Day, L. A., and Winter, W. T. (1982). *J. Mol. Biol.* **162**, 887–891.

Richards, F. M. (1974). *J. Mol. Biol.* **82**, 1–14.

Richmond, T. J., and Richards, F. M. (1978). *J. Mol. Biol.* **119**, 537–555.

Rodi, D. J. (1983). Thesis, Columbia Univ., New York.

Salstrom, J. S., and Pratt, D. (1971). *J. Mol. Biol.* **61**, 489–504.

Schaller, H. (1978). Cold Spring Harbor Symp. Quant. Biol. **43**, 401–408.

Simons, G. F. M., Konings, R. N. M., and Schoenmakers, J. G. G. (1981). *Proc. Natl. Acad. Sci.* **78**, 4194–4198.

Smilowitz, H., Carson, J., and Robbins, W. P. (1972). *J. Supramol. Struct.* **1,** 8–18.

Steckert, J. J., and Schuster, T. M. (1982). *Nature (London)* **299**, 32–36.

Stubbs, G., and Diamond, R. (1975). *Acta Crystallogr. A* **31**, 709–718.

Stubbs, G., and Makowski, L. (1982). *Acta Crystallogr. A* **38**, 417–425.

Stubbs, G., and Stauffacher, C. (1981). *J. Mol. Biol.* **152**, 387–396.

Stubbs, G., Warren, S., and Holmes, K. C. (1977). *Nature (London)* **267**, 216–221.

Thomas, G. J. (1976). *SPEX Speaker*, Vol. 4. SPEX Industries, Inc., Metuchen, New Jersey, pp. 1–11.

Thomas, G. J., and Day, L. A. (1981). *Proc. Natl. Acad. Sci. U.S.A.* **78**, 2962–2966.

Thomas, G. J., Prescott, B., and Day, L. A. (1983). *J. Mol. Biol.*, **165,** 321–356.

Torbet, J., and Maret, G. (1981). *Biopolymers* **20**, 2657–2669.

van Weczenbeek, P. F. M., Hulsebos, T. J. M., and Schoenmakers, J. G. G. (1980). *Gene* **11**, 129–148.

Wachtel, E. J., Marvin, F. J., and Marvin, D. A. (1976). *J. Mol. Biol.* **107**, 379–383.

Webster, R. E., Grant, R. A., and Hamilton, L. A. W. (1981). *J. Mol. Biol.*, **152**, 357–374.

Wickner, W. (1975). *Proc. Natl. Acad. Sci. U.S.A.* **72**, 4749–4753.

Wickner, W., (1976). *Proc. Natl. Acad. Sci. U.S.A.* **73**, 1159–1163.

Wiseman, R. L., Berkowitz, S. A., and Day, L. A. (1976). *J. Mol. Biol.* **102**, 549–561.

Zimmern, D. (1977). *Cell* **11**, 463–482.

Zimmern, D., and Butler, P. J. G. (1977). *Cell* **11**, 455–462.

7

Animal Virus Structure

IVAN RAYMENT
Rosenstiel Basic Medical Sciences Research Center
Brandeis University
Waltham, Massachusetts

CONTENTS

1. INTRODUCTION

Crystals of animal viruses have been known for many years. Indeed, crystals of poliovirus were first studied by Finch and Klug in 1959. However, it is only recently that serious attempts at determining the structure of an animal virus have been initiated. At this time the low resolution crystal structure of polyoma virus capsid is the only structural determination for an animal virus (Rayment et al., 1982).

It has taken many years to develop the tools necessary for solving a crystallographic problem the size of a virus. The monumental efforts that were required to solve the problems of data collection and processing required large amounts of stable virus. Plant viruses were the only choice for systems on which to develop the methodology. These efforts have resulted in the detailed structures of three plant viruses: tomato bushy stunt virus (Harrison et al., 1978), southern bean mosaic virus (Abad-Zapatero et al., 1980), and satellite tobacco necrosis virus (Liljas et al., 1982) which have been discussed in Chapters 1, 2, and 3.

The study of biologically interesting animal viruses has had to wait until new techniques for data collection and phase determination were developed. A second major problem in studying animal viruses is obtaining sufficient material in order to grow crystals. This generally involves cell culture which is expensive and time consuming. At this time the structures of poliovirus (Hogle, 1982), rhinovirus (Erickson et al. 1982) and SV40 (Lattman, 1980), in addition to polyoma virus, are actively being pursued. Many animal viruses are too large to study by current X-ray technology. Study of these systems is only possible by examining the individual components of the virus. This approach has been used to determine the molecular structure of influenza spike protein and the hexon protein of adenovirus which are described in detail in Chapters 8 and 9.

Animal viruses exhibit a wide variety of shapes and sizes which were, until now, accounted for by the principles of quasi-equivalence set forth by Caspar and Klug in 1962. However, the structure of polyoma virus described in this chapter does not follow those principles so it may be appropriate here to reexamine the theory of quasi-equivalence, which has been the cornerstone of structural virology for the last 20 years.

2. PHYSICAL PRINCIPLES OF VIRUS STRUCTURE: THEORY OF QUASI-EQUIVALENCE

The structure and assembly of spherical viruses have presented an array of questions. The first major step in understanding their construction was made in 1956 by Crick and Watson, who suggested that the virus coat could be built up from a number of small protein subunits arranged in a symmetrical way on the surface of the genetic material. They rationalized that this would be an efficient use of the coding potential of the viral RNA or DNA, which, in any event, would be insufficient to code for a single polypeptide chain that could enclose the genetic material. Crick and Watson noted that of all the types of symmetry arrangements possible only the cubic point groups were likely to lead to an isometric particle. This suggestion was confirmed by Caspar (1956), who demonstrated that the X-ray diffraction photographs from crystals of tomato bushy stunt virus showed intensity distributions that were characteristic of an icosahedral particle.

Icosahedral symmetry demands that the maximum number of subunits that can be arranged on the surface of an icosahedron, such that each has exactly the same environment, be 60. However, it soon became evident that most viruses have considerably more than 60 protein subunits in their protective coat. This presented the problem of how to form a closed shell using more than 60 chemically identical protein subunits. A rational answer to this problem was suggested by Caspar and Klug (1962) who proposed the idea of quasi-equivalence. They suggested that the protein subunits could be positioned on a triangulated icosahedral surface lattice such that the environments of the individual proteins would be similar though not identical. They postulated that such an arrangement would lead to a conservation of the subunit–subunit interactions.

The spherical icosahedron (the projection of an icosahedron onto a concentric sphere) partitions the surface of a sphere into the maximum number of 20 equivalent spherical triangles. Thus, for a virus particle built up from 60 subunits this means that three protein subunits will be associated with each triangular face. These are represented by commas in Figure 1. In this case each subunit is equivalent and has exactly the same chemical environment.

The size of the shell can be increased by partitioning the surface of the sphere into a larger number of triangles, but now the triangles will no longer all be in equivalent positions or have equivalent shapes. Thus, the three protein subunits assigned to each triangular face cannot occupy equivalent environments. Caspar and Klug noted that the departures from strict equivalence could be minimized if the spherical polyhedron so formed had icosahedral symmetry. They suggested that such an arrangement would represent an energy minimum for closed shells built of reg-

Figure 1. The orthogonal projection of a spherical icosahedron on which the 60 asymmetric units are represented by commas. Each triangle has an identical arrangement of three units related by a threefold axis. All 60 asymmetric units are in identical environments and are thus undistinguishable from each other. This represents the largest number of asymmetric objects that can be arranged with complete equivalence to form an isometric group.

ularly bonded identical structure units. Caspar and Klug (1962) enumerated all the possible icosahedral surface lattices and numbers of structure units involved in each case in terms of a triangulation number, T, where the number of triangles in the surface lattice is $20T$ and the number of structure units is $60T$. The possible values for T are given by the rule $T = H^2 + HK + K^2$ where H and K are any pair of integers. The significance of this rule is that it predicts that only certain numbers of subunits can be expected to form closed icosahedral shells.

The triangulated icosahedral surface lattices can be derived from a planar equilateral triangular net by converting some of the sixfold axes to fivefold axes. This useful conceptualization of the symmetry of icosahedral surface lattices is illustrated in Figure 2. The process has the effect of folding the plane into a closed surface. If three protein subunits are associated with each triangular segment of the lattice, then a close packed surface is formed. The formation of a shell in this way gives an

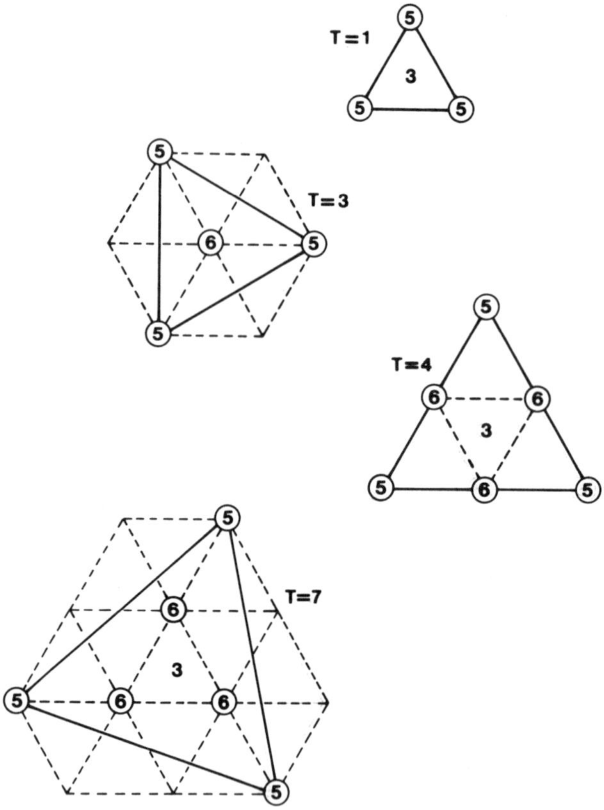

Figure 2. Construction of the first four icosahedral triangulated surface lattices by excising a triangular fragment from a planar hexagonal array. The surface can be folded into a closed container by converting the vertices of the triangular fragment into fivefold axes.

efficient packing scheme for protein subunits that have the ability to fill a triangular lattice. In the simplest case where all the sixfold axes of the planar hexagonal net are converted to fivefold axes a simple $T = 1$ icosahedral arrangement of subunits is formed. In this case all the subunits are equivalent and the shell is built from 60 subunits.

The size of the shell can be increased by excising a larger equilateral triangular portion of the hexagonal planar array to form the face of an icosahedron. Using this packing scheme the next largest shell that is predicted contains 3 times as many triangular facets and therefore 180 protein subunits (H and $K = 1$). This lattice, which is shown in Figure 3, is known as a $T = 3$ icosahedral surface lattice. This arrangement of protein subunits occurs for many small plant viruses.

In a $T = 3$ icosahedral surface lattice all 180 protein subunits cannot possess chemically identical environments. The subunits fall into three groups of 60, each of which occupies symmetrically distinct locations. Caspar and Klug suggested that the different bonding environment experienced by each class of subunit could be accommodated by deformation of the bonds between subunits where bonds in this context refer

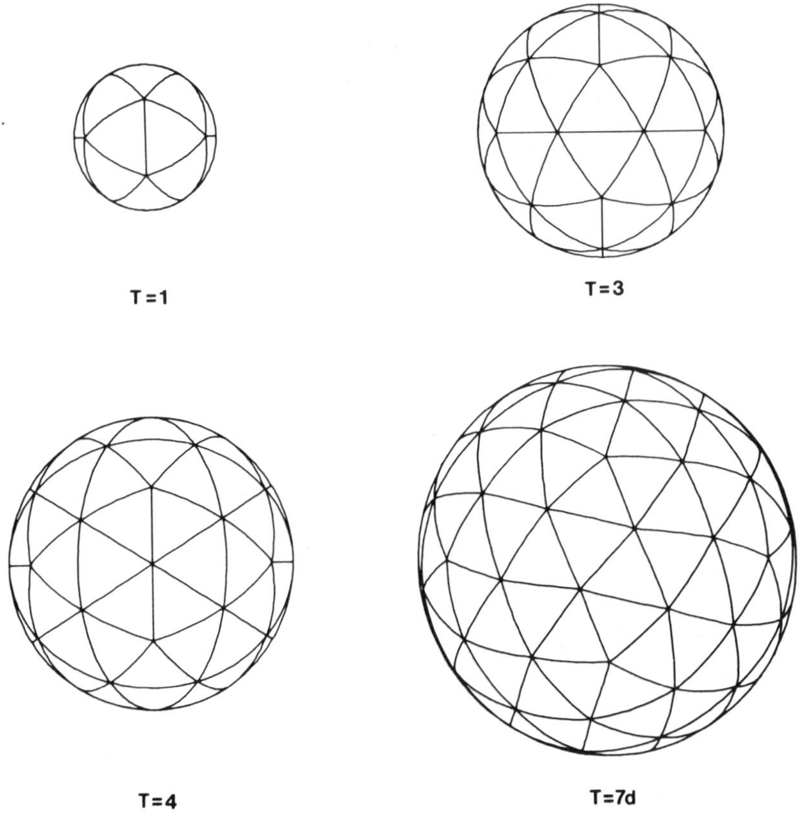

Figure 3. Orthogonal projection of the spherical triangulated icosahedral surface lattices for triangulation numbers 1, 3, 4, and 7*d*. The lattices are scaled such that the relative surface area of the sphere increases in direct proportion to the triangulation number. The relative increase in volume in going from $T = 1$ to $T = 7$ is 1:5.2:8.0:18.5. This illustrates that the use of quasi-equivalence to generate a larger spherical shell gives more efficient use of the genetic material within the virus than increasing the size of the protein subunit.

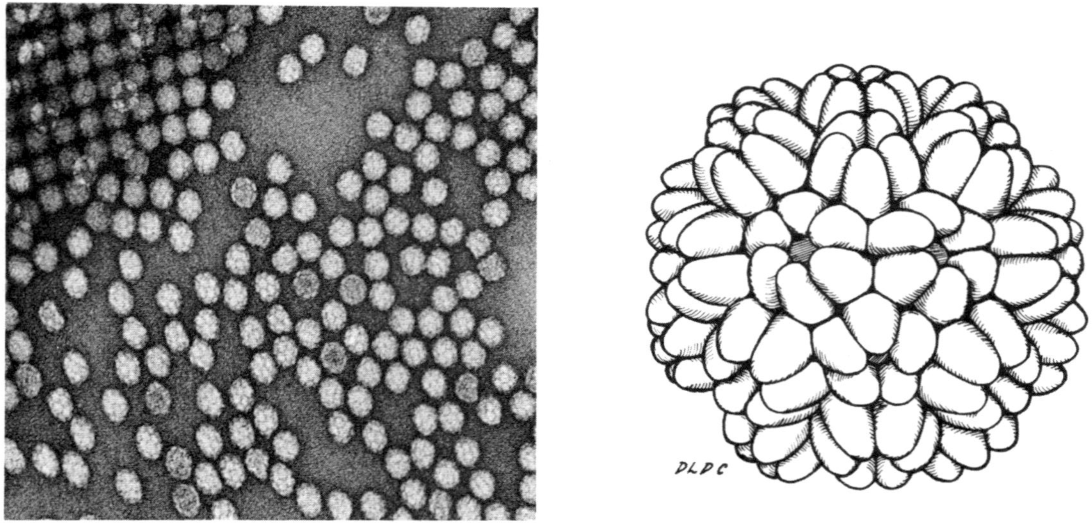

Figure 4. (*a*) Electron micrograph of negatively stained turnip yellow mosaic virus (magnification $\times 200{,}000$) showing the coarse surface features characteristic of pentamer hexamer clustering. Courtesy of Dr. T. S. Baker. (*b*) A drawing of the $T = 3$ protein shell of turnip yellow mosaic virus. The shell is built from 180 identical subunits in quasi-equivalent positions to illustrate the pentamer-hexamer clustering. Original drawing by Caspar.

to the interaction of a region of the surface of one protein molecule with a complementary region on another protein molecule. However, the high resolution structures of TBSV and SBMV show that quasi-equivalence is attained by the use of several chemically distinct interactions rather than by bond deformation.

The next possible triangulation numbers are $T = 4$ and $T = 7$ which are shown in Figure 3. For triangulation number $T = 7$ the spherical lattice no longer possesses inversion symmetry so that there are two enantiomorphic lattices possible.

The great success of this hypothesis was its ability to explain the vastly different surface appearances of viruses in terms of an underlying arrangement of the protein subunits. Caspar and Klug noted that if the protein subunit was to protrude close to a symmetry axis of the icosahedral surface lattice this would give rise to dimer, trimer, or pentamer-hexamer clustering. The resulting morphological units thus show characteristic features in negatively stained preparations even though the individual subunits may not be distinguished because the stain does not penetrate between them.

A good example is turnip yellow mosaic virus which exhibits pentamer-hexamer clustering. Electron micrographs of negatively stained particles show (Fig. 4*a*) that the surface of the virus is composed of 32 doughnut-shaped morphological units. The virus coat is built from 180 copies of the subunit protein arranged on a $T = 3$ icosahedral surface lattice. It appears that the protein subunit protrudes close to the 12 icosahedral fivefold and 20 local sixfold axes to give rise to the 32 morphological units seen in the electron micrographs. This is illustrated in a drawing shown in Figure 4*b*.

The theory of quasi-equivalence has been used to account for the surface morphology of all isometric viruses; however, except for the small plant viruses, these conclusions have rarely been unequivocally confirmed because of difficulty in determining the exact molecular weight of the virus capsid and that of its protein subunits. Because the theory is elegant and comprehensive, however, it has been widely assumed correct.

3. STRUCTURAL STUDIES ON POLIOVIRUS AND RHINOVIRUS

Poliovirus and rhinovirus belong to the picornavirus family. This class of viruses is among the smallest RNA containing animal viruses currently known and has been the subject of intensive biochemical and genetic study. Under suitable conditions, these viruses rapidly take over the synthetic machinery of the host cells in culture. Thus, it has been possible to investigate the synthesis and assembly of the viral macromolecules within the framework of injected cells in the virtual absence of background from host protein and RNA synthesis. As such, these viruses have proved to be excellent models for studying the chemistry and control of viral replication and assembly.

The picornaviruses consist of a single strand of RNA encapsidated by a protein coat which is ~ 300 Å in diameter for poliovirus and rhinovirus (Rueckert, 1976). The coat is composed of 60 copies each of four proteins, three of which have a similar weight and one of which is lighter. For poliovirus the molecular weights of the polypeptides are 34,000, 30,000, 26,000, and 7400 (Kitamura et al., 1981). These four subunits are derived from a larger precursor by proteolytic cleavage (Holland and Kiehr, 1968; Jacobsen and Baltimore, 1968; Maizel and Summers, 1968).

Small crystals of poliovirus were first observed in 1955 by Schaffer and Schwerdt. Large crystals of the type 1 Mahoney strain were observed to form during prolonged storage of concentrated stocks of the virus somewhat later (Steere and Schaffer, 1958). These crystals were characterized crystallographically by Finch and Klug (1959) and shown to diffract to very high resolution. Recent studies on poliovirus aimed at determining its three-dimensional structure have shown that the crystals diffract to at least 2.9 Å spacings and are suitable for high resolution data collection.

The Mahoney strain of poliovirus crystallized in space group $P2_12_12$ with $a = 324$ Å, $b = 359$ Å and $c = 381$ Å with half of a virus particle in the asymmetric unit. The diffraction patterns show strong pseudo I222 symmetry for which reflections $h + k + l = 2n + 1$ are systematically absent to about 35 Å resolution and systematically weak to approximately 25 Å resolution. Packing considerations require that the virus particle lie on the crystallographic twofold axis. The pseudo I222 extinctions imply that the virus particle lies close to $z = \frac{1}{4}$ on the twofold axis.

Rhinovirus also crystallizes in space group $P2_12_12$ and diffracts to high resolution (Erickson et al., 1982). It appears that the crystals are isomorphous to ~ 30 Å resolution with those observed for poliovirus even though there is *a priori* no reason for this. Correspondence may be an indication of a similarity in the organization of the polypeptide chains in the viral capsid. A comparison of the high resolution structures of these two viruses should provide new insight into the factors controlling viral assembly and the structural features responsible for the antigenic differences between these viruses.

4. STRUCTURAL STUDIES ON POLYOMA VIRUS

4.1. Background

Polyoma virus has been the subject of intensive biochemical and genetic study over the past 25 years. Polyoma virus induces tumors in newly born or immunologically deficient mice but is passive in adult mice. Early on it was recognized that this virus was an excellent system for studying the onset of tumor formation because the virus can be prepared and studied in cell culture and because the virus transforms cells *in vitro*. These studies have been reviewed recently by Tooze (1981).

Polyoma virus is a small icosahedral virus of mean diameter 490 Å. It contains a double strand of DNA of molecular weight 3.3×10^6. The DNA is associated with an approximately equal weight of cellular histones forming a circular piece of chromatin (Griffith, 1975), consisting of about 26 nucleosomes. The chromatin core is encapsidated in a protein shell composed primarily of the major coat protein of molecular weight 42,000. Electron micrographs of negatively stained particles show that the virus has a coarse surface (Fig. 5) composed of doughnut-shaped morphological units. Klug in 1965 showed that there are 72 morphological units on the surface of the virus and their locations are consistent with a $T = 7$ icosahedral surface lattice. Infected cells also produce nucleic acid-free empty capsids whose surface morphology is identical to that of the intact virion.

Complete virions contain seven different proteins. Only three of these are coded for by the viral DNA. The remaining four are cellular histones (H2A, H2B, H3, H4) associated with the nucleosomes. The three virus-coded proteins, which are usually denoted VP1, VP2, and VP3, have molecular weights of ~42,000, ~35,000 and ~23,000, respectively, where VP3 corresponds to the C terminal portion of VP2. Comparison of tryptic fragments for VP1 and VP2/VP3 shows no homology. VP1 is the major coat protein and constitutes 80% of the viral protein. The protruding morphological units must be composed entirely of VP1 since purified capsid

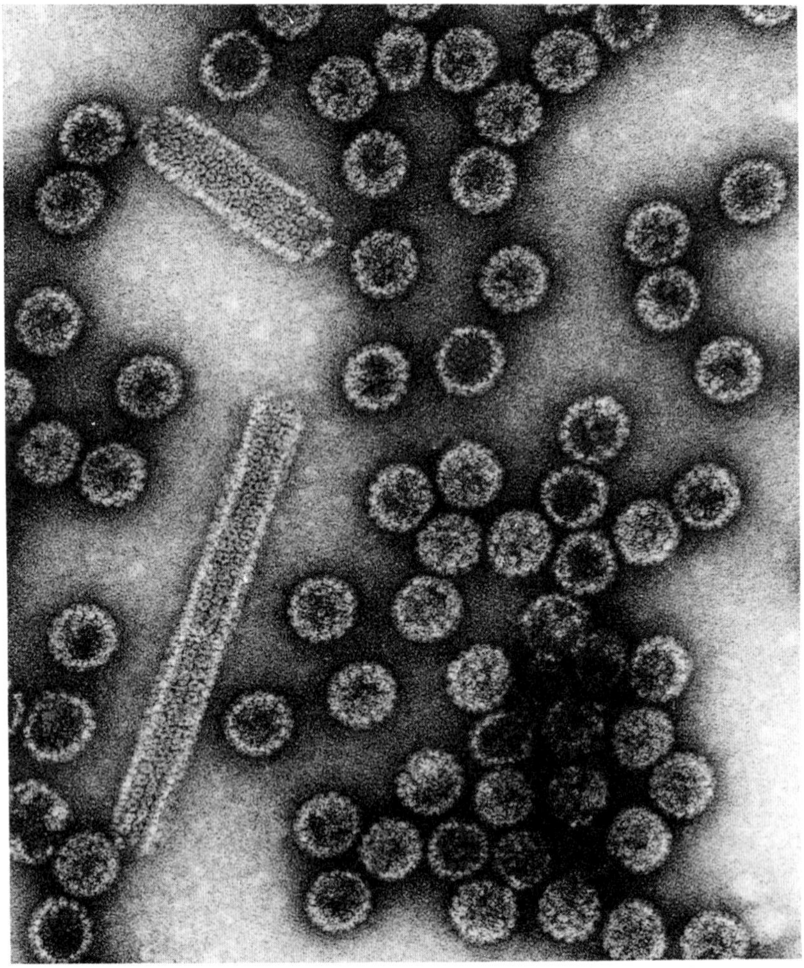

Figure 5. Electron micrograph of negatively stained polyoma virus particles (magnification ×200,000) showing the coarse surface morphology. The micrograph also shows two different examples of hexamer tubes which are polymorphic aggregates of the major coat protein. Micrograph courtesy of Dr. T. S. Baker.

preparations have been isolated that contain only VP1 and whose surface morphology is identical to that of normal virus preparations (Murakami and Schaffhausen, 1974). Consequently, the role of the minor proteins, VP2 and VP3, in the structure is unclear.

The complete nucleotide sequence for two strains of polyoma virus has been determined (Deninger et al. 1979; Soeda et al., 1980). This work has given rise to tentative amino acid sequences and molecular weights for the structural proteins VP1, VP2, and VP3. The polyoma genome is very compact as demonstrated by the partial overlap of the genes for VP1 and VP2/VP3. In addition to the three structural proteins present in the intact virus, the DNA is known to code for three other proteins that are involved in cell transformation and viral replication. These are usually designated large, middle, and small T antigens and are often termed early proteins since they are produced early in the lytic cycle.

4.2. Early Structural Studies on Polyoma and Papilloma Viruses

The early structural studies on polyoma and papilloma viruses played a key role in supporting the precepts of quasi-equivalence. These two groups show considerable structural and organizational similarity as indicated by a comparison of their physical characteristics summarized in Table 1. Both groups are DNA viruses that cause tumors. Polyoma viruses cause tumors that contain little or no detectable virus in nonpermissive hosts. By contrast papilloma viruses cause benign skin tumors in a wide variety of mammalian species in which the tumors themselves contain relatively large amounts of virus. For this reason papilloma viruses were among the first animal viruses to be studied by structural methods.

Information on the physical structure of polyoma and papilloma viruses has, until recently, been obtained primarily by electron microscopy. Figure 5 shows an electron micrograph of negatively stained polyoma virus. The field is composed mostly of isometric particles approximately 450 Å in diameter. Also present are tubular variants that do not contain any DNA. From the micrograph it is clear that the surface of the virus is built from a number of doughnut-shaped morphological units. The papilloma viruses have a very similar appearance and show the same types of tubular variants.

Electron micrographs of negatively stained polyoma virus were first published by Wildy et al. (1960). In their micrographs they could see that the surface consisted of close-packed large morphological units some of which were 6 coordinated with their neighbors and some 5 coordinated. They concluded that the virus particles had icosahedral symmetry and proposed a 42 morphological unit model for the surface structure. This

Table 1. Structure and Composition of the Papovaviruses

Feature	Polyoma Virus	SV40	Papilloma Viruses
Diameter[a]	450 Å	450 Å	550 Å
Symmetry	Icosahedral	Icosahedral	Icosahedral
Triangulation number	7*d*[b,c]	7*d*[d]	7*l* (rabbit)[e] 7*d* (human)[f]
Number of capsomeres	72	72	72
DNA (MW)	3.4×10^6[g]	3.4×10^6[h]	4.9×10^6[i]
DNA content (% w/w)	12.5%	12.5%	12.5%
Virion molecular weight	24.6×10^6	17.6×10^6[d]	40×10^6
Protein composition			
VP1	42,404[g]	40,168[h]	57,000 (human)[j]
VP2	34,761	38,531	Several minor proteins
VP3	22,866	26,965	Several minor proteins

[a] Diameter estimated from electron micrographs of negatively stained particles.
[b] From Klug (1965).
[c] From Finch (1974).
[d] From Anderer et al. (1967).
[e] From Finch and Klug (1965).
[f] From Klug and Finch (1965).
[g] Calculated from DNA sequence (Soeda et al., 1980).
[h] Calculated from SV40 DNA sequence (Reddy et al., 1978; Fiers et al., 1978).
[i] From Gissmann and Zur Hausen (1976).
[j] From Gissmann et al. (1977).

model was criticized by Caspar and Klug in 1962 since it was not based on a detailed analysis of the electron microscope images. The model was also questioned by Mattern (1962) who proposed an alternative 92 morphological unit model. Following the demonstration that the surface structure of human papilloma virus was composed of 72 morphological units (Klug and Finch, 1965) in an arrangement corresponding to the clustering of 420 structure units in hexamers and pentamers at the vertices of the $T = 7d$ icosahedral surface lattice, Klug (1965) showed that the published electron micrographs of Wildy et al. (1960) were consistent with a 72 unit arrangement on a $T = 7$ icosahedral surface lattice. Klug was unable to decide upon the hand of the $T = 7$ skew lattice.

These conclusions were confirmed in the image reconstruction of polyoma virus from electron micrographs of negatively stained particles

(Finch, 1974). In addition, from a series of tilting experiments, Finch was able to show that the skew surface lattice was right handed. The reconstruction was based on images of 10 particles. A stereo projection of the reconstruction is shown in Figure 6. The best images showed some correlation with icosahedral symmetry to spatial frequencies of about 25 Å. The reconstruction showed that the surface of the virus was built from 72 morphological units lying on a $T = 7d$ icosahedral surface lattice. The morphological units in the reconstruction were seen to extend from a radius of 180 Å to about 210 Å. At a radius of 190 Å the distances between the capsomeres were about 75–80 Å except for the distances between the hexavalent capsomeres immediately around the icosahedral fivefold axes which were about 93 Å. This distribution is similar to that found for human papilloma virus. The morphological units in the reconstruction were hollow and slightly conical with a diameter of about 50 Å although the size is very sensitive to the choice of contour level chosen as the boundary between stain and protein. It is of interest to note that no significant difference between the size of the pentavalent and hexavalent capsomeres could be detected although it would be difficult to discern a 20% increase in mass between a pentameric and hexameric capsomere at the resolution of the image reconstruction. No substructure in the morphological units

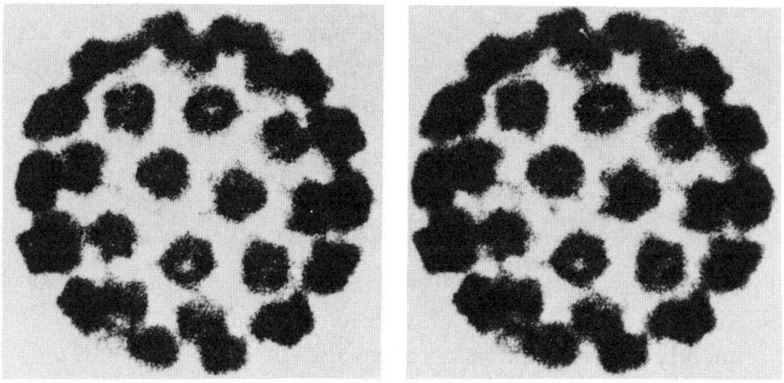

Figure 6. Stereo view of a reconstruction of polyoma virus, including data to a Fourier cutoff of 25 Å. The view is close to a twofold symmetry axis and only the top half of the reconstruction is shown. Two 5-coordinated morphological units can be seen clear of the periphery, toward the upper right and lower left. The path between them is two steps out and one to the right corresponding to the $T = 7d$ icosahedral lattice. Reproduced with permission from Finch (1974).

could be detected in the reconstruction. Likewise, no information about the inner protein shell could be seen in the image reconstruction.

On account of the similarity in size of the pentavalent and hexavalent morphological units, Finch suggested that the 12 pentameric capsomeres might be constructed from the minor virus proteins VP2 and VP3 whereas the 60 supposedly hexameric capsomeres could then be constructed from the major coat protein. This suggestion cannot be true since as noted earlier it is possible to isolate empty virus capsids that contain no VP2 or VP3 and yet have the same surface morphology as the complete virus.

The image reconstruction of polyoma virus was an important stepping-stone in the X-ray structure determination of polyoma virus. It provided a good model for the external envelope of the virus and location for the hexavalent morphological unit. It also provided a benchmark by which crystallographic studies could be measured.

4.3. Structure Determination of Polyoma Virus Capsid

The three-dimensional structure of the nucleic acid-free capsid of polyoma virus has been determined at 22.5 Å resolution by single crystal X-ray diffraction (Rayment et al., 1982). The structure was solved by using the fivefold noncrystallographic symmetry present in the diffraction data. The resultant electron density map shows that all 72 of the protruding morphological units on the surface of the virus are built from five copies of the major coat protein. This finding contradicts the theory of quasi-equivalence in virus particle design (Caspar and Klug, 1962) which predicts that the 60 hexavalent capsomeres in the $T = 7d$ icosahedral surface lattice should be hexamers whereas only the 12 pentavalent capsomeres should be pentamers.

The crystals of polyoma capsids diffract to at least 8 Å resolution on still photographs. In order to solve the structure to this resolution it was decided to collect and phase a low resolution data set that could be used subsequently to locate the heavy atom sites in an isomorphous derivative. This strategy was adopted because it is impossible to locate the heavy atoms directly from a Patterson function for a macromolecular assembly as large as polyoma virus.

The diffraction data to 22.5 Å resolution was collected on film using conventional screened precession photography. A total of 11 zero level and four upper level photographs were recorded. These films gave a total of 1481 independent reflections representing 94% of the data to 22.5 Å resolution. Great care was taken in collecting the data, as reflected in the low overall scaling R factor of 0.057. This measures the mean variation in intensity of symmetry-related reflections on different films. The quality of the data was an important factor in the determination of the structure and the clarity of the final result.

4.3.1. *Initial Low Resolution Phases*

The initial phases to 30 Å resolution were derived from a model that was built to fit the information available from electron microscopy and X-ray small-angle scattering. The model was built as an electron density map with the same symmetry and unit cell as the crystalline virus. The phases were obtained by Fourier transforming this model electron density map. The starting model consisted of 72 morphological units arranged on a $T = 7d$ icosahedral surface lattice (Fig. 7). The arrangement of the pro-

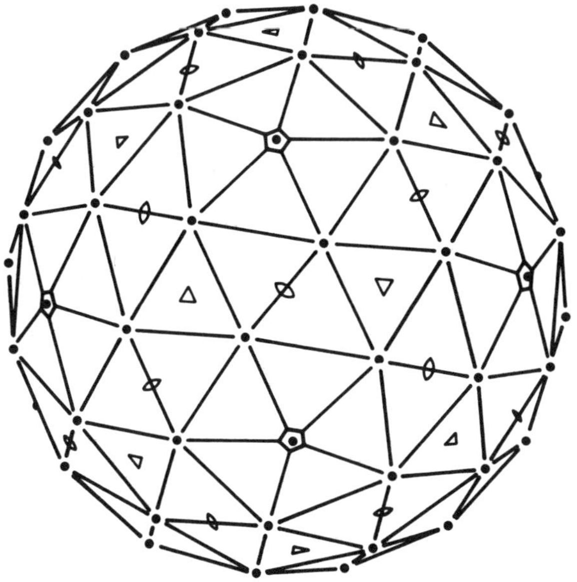

Figure 7. $T = 7d$ icosahedral surface lattice with five-, three-, and twofold axes marked. The drawing shows one side of the polyhedral surface which consists of 60 six-coordinated and 12 five-coordinated lattice points at the same radius. The location of the six-coordinated point is that determined for the hexavalent morphological unit in the polyoma capsid.

truding morphological units introduces an acentric component into the initial phases. This is essential to the success of the subsequent real-space phase refinement and extension.

The exact size and location of the 72 hollow cylindrical capsomeres were adjusted to give a minimum *R* factor between the observed and the calculated structure factors from the model. An example of this process is shown in Figure 8 where the *R* factor is plotted as a function of the location of hexavalent morphological units. The position of the minimum is close to the location of the capsomere observed in Finch's image reconstruction of polyoma virus (1974). The models which gave the best *R* factor were built from morphological units 40 Å in height and 80 Å in diameter with a 35 Å diameter axial hole resting on a shell extending from a radius of 180–200 Å. The *R* factor for this model to 30 Å was 50% where *R* is defined as

$$R = \frac{\Sigma \, ||F_o| - |F_c|| \times 100\%}{\Sigma \, | F_o |}$$

where $|F_o|$ and $|F_c|$ are the observed and calculated structure factor amplitudes, respectively. A projection of this "best" model down a twofold axis is shown in Figure 9*a*.

4.3.2. *Phase Refinement*

The calculated phases from the model structures were refined to 30 Å resolution against the observed structure factors by imposing the constraints of solvent flattening and the fivefold noncrystallographic symmetry (Bricogne, 1974, 1976; Johnson, 1978). An electron density map was calculated to 30 Å resolution using the model phases and observed amplitudes and averaged over the five noncrystallographic asymmetric units within the particle envelope.

The averaged solvent-flattened density was Fourier transformed to derive the calculated amplitudes and phases from the first cycle of refinement. The rms phase change and *R* factor between the observed and refined structure factors were evaluated. The new phases were combined with the measured amplitudes to initiate the next cycle of refinement. These calculations were iterated until the *R* factor had converged and there was no significant phase changes from one cycle to the next. After

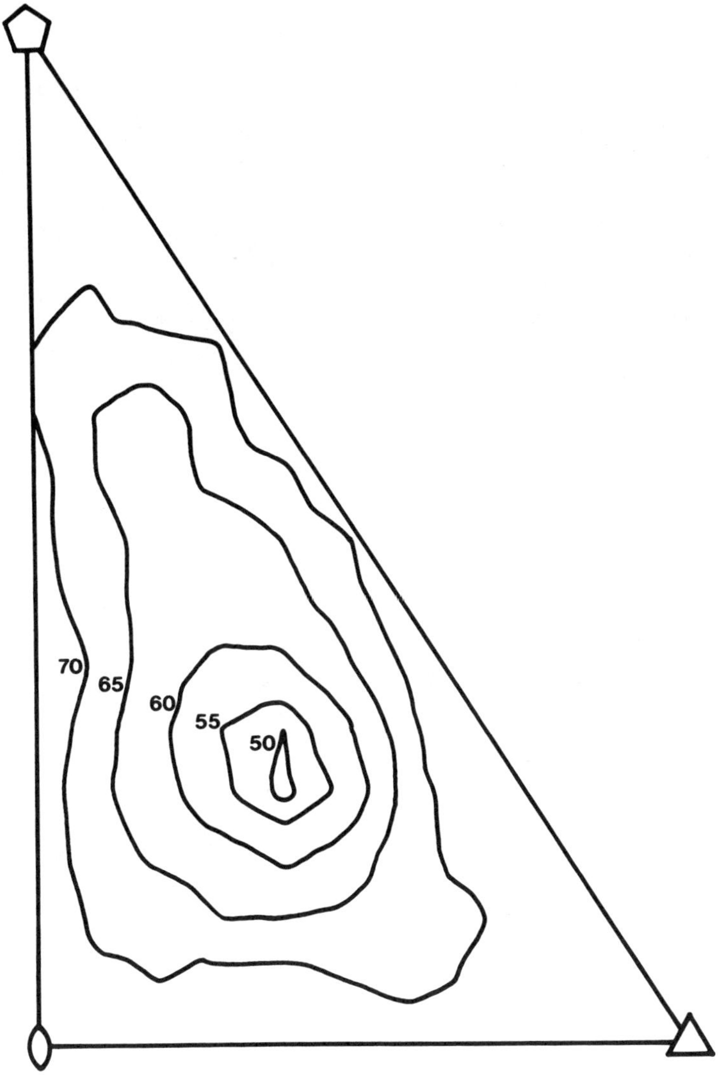

Figure 8. Location of the hexavalent unit in the icosahedral surface lattice. The *R* factor comparing the calculated and measured diffraction pattern is plotted as a function of the position of the hexavalent unit in the icosahedral cell defined by the five-, three-, and twofold axes (see Fig. 7). The model giving the smallest *R* factor is shown in Figure 9*a*.

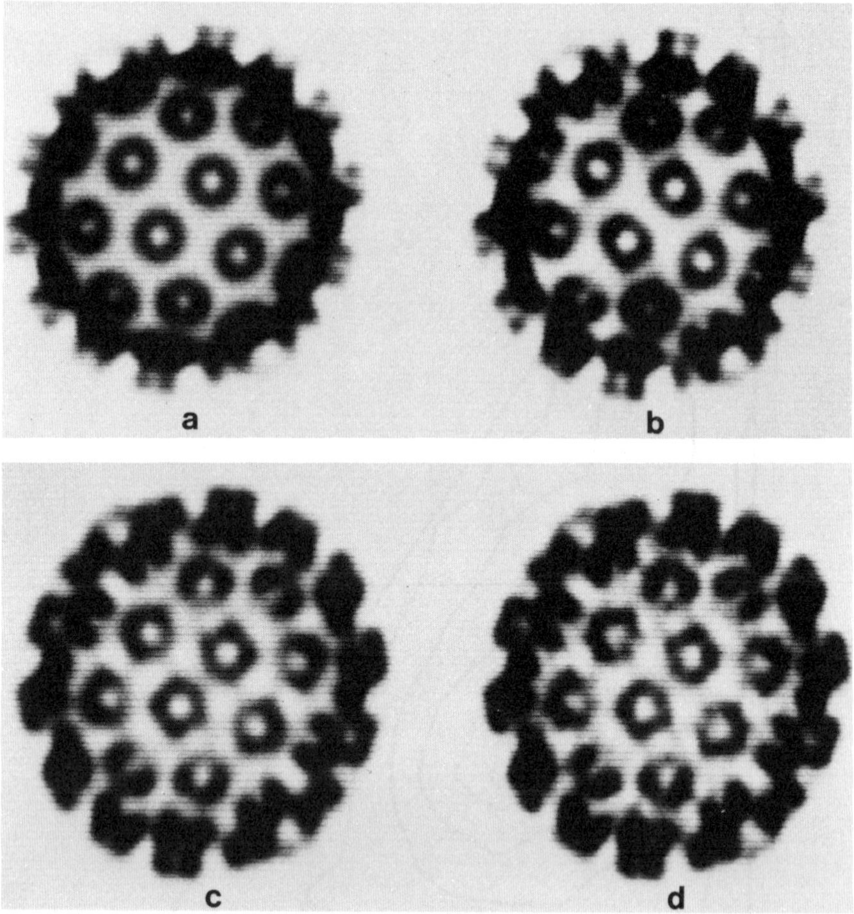

Figure 9. An illustration of the refinement process showing density maps of the top halves of the particles (diameter 495 Å projected down a twofold axis). (*a*) Model map at 30 Å resolution. (*b*) Map computed using the observed data and model phases to 30 Å resolution without refinement. (*c*) 30 Å resolution refined map. (*d*) Refined map at 22.5 Å resolution calculated by phase extension from 30 Å. Note how the substructure in the morphological units develops as the refinement proceeds. Even at 30 Å the hexavalent unit has a pentagonal appearance.

14 cycles of refinement the *R* factor between the observed and calculated structure factors was 10.5%.

There are two possible orientations for the icosahedral fivefold axis in the I23 unit cell. The orientation used was that indicated by the intensity spikes in the diffraction patterns (Adolph et al., 1979). If the alternative orientation of the fivefold axis was chosen the *R* factor between the refined and observed structure factors never dropped below 50%. The standard deviation in the density between the five icosahedral units in the crystallographic asymmetric unit for the correct orientation was one-seventh of that calculated for the alternative.

Great care was taken to measure as much of the data within the resolution limits as possible. Even so no reflections were measured with spacings >150 Å. These very low resolution terms are important in the phase refinement, because they include some of the most intense reflections in the data set. Their absence from the calculations severely restricts the phase refinement and generates an ambiguous electron density map (Rayment et al., 1983). This problem was overcome by including calculated values for the unmeasured data after the fourth cycle of refinement. This strategy has not been adopted in any previous structure determination using molecular replacement.

Inclusion of the calculated values for the missing data is necessary because phase refinement by symmetry averaging depends upon the continuity of the Fourier transform in reciprocal space. Ignoring unmeasured data introduces artifactual discontinuities in the transform. The refinement process smoothes out the perturbations thereby reducing the value for the terms close to the missing data and predicting an amplitude and a phase for the missing data.

4.3.3. *Phase Extension*

The polyoma diffraction data between 30 and 22.5 Å resolution were phased by extension from the refined 30 Å phases. This was necessary because the model phases were only reliable to ~30 Å resolution. Attempts to refine the data directly to 22.5 Å resolution did not converge satisfactorily.

Phase extension is possible because solvent flattening generates phase information for reflections just beyond the resolution of those used to calculate the map. Thus predicted phases for structure factors within one reciprocal lattice distance $\frac{1}{572}$ Å of the previous resolution were included in alternate cycles. After 18 cycles of phase extension and refinement all the terms between 30 and 22.5 Å resolution were incorporated in the calculation. The final *R* factor between the observed and calculated structure factor amplitudes was 14.2%. A projection down a twofold axis of half a virus particle excised from the refined map at 22.5 Å resolution is shown in Figure 9*d*.

4.3.4. *Tests of the Refinement Method*

The method used to solve the structure has been investigated in detail using model data in place of the observed data (Rayment, 1983). This was necessary in order to establish that the pentameric nature of the 60 hexavalent morphological units was not an artifact of the refinement procedure. There is no other way to establish a level of confidence in the low resolution electron density map since the normal criteria of stereochemical consistancy or heavy atom refinement do not apply at this resolution.

The model calculations were used to test all aspects of the refinement procedure. The most significant test however was the demonstration that the phases refined toward the true phase solution for the amplitudes and that a low R factor between the calculated and observed amplitudes was indicative of the correct phase solution. For example, in one set of calculations structure factors were generated from a model that exhibited true $T = 7d$ icosahedral symmetry.

The model consisted of 72 cylindrical protrusions 42 Å tall with an external and internal diameter of 70 and 34 Å, respectively. These morphological units were arranged on a spherical shell extending from a radius of 190–200 Å. The locations of the 60 hexavalent units were close to that determined from the image reconstruction of polyoma virus (Finch, 1974). Additional features were added to this simple model by imbedding small cylinders of density 12 Å in diameter in the walls of the morphological units. Five small cylinders of density were placed in each of the 12 pentavalent units whereas six small cylinders were placed in each of the 60 hexavalent units. In this way the substructure showed a $T = 7d$ distribution in which the small cylinders represented the expected 420 quasi-equivalently related protein subunits. A projection of this model electron density map at 22.5 Å resolution down an icosahedral fivefold axis and through a hexavalent unit is shown in Figure 10*a* and *b*.

The structure factors from this 420 subunit model were used in place of the observed polyoma data in order to test the refinement method. In one experiment phases were taken from the 360 subunit model in which all of the morphological units contained strong fivefold substructure. These phases were combined at 30 Å resolution with the amplitudes from the 420 subunit model and refined using the constraints of solvent flattening and noncrystallographic averaging. The initial R factor between

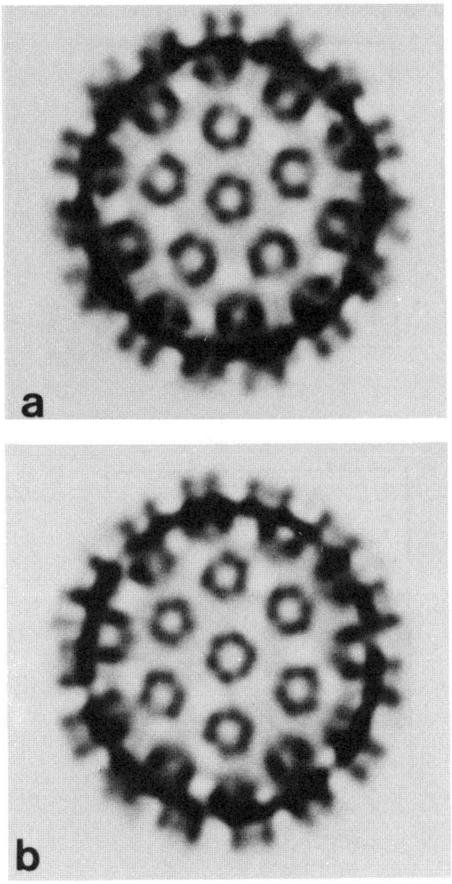

Figure 10. Density maps of half of the 420 subunit model at 22.5 Å resolution projected (*a*) down the icosahedral five-fold axis and (b) through the hexavalent morphological unit.

the two data sets was 50%. After 15 cycles of refinement the *R* factor fell to 4.5%. This refinement experiment is shown graphically in Figure 11*a* through *d*, which shows a section through the hexavalent morphological unit of (1) the 420 subunit model from which the amplitudes were taken, (2) the 360 subunit model from which the phases were taken, (3) the initial combination of amplitudes and phases which shows that at this point the phasing model dominates the electron density map, and (4) the section through the refined electron density map that is indistinguishable from the 420 subunit map at 30 Å resolution. This clearly demonstrates that the refinement converges on the information present in the amplitudes and does not contain a bias toward the phasing model.

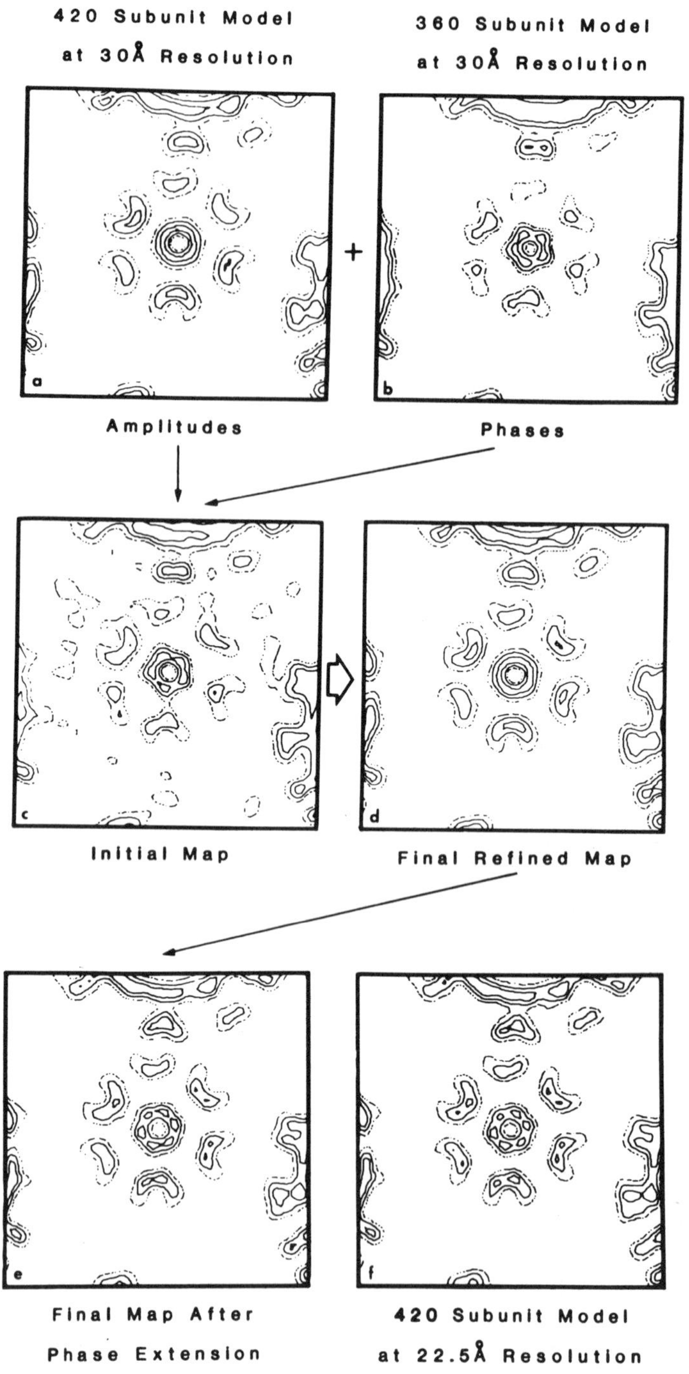
420 Subunit Model
at 30Å Resolution
360 Subunit Model
at 30Å Resolution
+
a
b
Amplitudes
Phases
c
d
Initial Map
Final Refined Map
e
f
Final Map After
Phase Extension
420 Subunit Model
at 22.5Å Resolution

The validity of phase extension was tested using the amplitudes from the 420 subunit model and the refined phases to 30 Å resolution from the previous experiment. This process is illustrated in Figure 11, where Figure 11*d* shows one section through the initial electron density map at 30 Å resolution, Figure 11*e* shows the phase extended and refined electron density map at 22.5 Å resolution, and Figure 11*f* shows the correct 22.5 Å resolution map. These figures clearly show that phase extension is valid, and that the refinement strategy is capable of distinguishing between pentameric and hexameric substructure in the hexavalent morphological units.

4.3.5. Structure of Polyoma Virus Capsid

In Figure 12 a view of half of a virus particle projected down the icosahedral fivefold axis is compared with that down the axis of a hexavalent morphological unit. This reveals a striking similarity between the penta- and hexavalent morphological units. Indeed the hexavalent morphological unit shows a distinctive fivefold character. This finding is inconsistent with the principles of quasi-equivalence which predicted that the 60 hexavalent morphological units of the $T = 7d$ icosahedral surface lattice should show sixfold substructure.

The electron density map shows that the 12 pentavalent morphological units and 60 hexavalent units have the same diameter and height, and also contain the same total electron density. This finding is shown quantitatively in Figure 13, which shows a power spectrum analysis of the electron density in the pentavalent and hexavalent morphological units. The power spectrum was determined for a cylindrical portion of the density 93 Å in diameter centered on the pentavalent and hexavalent unit. These cylinders touch at a radius of 170 Å. The correspondence in the rotational power spectra distribution of the pentavalent and hexavalent units indicates a close similarity in their structures and a small difference in their radial positions. The order 5 harmonic for the hexavalent capsomere is always much stronger than that of the order 6 or indeed any other nonzero harmonic. The relative strength of the order 5 harmonic

Figure 11. (*Opposite*) Refinement and extension of phases from a 360 subunit model against amplitudes from a 420 subunit model. The maps are perpendicular to the axis of a hexavalent morphological unit sectioned 220 Å from the particle center.

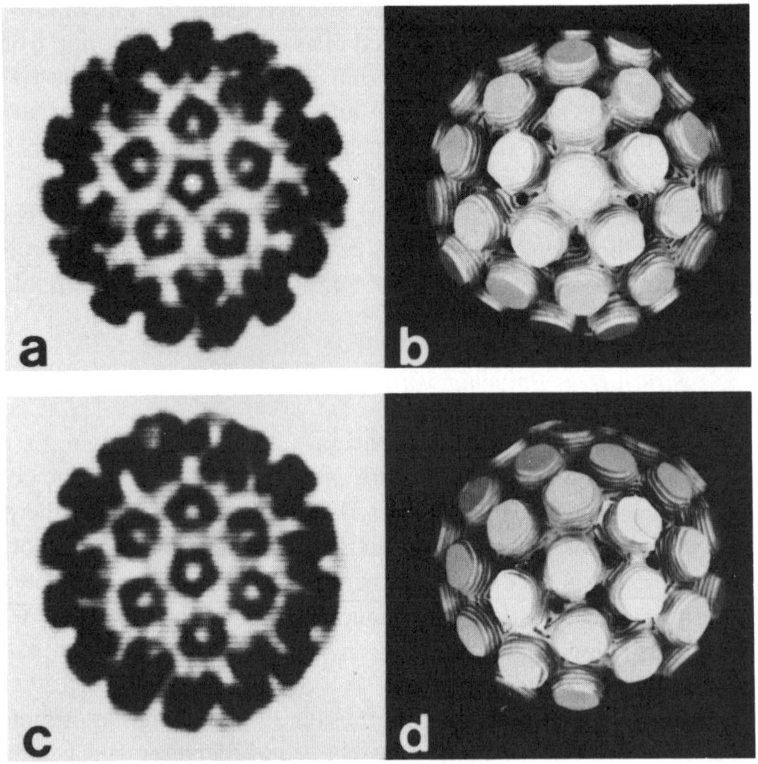

Figure 12. Views of half the capsid (495 Å diameter) (*a* and *b*) down the fivefold axis and (*c* and *d*) down the axis of a hexavalent unit. (*a* and *c*) Computer graphics projections of the electron density map at 22.5 Å resolution and (b) and (c) are photographs of a model, of stacked sections of the map at 30 Å resolution with the contour level set at 0.2 of the maximum density. The half-model was built symmetric about the fivefold axis (*b*), thus when it is tilted to view down the hexavalent unit (*d*) the bottom portion of the image is incomplete.

for the outer portion of both the pentavalent and hexavalent capsomere is reduced because the cylindrically symmetric component, which is the dominant term in P_t, is very strong for this part of the structure. It is important to note that the icosahedral symmetry requires the 12 identical pentavalent units to have fivefold symmetry but that no constraints on the size, shape, or location of symmetry have been imposed on the hexavalent unit. Thus, the similarity of the pentavalent and hexavalent capsomeres seen in the electron density map cannot be an artifact of the imposition of the noncrystallographic symmetry in the phase refinement.

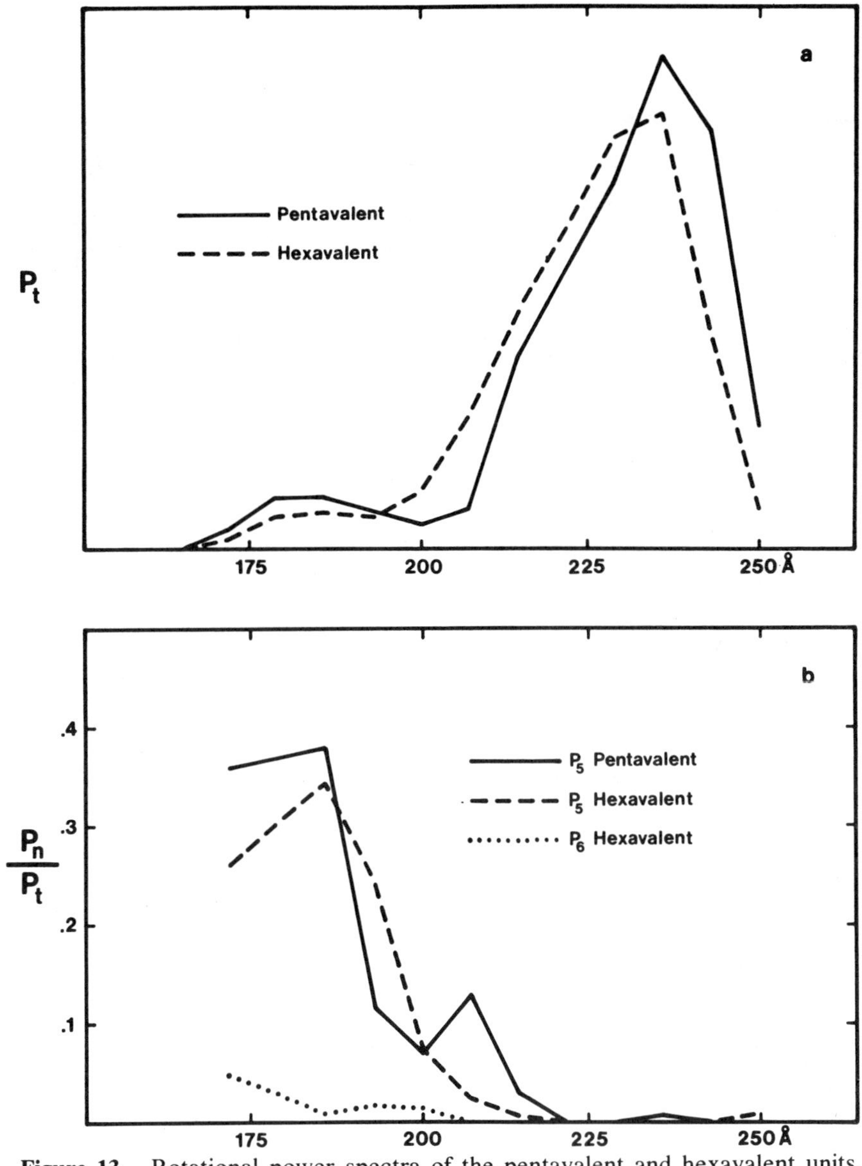

Figure 13. Rotational power spectra of the pentavalent and hexavalent units. (*a*) P_t is the sum of all cylindrical harmonics in sections of the map through the capsomeres normal to the fivefold axis and the "best" axis of the hexavalent units, plotted as a function of the distance of the section from the center of the particle. (*b*) Relative strength of the fivefold harmonic of the pentavalent unit and the five- and sixfold harmonic of the hexavalent unit. The correspondence in the rotational power spectra distribution of the pentavalent and hexavalent units indicates a close similarity in their structure and a small difference in their radial positions.

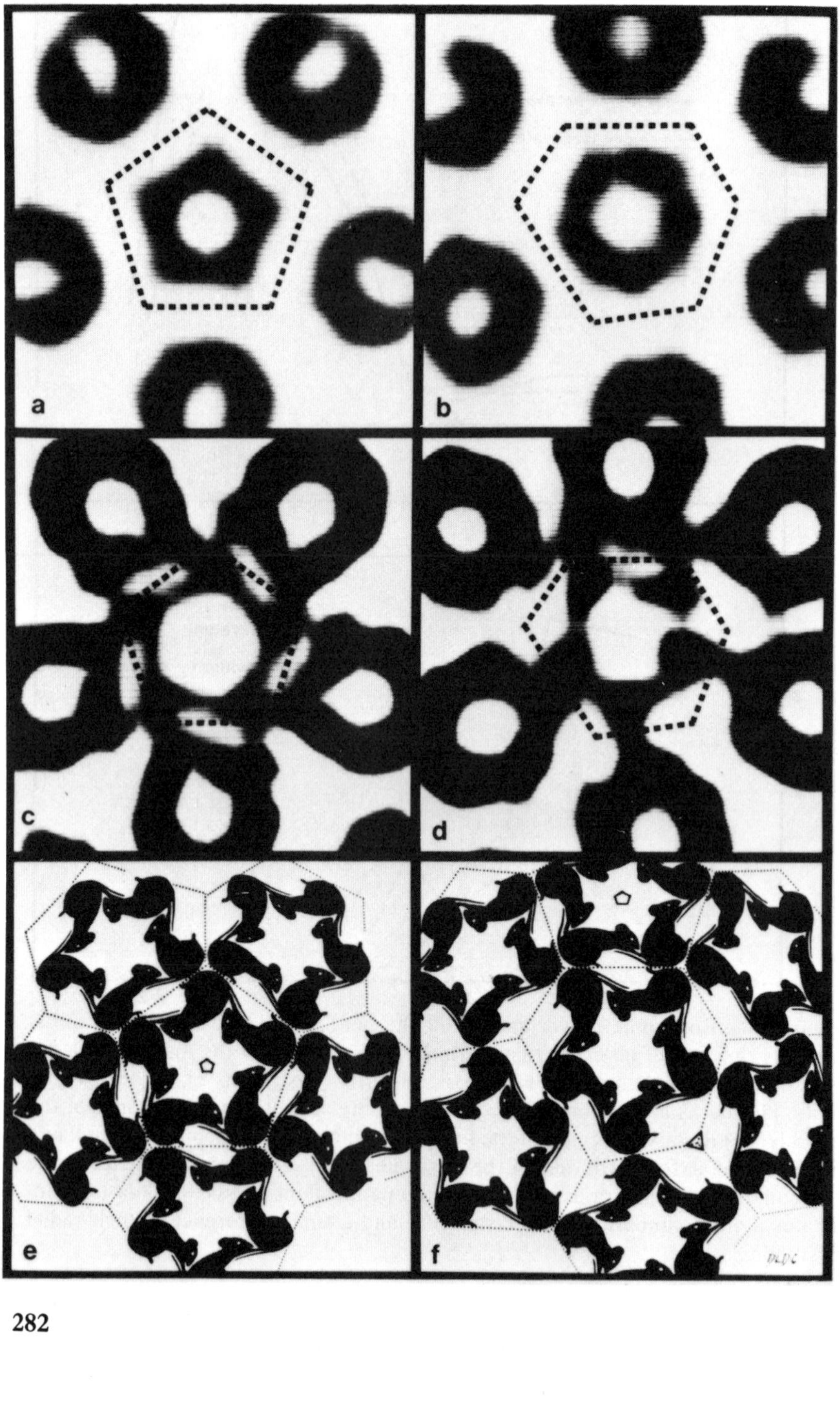
a
b
c
d
e
f

These results indicate that all 72 capsomeres are pentamers of identical or very similar subunits.

The electron density map may be divided into two regions: the protruding pentagonal caps and the basal parts connecting the capsomeres. The contacts between capsomeres extend inward from a radius of ~ 210 Å. The protruding caps are 85 Å in diameter and extend out ~ 40 Å. The central hole in the capsomere has a diameter of 40 Å at a radius of 210 Å and tapers shut ~ 15 Å from the surface, leaving a dimple on the top of the cap. The basal parts of the capsomeres extend in ~ 40 Å. A section through the protruding cap and basal regions of the pentavalent and hexavalent morphological units is shown in Figure 14. Figure 14*a* and *b* show the electron density distribution in the protruding cap at a height at which there is no contact with the neighboring capsomeres. Figure 14*c* and *d* show the electron density distribution in the shell of the virus where the subunits of the capsomeres interact with one another. Figure 14 *a* and *b* show that the protruding portion of the pentavalent and hexavalent capsomeres have very similar substructure. This suggests that the same sub-

Figure 14. (*Opposite*) Electron density sections. Sections of the electron density map showing the substructure of the capsomeres (*a–d*) and drawings on the polyhedral surface (*e–f*) which illustrate an inferred packing relationship of structure units. Left-hand column, views down the fivefold axis: righthand column, view through a hexavalent capsomere. The width of the section displayed is 215 Å. (*a*) and (*b*) are sections 215 Å above the center of the capsid at a level where there is little contact between neighboring capsomeres. (*c*) and (*d*) show sections 195 Å above the center. At this level and below, the five subunits of each capsomere splay out and contact the basal parts of the neighboring subunits. The section planes are perpendicular to the axis of the central capsomere and slice through the neighboring capsomeres obliquely. In the obliquely sectioned parts, the substructure is obscured. The polygons delineating the domain of the central capsomere are marked by dotted lines on the density maps. In (*e*) and (*f*), the drawings (constructed according to polyhedral surfaces made up of pentagonal and hexagonal facets) illustrate that the interactions between subunits (represented by mice) are not conserved. If the structure units in the hexavalent pentamer are arranged with quasi-fivefold symmetry resembling that seen in the pentagonally symmetric pentamer (center of *e*), this gives rise to three types of contact between structure units of neighboring pentamers: (1) the two structure units at the top of the central pentamer in (*f*) each contact two neighbors (one from the pentavalent and one from another hexavalent pentamer) related by a quasi-threefold axis; (2) the two structure units below and right of center of (*f*) make quasi-twofold contacts with neighbors related by an icosahedral threefold axis; (3) the one structure unit left of center of (*f*) makes a close contact with a unit related by an icosahedral twofold axis.

unit–subunit interaction occurs in both types of capsomeres for this part of the capsid structure. Conversely, where the polyoma capsomeres interact in the shell of the virus, the subunit–subunit contact and its six neighbors cannot show fivefold symmetry. At 22.5 Å resolution it is impossible to observe how a pentameric capsomere can interact with six neighbors. Each subunit in the hexavalent capsomere occupies a symmetrically distinct location in the capsid. Figure 14*c* and *d* suggest, however, that the interactions between the subunits in the shell can be rationalized by dimer and trimer associations. This is shown diagramatically in Figure 14*e* and *f* in which the structure units have been represented by a mouse motif. The structure units in the hexavalent pentamer have been arranged with quasi-fivefold symmetry similar to that found in the pentavalent pentamer. This results in three types of contact between the six structure units in the iscosahedral asymmetric unit. Three units are connected in one quasi-trimer, two in one quasi-dimer, and one in half of an exact dimer. The drawing demonstrates that the bonding between the subunits cannot be conserved but it also shows that a closed shell can be formed by a pentameric aggregate.

4.3.6. *Molecular Weight Measurements*

The molecular weight determinations for both isolated major coat protein and the complete capsid are summarized in Tables 2 and 3. All the measurements are consistent with a capsid built from 360 subunits. The best estimates for the molecular weight of VP1 (42,834 and 42,458) are those deduced from the DNA sequence for two strains of polyoma virus. The estimates for the molecular weight derived from SDS polyacrylamide gel

Table 2. Molecular Weight Determinations for Polyoma Virus Major Coat Protein, VP1

Method	Molecular Weight
DNA sequence	
A2 strain	42,404[a]
Large plaque strain 3	42,834[b]
SDS-polyacrylamide gel electrophoresis	42,000–50,000[c]
Gel exclusion chromatography on guanidine HCl dissociated VP1	41,400[c]

[a] From Soeda et al. (1980).
[b] From Deninger et al. (1979).
[c] From Murakami and Schaffhausen (1974).

Table 3. Molecular Weight of VP1 Deduced from Particle and Capsomere Molecular Weight

	Molecular weight of VP1 assuming	
	360 Subunit Capsid	420 Subunit Capsid
Molecular weight of empty capsid = $15 \pm 1 \times 10^{6a}$ (equilibrium sedimentation)	41,700 ± 360	35,700 ± 7%
	Pentameric Capsomere	Hexameric Capsomere
Molecular weight of dissociated capsomeres = 221,000[b] (equilibrium sedimentation)	44,200	36,833

[a] From Murakami and Schaffhausen (1974).
[b] From Friedman (1971).

electrophoresis which range from 42,000 to 50,000 are not very reliable because of a dependence of the mobility of the protein with gel concentration (Murakami and Schaffhausen, 1974).

The molecular weight of the empty capsid as determined by equilibrium sedimentation is $15 \pm 1 \times 10^6$ (Murakami and Schaffhausen, 1974). This value agrees with that obtained from the molecular weight of the complete virus (24×10^6) after accounting for the contribution of the mini chromosome and minor protein. Thus, the molecular weight of VP1 for a 360 or a 420 subunit capsid should be 41,700 ± 7% or 35,700 ± 7%. Clearly the capsid molecular weight is consistent with a 360 subunit capsid. Unfortunately none of the molecular weight measurements were undertaken with the aim of establishing the number of subunits in the capsid and as a consequence individually they cannot be taken as establishing categorically the capsid composition. As a whole however these measurements support a 360 subunit capsid.

5. SIMILARITIES BETWEEN POLYOMA VIRUS AND SV40

Polyoma virus and SV40 show remarkable structural and genetic similarities. The viruses have the same diameter and contain three viral proteins of similar molecular weight. Furthermore the surface appearance of polyoma and SV40 on negatively stained electron micrographs is almost identical. The capsids of both viruses appear to be composed of 72 morphological units arranged on the vertices of a $T = 7d$ icosahedral surface lattice.

The similarity is also expressed in the genetic organization. Both viruses transform cells using a similar mechanism of transformation. The DNA sequences of the genomes of both SV40 and polyoma have been determined. A comparison of the early and late regions show considerable homology in the DNA sequences (Soeda et al., 1980) of these two viruses. The early region that codes for the proteins involved in cell transformation and DNA replication shows a 36% nucleotide homology. In the late region that codes for the structural proteins there is a 53% and 38% nucleotide sequence homology for VP1 and VP2, respectively. The high nucleotide homology strongly implies that there is a close structural similarity in the organization of the protein coat. It is fair to suggest that polyoma virus and SV40 are likely to have the same all-pentameric arrangement of capsomeres on their surfaces.

6. STRUCTURAL STUDIES ON SV40

Small crystals of SV40 have been grown which are similar in morphology to the crystals of polyoma virus (Lattman, 1980). Preliminary investigations suggest that the packing of SV40 particles in the crystalline lattice may be identical to that observed for polyoma. In light of the great similarity in the genetic organization of these viruses and their similar appearance on negatively stained electron micrographs, a comparative study of the crystal structure of SV40 would be of considerable interest. This could independently confirm that the organization of the subunits in the capsid of polyoma is not an anomaly.

An additional reason for pursuing the structure of SV40 is the hope that the crystals might diffract to high resolution. At this time crystals of complete infectious polyoma virus diffract to only 5.5 Å resolution. This would be adequate to determine domain boundaries but insufficient to resolve the structural features responsible for the final viral assembly. In addition SV40 is more readily available than polyoma virus so that it would be possible to rigorously investigate the crystallization conditions.

7. POLYMORPHIC ASSEMBLIES OF POLYOMA AND PAPILLOMA VIRUSES

Unfractionated preparations of polyoma virus and the papilloma viruses normally contain tubular particles that appear to be hollow and often have rounded ends. The tubes are similar in their surface appearance to that of the virus particles in that they also appear to be composed of morphological units packed approximately hexagonally about 100 Å apart. These tubular variants are interesting because analysis of the errors in viral assembly provides insight into the ways in which the assembly of the complete particle is controlled.

There are two main categories of hollow tube found for both the polyoma and papilloma viruses: wide and narrow. The first category has a diameter of about 450–550 Å, close to that of the normal virus particle. The second category is narrower with a diameter of about 300 Å.

Detailed analysis of the tubular variants and of rabbit and human papilloma virus by means of electron microscopy was undertaken by Kiselev and Klug (1969). They showed that the tubes consisted of capsomeres arranged on a hexagonal lattice. These were presumed to be a tubular variant built from hexameric aggregates of the major coat protein and in that respect this polymorphic assembly did not seem surprising. This type of mistake in self-assembly had previously been predicted by Caspar and Klug (1962). The narrow tubes, however, show a novel arrangement of capsomeres. A detailed examination of the optical diffraction patterns from electron micrographs of the narrow tubes shows that the capsomeres are arranged with a particular type of pentagonal tessellation which only roughly approximates an ordinary hexagonal lattice. This structural organization was not expected but was consistent with the precepts of quasi-equivalence. The existence of these two types of tubular variants was taken as confirming evidence for pentamers and hexamers of the major coat protein and lent considerable support to the theory of quasi-equivalence.

The all-pentamer structure observed for polyoma virus raises questions about the nature of the hexamer tubes. Conversely, it is now straightforward to explain the pentamer tubes if all the capsomeres are pentameric. A recent reexamination of electron micrographs of the wide hexagonal tubes has shown that these are also made of pentamers (Baker et al., 1983). This is not unreasonable or inconsistent with the observed structure for the hexavalent morphological unit of polyoma since in the latter, a pentameric aggregate of the major coat protein, has interactions with six neighbors.

8. ASSEMBLY OF POLYOMA VIRUS AND SV40

Polyoma virus and SV40 appear to have similar assembly pathways. The synthesis of the viral proteins occurs in the cytoplasm and the proteins are then transported through the nuclear membrane into the nucleus where assembly occurs. Study of viral assembly is difficult because of experimental problems in isolating the cell nuclei without perturbing viral intermediates.

More research regarding the assembly of SV40 has been described than for polyoma; hence, it is convenient to discuss the problems of assembly with respect to these studies. All the findings, however, are essentially consistent with the investigations on polyoma virus (Murakami, personal communication). The assembly pathway for SV40 is summarized in Figure 15 and is taken from the work of Garber et al. (1980). The assembly process involves adding the viral proteins to a nucleoprotein complex to build up the final virus. This scheme differs from that found for bacteriophage which involves a prohead precursor into which the genetic material is injected. Early studies of SV40 and polyoma using pulse-chase labeling of viral capsid proteins indicated that the empty shells were precursors of the virions and that assembly occurs by encapsidation of the viral genome by the preformed shell (Ozer and Tegtmeyer, 1972). However, these investigations used CsCl gradient analysis of infected cell lysates, a procedure known to disrupt many nucleoprotein complexes (Coca-Prados and Hsu, 1979). Thus, it is not certain whether the empty capsids observed in the CsCl gradients were independently synthesized or whether they were derived from CsCl-destabilized assembly intermediates. Studies on the intermediate nucleoprotein complex of polyoma virus indicate that the complex is dissociated by high ionic strength to give free DNA, cellular histones, and an aggregate of viral proteins that sediment slightly slower than the virus capsid (Murakami, personal communication).

Assembly is thought to begin at the 75S compact nucleoprotein complex. This is composed of closed circular DNA, nucleosomal histones,

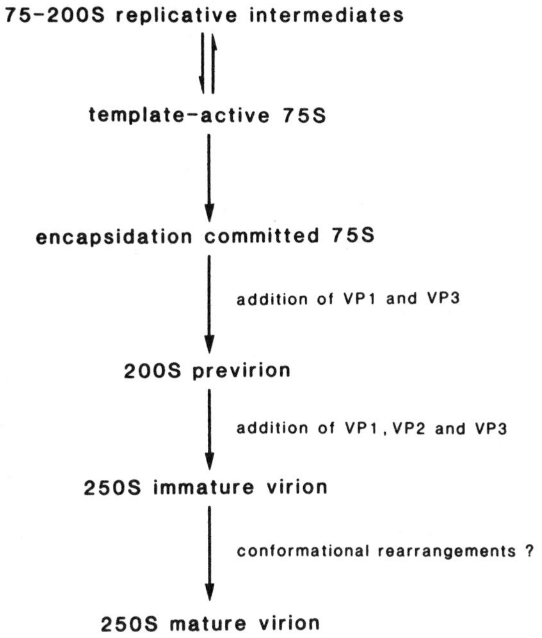

Figure 15. Model for SV40 virus assembly adapted from Garber et al. (1980).

and low levels of capsid proteins, and about 10 copies of VP1 and one copy of VP3 per nucleoprotein complex. The viral proteins do not appear to be essential for maintenance of the compact structure because temperature sensitive mutants that lack VP1 and VP3 produce nucleoprotein complexes that have the same properties as the wild-type SV40 but lack VP1 and VP3 (Garber et al., 1980).

The 75S complex can function as an active template for DNA replication giving rise to a heterogeneous population of faster sedimenting replicative intermediates where the faster sedimenting forms contain the more extensively replicated molecules. Upon completion of replication and segregation of the daughter strands a 75S gapped circular intermediate (Laipis et al., 1975) is produced which is then converted to a 75S nucleoprotein complex containing closed circular DNA. Some of these templates are chosen for another round of replication. Other complexes become assembly committed. It is not known what controls commitment. Addition of VP1 and VP3 to the 75S complex produces a 200S intermediate previrion. Further addition of VP1 and VP3 together with VP2 forms a high salt unstable 250S intermediate (Seidman et al., 1979; Baumgartner et al., 1979; Jakobovits and Aloni, 1980). The steps involved in final maturation are unknown. It has been suggested that for SV40 this could involve intersubunit sulfhydryl crossbridges (Etchison and Walter, 1977).

9. IMPLICATIONS OF THE POLYOMA VIRUS STRUCTURE FOR VIRAL ASSEMBLY

The structure of the polyoma virus was unexpected. After the comprehensive theory of quasi-equivalence by Caspar and Klug it was presumed that the structure and assembly of viruses were, in principle at least, solved. Implicit in these ideas was the concept of self-assembly by the protein subunit into a viral capsid of proper size. This concept has required modification in order to encompass the complex assembly pathways of the bacteriophages so that "self-assembly" is more appropriately replaced by "regulated assembly" (Wood and King, 1979). However, the final organization of the proteins in the viral capsid was still expected to adopt a close-packed arrangement on a triangulated icosahedral surface lattice. The structure of the polyoma virus capsid now raises questions not only about the assembly but the logic underlying the organization of the subunits.

Given the structure of the polyoma virus capsid it is perhaps appropriate to reexamine the theory of quasi-equivalence. In an earlier section it was noted that on a triangulated icosahedral surface lattice the protein subunits fall into "T" groups of 60 where T is the triangulation number for the lattice. Each of these groups has a chemically distinct environment on the surface of the icosahedron. In the original formulation of the theory of quasi-equivalence, the reason for postulating a triangulated icosahedral arrangement of protein subunits was that it would allow a subunit to experience an environment that was similar to that sensed by other groups of nonequivalent subunits. This arrangement it was felt would lead to a conservation of bonding specificity in which the interactions between molecules would be similar, representing perhaps small deviations from one another. The extent of the differences in environment is hard to assess since it depends on the thickness of the protein subunits that constitute the shell. Caspar and Klug noted that a quasi-equivalent arrangement of subunits has a lower energy than the related $T = 1$ arrangement of subunits and that this latter arrangement is not seen to coexist with the larger shell. They postulated that "bond" deformation lowered the energy of the larger aggregate.

The studies on the small $T = 3$ plant viruses show that the different environment of the three groups of protein subunits is accomplished by using several chemically distinct subunit–subunit interactions rather than deformation of a more general interaction (Rossmann et al., 1983). Some interactions are conserved for all three subunits, for example, the bonding around the local threefold axis, whereas others involved with the icosahedral and local twofold axes are essentially different. These results might appear to be an extension of the original suggestion of Caspar and Klug; however, in reality they imply that the organization is not specified by the principles of quasi-equivalence. The existence of distinctly different interactions between identical protein subunits removes the necessity for proteins to be arranged on a triangulated icosahedral surface lattice. However, it must be noted that the quasi-equivalent arrangement of protein subunits is still an efficient way of building a protein shell. The parameters that dictate the final organization must be the regulation of the assembly pathway and packing of the subunits to form a closed shell.

It is clear why distinct intermolecular subunit interactions are observed in the high resolution virus structures in place of bond deformations. The interactions between subunits consist of hydrogen bonds and salt bridges for the hydrophilic regions of the protein and hydrophobic interactions for the nonpolar portions of the subunit surface. Such interactions have a specific directionality, organization, and distinct energy minimum. Deformation of this set of interactions would lead to an increased energy of the aggregate. What is observed in place of this deformation is a conformational switching between alternative sets of interactions representing distinct energy minima. The concept of conformational switching is not new. It was first introduced by Caspar and Cohen (1969) to account for the polymorphism of tropomyosin. It was defined more explicitly to account for the structure and assembly of bacterial flagella (Asakura, 1970) and tail sheath contraction of bacteriophage T4 (Moody, 1973). The importance of conformational switching in self-assembly of macromolecular aggregates has been reviewed by Caspar (1976).

The importance of regulation in the assembly of viruses has been elegantly shown by studies on the small bacteriophages such as T4, λ, or P22 (Wood and King, 1979). It is also intuitively clear from study of the small plant virus structures. In the $T = 3$ plant viruses the 180 proteins fall into three groups of 60. Each group occupies a chemically distinct environment on the surface of the virus. The viral protein itself does not assemble to form a closed shell in the absence of RNA. The steps or sequence involved in the formation of the viral shell are not known; however, in the assembly process the protein subunits must be able to distinguish which set of interactions to use, or, in other words, the information necessary for assembly must be carried by the protein itself.

In the case of the larger bacteriophages such as P22 it is known that assembly occurs via a precursor prohead composed of the major coat protein and a scaffold protein. In the maturation process the DNA is inserted into the prohead with concurrent loss of the scaffold protein (King and Casjens, 1974). This process also involves a substantial structural rearrangement since the prohead expands from a diameter of 550–600 Å on loss of the scaffold protein and injection of the DNA (Earnshaw et al., 1976). The prohead is known to contain about 420 copies of the major coat protein and about 240 copies of the scaffold. The surface of the bacteriophage shows morphological features that are consistent with $T = 7$ icosahedral symmetry. The assembly pathway for the prohead is unknown. It is clear that the protein must occupy seven chemically distinct sites if a true $T = 7$ icosahedral surface lattice is formed. The coat protein by itself may not carry enough information to correctly choose which conformation to adopt. The purpose of the scaffold protein may be to cause the coat protein to aggregate in a different conformation from its normal state producing a structural differentiation between the coat protein subunits in the shell. In this way the scaffold protein would simplify the assembly pathway by dictating the conformation of its associated coat protein.

In the case of polyoma virus it appears that the close packing offered by the quasi-equivalent arrangement has not been adopted. This may have occurred in order to simplify the assembly process or to allow self-assembly to occur at all. There are two factors that may influence the structural motif adopted by polyoma. First, polyoma contains a very limited amount of genetic material; second, the major coat protein appears to exist in solution as tightly associated pentameric capsomeres.

The polyoma genome is very compact and therefore does not carry enough genetic material to code for a complex assembly pathway. The DNA codes for only three viral proteins involved in virus assembly. This is in contrast to the assembly pathways observed for the bacteriophages. For example, in bacteriophage P22 there are at least 11 proteins involved in capsid assembly, DNA injection, and phage maturation (Poteete and King, 1977). As noted earlier, in order to construct a true $T = 7$ triangulated surface lattice it is necessary to specify seven distinct modes of interaction for the protein subunit. For polyoma there may have been insufficient genetic information present to have allowed this pathway to develop.

The aggregation state of the viral protein in solution is very important because in order to form a $T = 7$ lattice it is necessary to construct both pentamers and hexamers of the major coat protein. This necessitates, at some point in the assembly, the existence of small monomeric or dimeric states of the major coat protein. In solution there is no evidence for the existence of monomers of VP1 in dissociated capsids or virions. This evidence supports an assembly process that utilizes preformed pentameric aggregates of the major coat protein which are added to a nucleoprotein complex. The structure itself favors this type of arrangement. As noted earlier, the electron density distribution is consistent with a two domain protein. One domain could form the protruding part of the morphological unit while the other could participate in interactions in the shell. In this respect the domain arrangement is similar to that seen in the high resolution structure of tomato bushy stunt virus (Harrison et al., 1978) in which the interactions between the protruding domains is conserved for all subunits. A flexible hinge connects these domains to the inner shell, so that there are no rigid domain–domain interactions between the protruding and shell domains. The interactions between the shell domains are modified by the variable orientation of the N-terminal portion of the polypeptide chain. This enables the protein subunits to have three stable but distinct chemical environments. The key features of the protruding domains in TBSV are the conserved interactions between twofold related domains and the flexible hinge region.

It is possible that the protruding domain in polyoma virus participates in a conserved interaction with four other protruding domains to form a fivefold symmetric pentameric aggregate. The protruding domain itself could then be connected to the shell domain by a flexible hinge. In this case the nonequivalent interactions in the shell would only involve the shell domains. Since nonequivalence is a necessity in the formation of a shell from more than 60 subunits, the use of a conserved interaction between the subunits could simplify the assembly pathway and result in fewer nonequivalent types of subunit–subunit interactions.

It might be expected that wherever a domain protrudes from the surface and interacts with the shell through a flexible hinge, the interactions between the protruding units will be conserved. A protruding "free-standing" domain connected only by a single polypeptide chain to the inner protein shell cannot be expected to associate with different rotational symmetries unless there is some mechanism for stabilizing each state. A change in the rotational association of five subunits to six subunits must entail some compensatory chain movement, which would not be possible if there were only one linkage between the shell and protruding domain.

The structural arrangement of the major coat protein on the surface of polyoma virus may be the result of minimizing the number of different types of interactions within the shell. At this stage nothing is known about the differentiation between the six structurally independent copies of the protein subunit. It is of interest to note that for both SV40 and polyoma there is microheterogeneity in the major coat protein. This is not caused

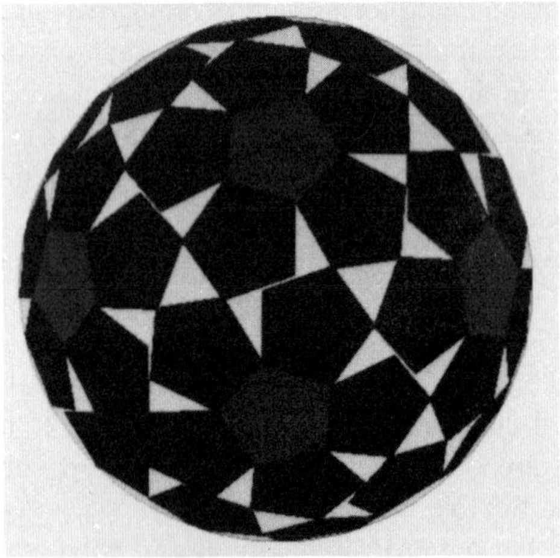

Figure 16. A spherical shell covered by pentagons which shows that a framework of 60 pentamers can be obtained by allowing distinct interactions between neighboring pentamers across the icosahedral twofold axes and around the icosahedral threefold axes.

by proteolytic cleavage but could be the result of phosphorylation or glycosylation. For SV40 it was found that the extent of microheterogeneity was the same in the previrion, immature, and mature virion. From this it was suggested that the small differences in the major coat protein were not important in the assembly. However, if the virus does assemble by adding pentameric aggregates to the nucleoprotein complex the heterogeneity in the major coat protein may be introduced to generate the correct asymmetry in the pentamer thus allowing an icosahedral shell to form. Figure 16 shows a spherical shell covered by pentagons. This illustrates that the interactions necessary for the self-assembly of such an arrangement are one twofold and one threefold interaction for each of the 60 hexavalent pentamers. Thus, it is possible that the role of the microheterogeneity is to specify the correct recognition and orientation of the assembling pentamers.

10. CONCLUSION

The structural studies on animal viruses by X-ray diffraction, although preliminary, illustrate the many unanswered questions concerning the structure and assembly of large macromolecular systems. The results for polyoma virus suggest that it may be important to reexamine the accepted principles of virus design. These questions can only be answered by experimentation. The studies currently in progress suggest that the next few years should be a particularly exciting era in our understanding of virus structure and assembly.

Acknowledgments. I wish to thank Dr. D. L. D. Caspar for continued support and encouragement. I also thank Drs. D. DeRosier, T. S. Baker, and W. T. Murakami for helpful discussions. This work was supported by a Young Investigators Research Grant CA27260 and NCI Grant CA15468 to D. L. D. Caspar.

REFERENCES

Abad-Zapatero, C., Abdel-Meguid, S., Johnson, J. E., Leslie, A. G. W., Rayment, I., Rossmann, M. G., Suck, D., and Tsukihara, T. (1980). *Nature (London)* **286,** 33–39.

Adolph, K. W., Caspar, D. L. D., Hollingshead, C. J., Lattman, E. E., Phillips, W. C. and Murakami, W. T. (1979). *Science* **203,** 1117–1119.

Anderer, F. A., Schlumberger, H. D., Koch, M. A., Frank, H., and Eggers, H. J. (1967). *Virology* **32,** 511–523.

Asakura, S. (1970). *Adv. Biophys.* **1,** 99–155.

Baker, T. S., Caspar, D. L. D., and Murakami, W. T. (1983). Nature (London) **303,** 446–448.

Baumgartner, I., Kuhn, C., and Fanning, E. (1979). *Virology* **96,** 54–63.

Bricogne, G. (1974). *Acta Crystallogr. A* **30,** 395–405.

Bricogne, G. (1976). *Acta Crystallogr. A* **32,** 832–847.

Caspar, D. L. D. (1956). *Nature (London)* **177,** 475–477.

Caspar, D. L. D. (1976). Proc. 3rd *John Innes Symp.,* pp. 85–99.

Caspar, D. L. D., and Cohen, C. (1969). *Symmetry and Function of Biological Systems at the Macromolecular Level,* A. Engström and B. Strandberg, Eds. Wiley, New York. *Nobel Symp.* **11,** 393–414.

Caspar, D. L. D., and Klug, A. (1962). *Cold Spring Harbor Symp. Quant. Biol.* **27,** 1–24.

Coca-Prados, M., and Hsu M. T. (1979). *J. Virol.* **31,** 199–208.

Crick, F. H. C., and Watson, J. D. (1956). *Nature (London)* **177,** 473–475.

Deininger, P., Esty, A., LaPorte, P., and Friedmann, T. (1979). *Cell* **18,** 771–779.

Earnshaw, E., Casjens, S., and Harrison, S. C. (1976). *J. Mol. Biol.* **104,** 387–410.

Erickson, J. W., Frankenberger, E. A., Rossmann, M. G., Fout, G. S., Medappa, K. C., and Ruekert, R. R. (1983). *Proc. Natl. Acad. Sci. U.S.A.,* **80,** 931–934.

Etchison, D., and Walter, G. (1977). *Virology* **77,** 783–796.

Fiers, W., Contreras, R., Haegeman, G., Rogiers, R., Van de Voorde, A., Van Heuverswyn, H., Van Herreweghe, J., Volckaert, G., and Ysebaert, M. (1978). *Nature (London)* **273,** 113–120.

Finch, J. T. (1974). *J. Gen. Virol.* **24,** 359–364.

Finch, J. T., and Klug, A. (1959). *Nature (London)* **183,** 1708–1714.

Finch, J. T., and Klug, A. (1965). *J. Mol. Biol.* **13,** 1–12.

Friedmann, T. (1971). *Proc. Natl. Acad. Sci. U.S.A.* **68,** 2574–2578.

Garber, E. A., Seidman, M. M., and Levine, A. J. (1980). *Virology* **107,** 389–401.

Gissmann, L., and Zur Hausen, H. (1976). *Proc. Natl. Acad. Sci. U.S.A.* **73,** 1310–1313.

Gissmann, L., Pfister, H., and Zur Hausen, H. (1977). *Virology* **76,** 569–580.

Griffith, J. D. (1975). *Science* **187,** 1202.

Harrison, S. C., Olson, A. J., Schutt, C. E., Winkler, F. K., and Bricogne, G. (1978). *Nature (London)* **276,** 368–373.

Hogle, J. M. (1982). *J. Mol. Biol.* **160,** 663–668.

Holland, J. J., and Kiehr, E. D. (1968). *Proc. Natl. Acad. Sci. U.S.A.* **60,** 1015–1022.

Jacobsen, M. F., and Baltimore, D. (1968). *Proc. Natl. Acad. Sci. U.S.A.* **61,** 77–84.

Jakobovits, E. B., and Aloni, Y. (1980). *Virology* **102,** 107–118.

Johnson, J. E. (1978). *Acta Crystallogr. B* **34,** 576–577.

King, J., and Casjens, S. (1974). *Nature (London)* **251,** 112.

Kiselev, N. A., and Klug, A. (1969). *J. Mol. Biol.* **40,** 155–171.

Kitamura, N., Semler, B. L., Rothberg, P. G., Larsen, G. R., Adler, C. J., Dorner, A. J., Emini, E. A., Hanecak, R., Lee, J. J., van der Werf, S., Anderson, C. W., and Wimmer, E. (1981). *Nature (London)* **291,** 547–553.

Klug, A. (1965). *J. Mol. Biol.* **11,** 424–431.

Klug, A., and Finch, J. T. (1965). *J. Mol. Biol.* **11,** 403–423.

Laipis, P. J., Sen, A., Levine, A. J., and Mulder, C. (1975). *Virology* **68,** 115–123.

Lattman, E. E. (1980). *Science* **208,** 1048–1050.

Liljas, L., Unge, T., Jones, T. A., Fridborg, K., Lovgren, S., Skoglund, U., and Strandberg, B. (1982). *J. Mol. Biol.* **159,** 93–108.

Maizel, J. V., and Summers, D. F. (1968). *Virology* **36,** 48–54.

Mattern, C. F. T. (1962). *Science* **137,** 612–613.

Moody, M. F. (1973). *J. Mol. Biol.* **80,** 613–635.

Murakami, W. T., and Schaffhausen, B. S. (1974). "Molecular Studies in Viral Neoplasia," 25th Annual Symposium on Fundamental Cancer Research, 1972, at the University of Texas, M. D. Anderson Hospital and Tumor Institute, Williams and Wilkins, Baltimore, Maryland, pp. 43–62.

Ozer, H. L., and Tegtmeyer, P. (1972). *J. Virol.* **9,** 52–60.

Poteete, A. R., and King, J. (1977). *Virology* **76,** 725–739.

Rayment, I. (1983). *Acta Crystallogr. A* **39,** 102–116.

Rayment, I., Baker, T. S., Caspar, D. L. D., and Murakami, W. T. (1982) *Nature (London)* **295,** 110–115.

Rayment, I., Baker, T. S., and Caspar, D. L. D. (1983). *Acta Crystallogr.* B 39, 505–516.

Reddy, V. B., Thimmappaya, B., Dhar, R., Subramanian, K. N., Zain, B. S., Pan, J., Ghosh, P. K., Celma, M. L., and Weisman, S. (1978). *Science* **200,** 494–502.

Rossmann, M. G., Abad-Zapatero, C., Hermodson, M. A., and Erickson, J. W. (1983). *J. Mol. Biol.,* **166,** 37–83.

Rueckert, R. R. (1976). *Comp. Virol.* **6,** 131–213.

Schaffer, F. L., and Schwerdt, C. E. (1955). *Proc. Natl. Acad. Sci. U.S.A.* **41,** 1020–1023.

Seidman, M., Garber, E., and Levine, A. J. (1979). *Virology* **95,** 256–259.

Soeda, E., Arrand, J. R., Smolar, N., Walsh, J. E., and Griffin, B. E. (1980). *Nature (London)* **283,** 445–453.

Steere, R. L., and Schaffer, F. L. (1958). *Biochim. Biophys. Acta* **28,** 241–246.

Tooze, J. (1981). *Biology of Tumor Viruses Part 2/Revised: DNA Tumor Viruses.* Cold Spring Harbor Laboratory, New York.

Wildy, P., Stoker, M. G. P., Macpherson, I. A., and Horne, R. W. (1960). *Virology* **11,** 444–457.

Wood, W. B., and King, J. (1979). *Comprehensive Virol.* **13,** 581–633.

8

The Hemagglutinin Membrane Glycoprotein of Influenza Virus

D. C. WILEY
Department of Biochemistry and Molecular Biology
Harvard University
Cambridge, Massachusetts

I. A. WILSON
Department of Molecular Biology
Research Institute of Scripps Clinic
La Jolla, California

J. J. SKEHEL
National Institute for Medical Research
London, England

CONTENTS

1. INTRODUCTION

Influenza virus, like a number of other enveloped viruses, contains a lipid membrane that it obtains during maturation by budding from the plasma membrane of an infected cell (Fig. 1). The membrane contains cellular lipids, but the membrane proteins are encoded by the virus. These pro-

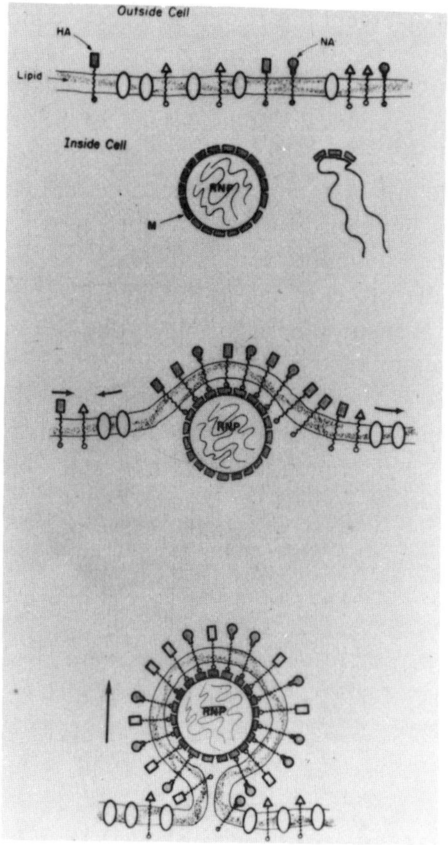

Figure 1. Virus budding. Viral glycoproteins are inserted into the cellular membrane by cellular compartmentalization mechanisms. Budding is envisioned as the transmembrane recognition of the nucleocapsid by the cytoplasmic portion of the membrane glycoprotein's anchor peptide. This specific transmembrane recognition and/or protein–protein contact in the external glycoprotein could drive the envelopment of the nucleocapsid by the membrane.

teins are inserted in the cellular membrane by the same "signal peptide" mediated process employed for compartmentalizing cellular membrane and secretory proteins. In the case of influenza virus two glycoproteins are present in the membrane: a hemagglutinin and a neuraminidase. The hemagglutinin (HA) binds to a sialic acid containing receptor on target cells to initiate infection and mediates the membrane fusion event required to transfer the viral nucleocapsid into the cytoplasm. Recurrent epidemics of respiratory disease in humans are occasioned by changes in the antigenicity of the influenza hemagglutinin. With the aim of establishing mechanisms for these activities, the structure of the hemagglutinin from the 1968 Aichi (68) Hong Kong influenza virus was determined.

The HA molecule is an integral membrane protein, which can be released intact from the lipid bilayer only by detergent solubilization of the membrane. In excess nonionic detergent this results in an active molecule embedded not in a lipid bilayer but instead in a detergent micelle (for detergent review, see Helenius and Tanford, 1978). The HA–detergent complex is stable, monodisperse, and water soluble (Fig. 2*a*). Efforts to crystallize that complex have been unsuccessful, although the recent success in crystallizing protein–detergent complexes (Garavito and Rosenbusch, 1980; Michel, 1982) has stimulated further attempts. Removal of detergent from the HA–detergent complex results in reassembly of lipid vesicles and reinsertion of the proteins unless the lipid has been removed, in which case protein–protein micelles result by the self-aggregation of the hydrophobic proteins of the HA molecules (Fib. 2*b*). These protein–protein micelles are heterogeneous aggregates unsuitable for crystallization. Brand and Skehel (1972) demonstrated that protease (bromelain) treatment of influenza membranes resulted in the release of

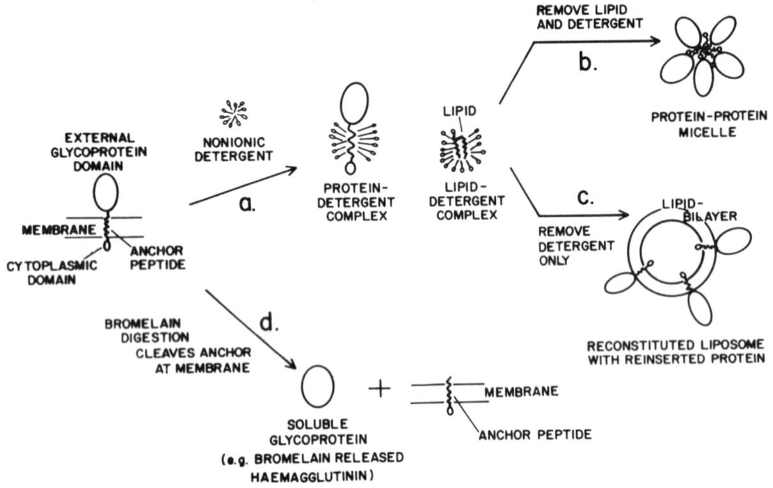

Figure 2. The isolation of membrane proteins. (*a*) Excess nonionic detergent solubilizes membrane proteins by interacting with the hydrophobic (wavy line) parts of the proteins and lipids. (*b*) Removal of the detergent in the absence of lipid results in protein–protein micelles, formed by association of the hydrophobic anchor peptides. (*c*) Removal of the detergent in the presence of lipid results in the reconstitution of lipid bilayer vesicles and the reinsertion of the proteins into the artificial membrane. (*d*) In some cases treatment of membrane proteins with proteases removes the hydrophobic anchor peptide near the membrane interface, releasing a soluble and active fragment of the glycoprotein. The case of the influenza virus hemagglutinin released by bromelain is illustrated.

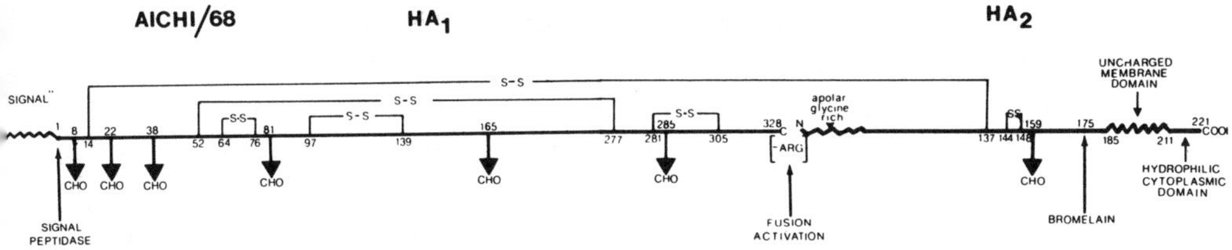

Figure 3. Schematic drawing of the 1968 Hong Kong influenza virus hemagglutinin showing the external domain (ends 185 HA_2), the membrane anchoring domain (185–211), and the cytoplasmic domain (212–221). The signal sequences, S–S bridges, CHO-attachment sites, and "fusion peptide" (see text) are shown.

a soluble glycoprotein (BHA), lacking a 47 amino acid C-terminal peptide, which anchors the HA to the membrane (Fig. 2*d*). This establishes the HA as a member of a class of glycoproteins that have terminal hydrophobic peptide anchors, which by spanning the membrane hold the protein in the lipid bilayer. The class contains, for example, all known viral membrane glycoproteins, glycophorin of the erythrocyte, and the histocompatibility antigens. These molecules all display surface enzyme activities or take part in recognition events at membranes. The second class of membrane proteins, that which forms the transport "pumps" and channels in membranes, by contrast, has a large domain embedded in the lipid membrane. In one case, bacteriorhodopsin, that domain has been shown to contain seven α-helices spanning alternately back and forth across the lipid (Henderson and Unwin, 1975).

The amino acid sequences of the hemagglutinin has been determined or deduced for many different strains and subtypes of virus by a number of laboratories. HA from 27 strains of virus have been sequenced, by protein and nucleic acid methods since the first sequence by Porter et al. in 1979. The HA from the 1968 Hong Kong strain of human influenza is synthesized as a single polypeptide chain of 550 amino acids which is post translationally cleaved by the removal of arginine 329 into two disulfide linked chains, HA_1 and HA_2 (Fig. 3). HA_1 begins with a hydrophobic signal peptide, which is cleaved off by signal peptidase during membrane insertion. HA_2 also begins with a hydrophobic sequence, which is the longest stretch of highly conserved residues. Membrane fusion requires this peptide and the cleavage between HA_1 and HA_2.

Twenty-seven residues near the C terminus of HA_2, 185–211, are uncharged and primarily hydrophobic. They form the hydrophobic anchor that spans the viral membrane. This segment is probably an α-helix for two reasons: (1) 27 amino acids of an α-helix would traverse a lipid bilayer, and (2) an α-helix is a structure that a single chain can adopt that satisfies all the hydrogen bonding capacity of the carbonyl oxygens and amide nitrogens in the main chain and therefore would not place unpaired polar atoms in the hydrophobic environment of the lipid bilayer. The oxygen atoms of serine and threonine residues found in anchor segments would also be expected to form hydrogen bonds to main chain atoms, as is often observed in the α-helices of soluble proteins. Residues 212–221 are on the cytoplasmic side of the membrane and presumably involved in the budding process (Fig. 1). The polypeptide molecular weight of HA_1 is 36,334, and HA_2 is 25,750, with an additional 13,000 daltons (~17%) of carbohydrate. Seven Asn-linked oligosaccharide attachment sites (Asn-X-Ser/Thr) are distributed along the extracellular portion of the sequence. Molecular weight measurements and chemical cross-linking have established the oligomeric structure of the hemagglutinin as a trimer of HA_1 + HA_2 units (Wiley et al., 1977b) in all environments including, on cellular membranes, the virus membrane, isolated in detergent, or bromelain released.

The soluble BHA molecule was chosen for structural study to circumvent difficulties with nonspecific aggregation encountered with the intact HA. It was argued that a protein like the HA which exists in three distinct environments would be folded into three independently stable "domains," one in the extracellular space, one in the lipid bilayer, and one in the cytoplasm (Wiley and Skehel, 1977). It should therefore be possible to cut the protein into these "domains" and study their structures separately. Circular dichroism (Wiley et al., 1977a), antibody binding (Brand and Skehel, 1972) and receptor binding (Skehel et al., 1982) studies confirm the assumptions that the external (BHA) "domain" is independently stable, active, and indistinguishable from that portion of the intact hemagglutinin. The proteolytic dissection of molecules into "stable domains" has been used previously in structural studies: two examples being the immunoglobulin structure (reviewed by Amzel and Poljak, 1979) and the inner membrane protein cytochrome b_5 (Mathews et al., 1971).

2. CRYSTALLIZATION

BHA crystals were grown from a variety of conditions (Wiley et al., 1977a) from HA prepared from influenza virus grown in embryonated chicken eggs. 1 mm × 1 mm × 1 mm crystals were routinely prepared by microdialysis (50 μl) of 40 mg/ml protein solution in 0.15 *M* sodium chloride against 1.32 *M* sodium citrate, 0.1% sodium azide, pH 7.5. The space group is $P4_1$, with $a = b = 163.2$ Å and $c = 177.4$ Å (Wiley and Skehel, 1977). The crystals are unusual in containing only 22% by weight protein. The asymmetric unit contains one BHA trimer of 208,422 daltons. A single heavy atom derivative was used in this study: a mercury adduct of the arginine reagent phenyl glyoxal synthesized by Monaco (1978). This reagent was expected to covalently label arginines with two phenyl glyoxal moieties (Takahashi, 1977) but was found after the structure was solved to be bound at single sites remote from arginine residues.

3. X-RAY DATA COLLECTION

X-ray data were collected on 1° oscillation photographs exposed for 12 hr at 4°C [Elliot GX-6 rotating anode, 100 μ^2 focus, Franks mirror optics (Harrison, 1968), C. Supper Co. automatic oscillation camera]. Two successive pictures were recorded from one location on the crystal after which the crystal was moved to expose a fresh volume. In the best cases ten-twenty 1° ranges could be collected from one crystal. Photographs were obtained from two settings of the crystal: (1) 110 into the beam and 110 on the oscillation axis, oscillation range 0°–30° and 70°–90°; (2) 110 into the beam and 001 on the the oscillation axis, oscillation range 0° – ±30°. This data collection strategy was partially dictated by the octahedral shape of the crystals with 110 and 001 type axes as vertices that masked the direction of the unique axis. By chance, two of three crystals mounted would be setting (1), and one of the three, setting (2). A 50° range of setting (1) is sufficient to record all the unique data. Approximately one-half as many photographs which resulted from the setting (2) were used to scale the data and provide accurate measurements for reflections near the oscillation axis on setting (1). A computer program written by C. Schutt assisted in the development of this strategy. A total of 408,000 native (78 films) and 338,000 (62 films) derivative reflections were collected to 3 Å resolution yielding 83,000 (of 93,000) and 78,000 unique reflections, respectively. Symmetry *R* factors were 0.11 and 0.13 on fully recorded reflections. Partially recorded reflections (>50%) were scaled into the data by a post refinement procedure (Winkler et al., 1979) yielding final *R* factors of 0.18 and 0.20 for the 3 Å data set.

4. HEAVY ATOM LOCATION

A 3 Å resolution difference Patterson was solved for six nearly equally occupied sites by automated symmetry minimum function searches of the Harker sections for potential single sites and automated cross vector searches for all possible pairs of single sites. The resulting six atom solution was confirmed by a higher order image seeking function using one known atom, by a difference Fourier, and by SIR refinement (Table 1). The six mercury atoms formed two triangles, which established the directions and location of the noncrystallographic threefold symmetry axis. Rotation function calculations were initially misleading as the major peak resulted from packing vectors. A smaller but significant peak was consistent with the noncrystallographic threefold symmetry indicated by the heavy atom positions (Wilson et al., 1980).

5. NONCRYSTALLOGRAPHIC SYMMETRY PHASE REFINEMENT

A 3 Å SIR phased Fourier map based on 59,669 reflections was calculated and threefold averaged about the noncrystallographic symmetry axis (Fig. 4*a* and 4*b*). The averaged map revealed the molecular envelope, which was traced from the electron density map on an Evans and Sutherland Picture System (with a program written by Craig Steele and Paul Kuttner). Noncrystallographic phase averaging using the algorithm and programs of Bricogne (1976) converged after 11 cycles, during which the R factor dropped from 47 to 27% (Table 2, Fig. 4*c*). Difference electron density

Table 1. Data Processing Statistics[a]

Data Set	No. Films	Total Reflections Measured	Unique Reflections	R_{sym}	R_{whole}	R_{asym}
Native	78	408,672	83,015	0.12	0.11	0.18
Derivative	62	338,738	78,077	0.13	0.13	0.20

[a] $R = \sum | s_i I_{hi} - \bar{I}_h | / \sum \bar{I}_n$ where s_i = scale factor for crystal i; I_{hi} = intensity of reflection h of crystal i; $\bar{I}_h$ = average intensity of reflections. R_{sym}, for symmetry related reflections on the same film. R_{whole}, for fully recorded reflections from different films. R_{asym}, for all reflections (≥50% recorded) on different films, final scaling.

(a) 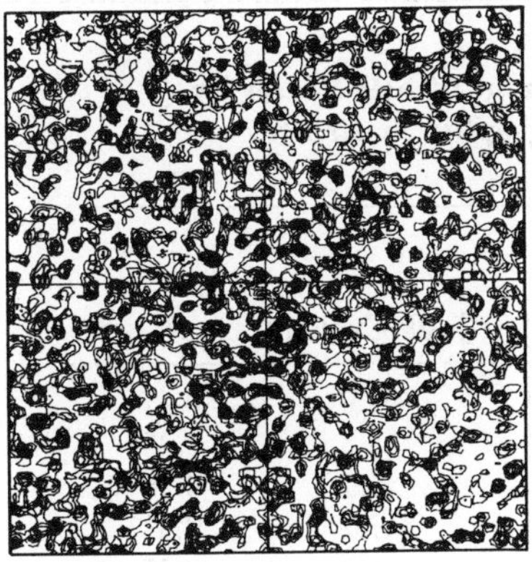

CYCLE 0 UNAVERAGED

(b) 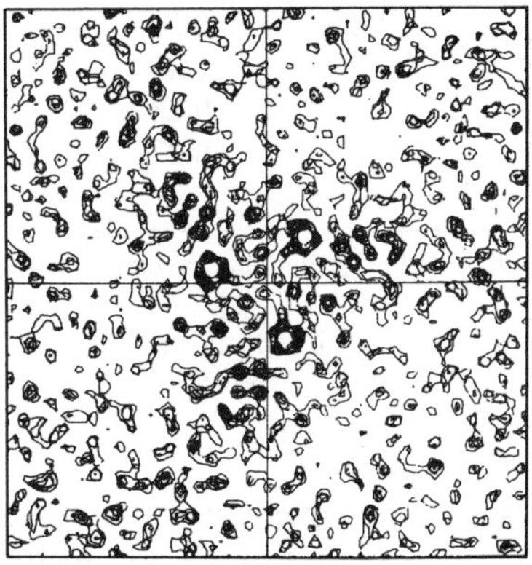

CYCLE 1 AVERAGED

(c) 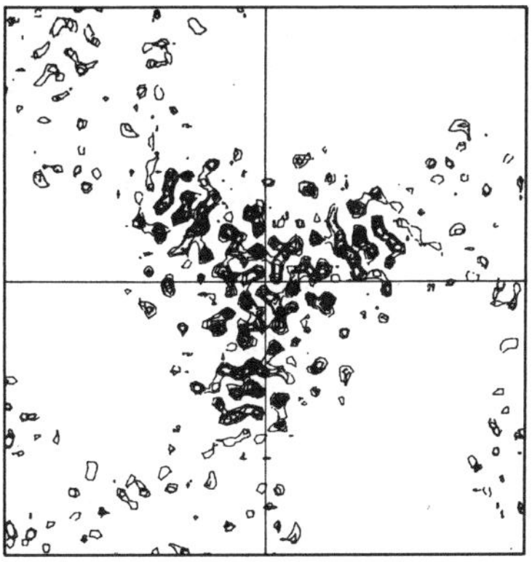

CYCLE 11 FINAL

Figure 4. Electron density maps. All maps are viewed down the molecular threefold axis and contoured at comparable levels. (*a*) SIR phased unaveraged electron density map. (*b*) SIR map real space averaged about the molecular threefold symmetry axis from which an approximate molecular envelope is evident. (*c*) Electron density map after eleven cycles (see Table 2) of real space phase averaging using noncrystallographic symmetry and solvent flattening.

Table 2. Phase Refinement of SIR Phases by Threefold Noncrystallographic Symmetry Averaging (Bricogne)

Cycle No. at 3 Å Resolution[a]	⟨| ΔΦ Overall |⟩[b]	⟨| ΔΦ Cycle |⟩[b]	R^c
+1	67.0	67.0	0.47
+2*	69.1	15.5	0.40
3	70.4	9.7	0.36
4	71.5	7.1	0.33
5	72.4	5.6	0.32
6	73.1	4.6	0.31
7*	73.6	5.4	0.30
8	74.1	3.6	0.29
9*	74.5	7.4	0.28
0	74.9	3.7	0.28
11	74.9	2.9	0.27
1			

[a] +, heavy atom cancel records included; *, molecular envelope recalculated.

[b] ⟨ΔΦ⟩ = mean phase change in degrees for all reflections.

[c] $R = \frac{\sum \left| |F_{obs}| - |F_{calc}| \right|}{\sum |F_{obs}|}$

maps $(F_o - F_c)\, e^{i\phi c}$ were examined during the refinement and revealed the need to redraw the envelope at cycles 2, 7, and 9 to include protruding regions of electron density and minimize molecular overlap. The SIR phases were dropped after the first cycle in favor of the "averaged" phases. For the first two cycles the electron density near heavy atom sites was modified to the local average value to "fill in" the perturbations around the heavy atom sites due to inadequate scaling and phasing. In the late stages of refinement native amplitudes not previously phased by SIR were included and phased.

It seems clear from the refinement statistics (Table 2) that convergence would have been reached in only a few cycles if the accurate final envelope could have been determined from the initial averaged SIR maps. The final map based on 74,194 reflections to 3 Å resolution was interpreted on a minimap of 4 Å/cm with the use of the amino acid sequence kindly provided by Ward and Dopheide (1980) and Verhoeyen et al. (1980). The atomic coordinates were determined including oligosaccharide chains for the entire monomer by molecular model building on the Evans and Sutherland Picture System II, with the graphics program BILDER (R. Diamond) which was extensively modified by R. Ladner.

6. DESCRIPTION OF THE STRUCTURE

The influenza HA trimer is 135 Å long (from membrane to tip) and approximately triangular in cross section varying in radius from 15 to 40 Å (Fig. 5). In the monomer (Fig. 6) there is a long, fibrous "stemlike" region

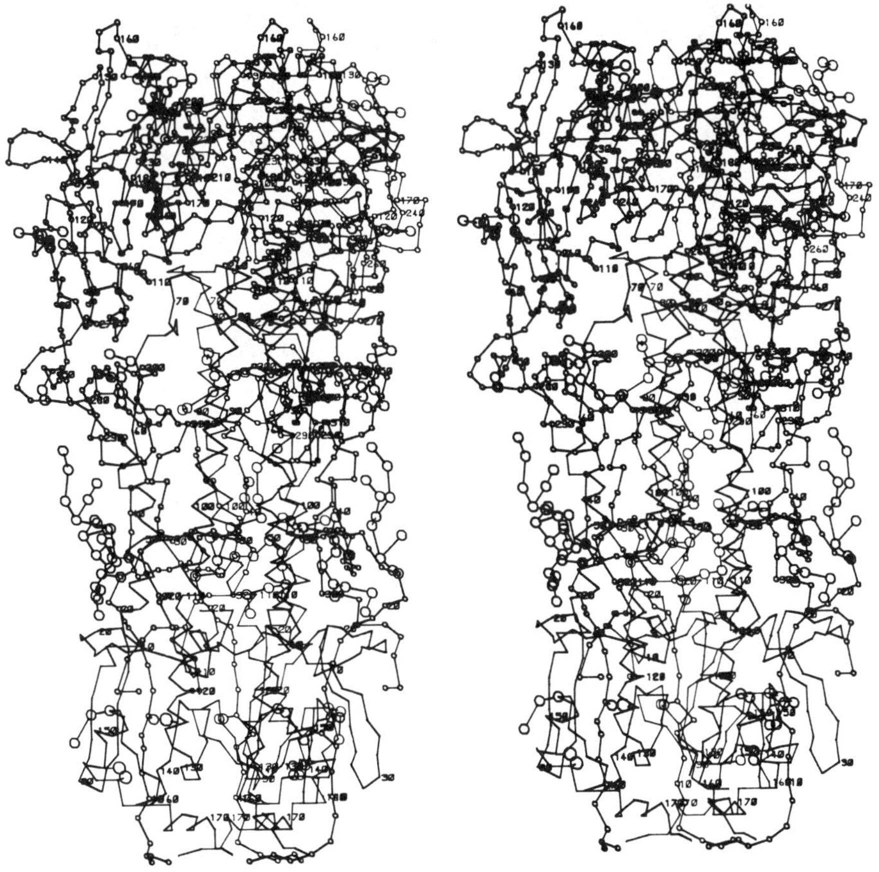

Figure 5. Stereo view of a carbon α-chain tracing of the hemagglutinin trimer. The membrane end of the molecule is at the bottom.

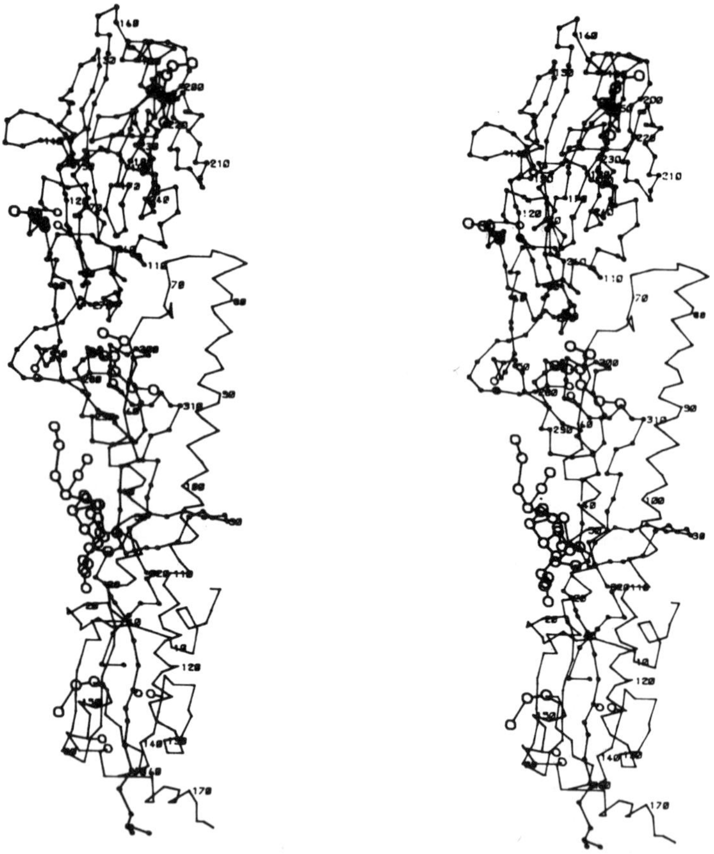

Figure 6. Stereo drawing of the α-chain tracing of an influenza hemagglutinin monomer: ●, HA_1; −, HA_2, ○, hexose. From Wilson et al. (1981).

extending outward from the membrane with a globular region at its tip. The globular region contains only residues of HA_1 while the fibrous "stem" contains residues from HA_1 and all of HA_2.

The unusually extended conformation of the molecule can be appreciated by following the schematic drawing of its chain tracing in Fig. 7. The amino terminus (HA_1) is found at the extreme membrane end of the molecule. The first 63 amino acids reach in a nearly extended structure 96 Å along the length of the molecule before the first compact fold (Fig. 7*a*). The globular region at the apex is an eight-stranded antiparallel β-sheet from residue 116 to 261 (Fig. 7*c* and *d*; also Fig. 8*b*). Similar "Swiss

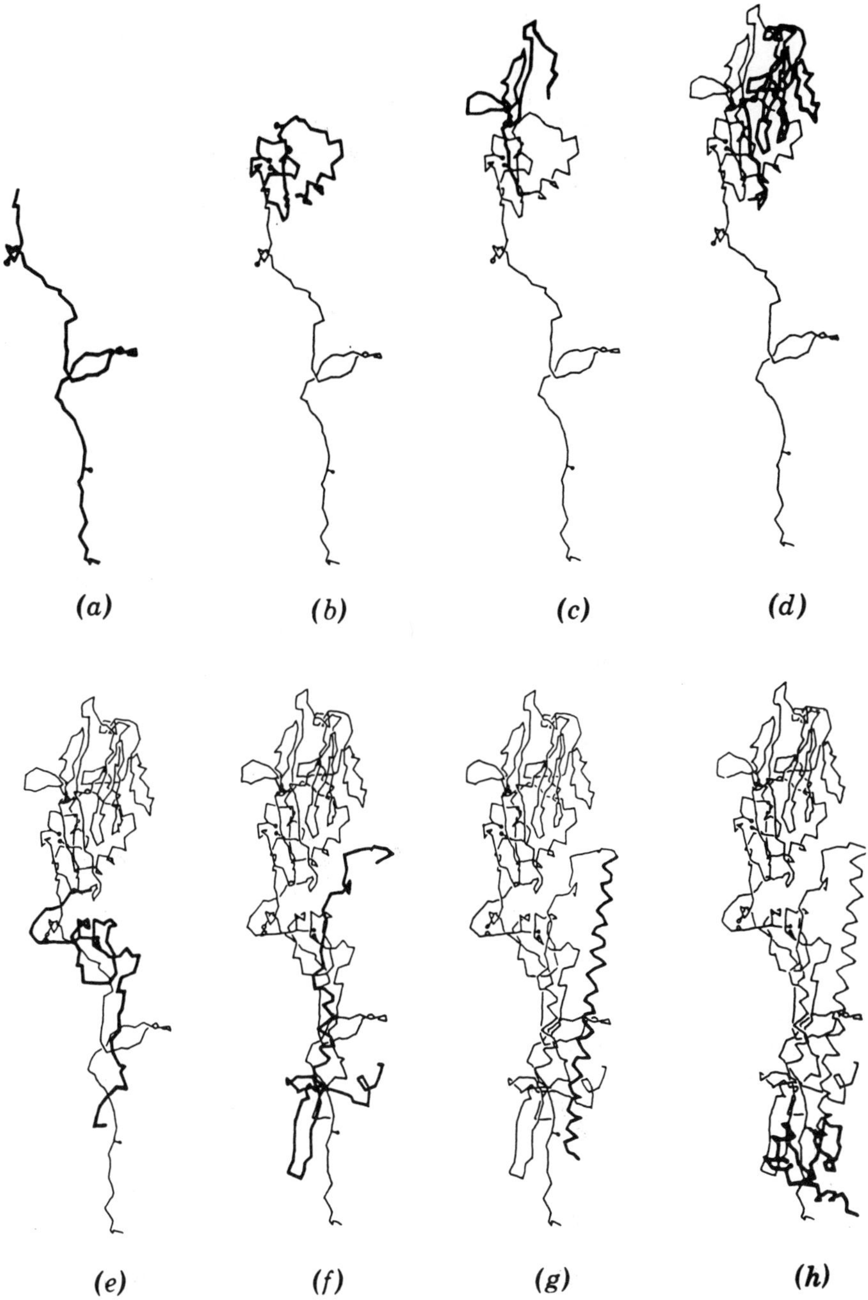

Figure 7. The path of the polypeptide chain of a hemagglutinin monomer (HA_1 and HA_2) traced from the N terminus of HA_1 to the C terminus of HA_2. From Wilson et al. (1981).

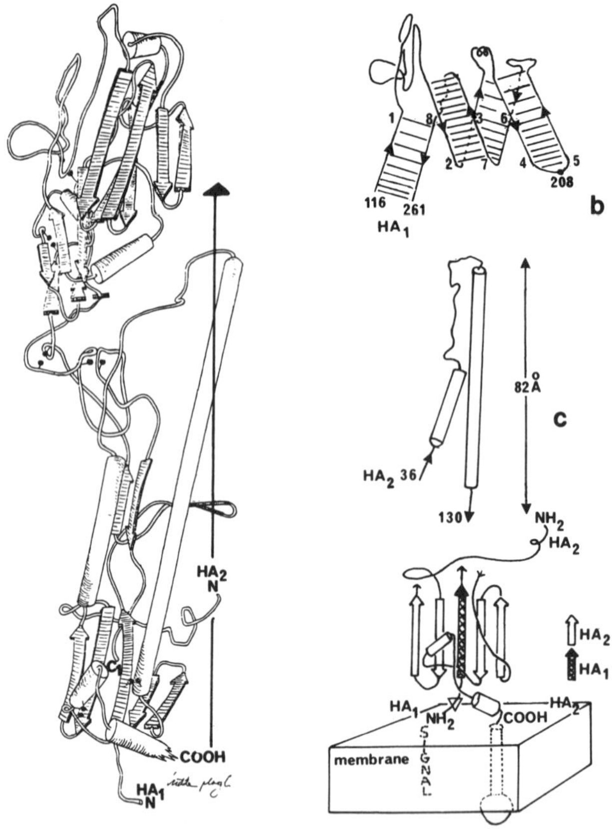

Figure 8. (*a*) Drawing of the three-dimensional structure of the influenza virus hemagglutinin. The terminal residues of HA_1 and HA_2 are labeled. The threefold symmetry axis relating molecules in the trimer is shown: the trimer is stabilized principally by packing of the long α-helices (cylinders). Note that the amino terminus of HA_2, the "fusion peptide," tucks into the threefold contact. From Wilson et al. (1981). (*b*) The eight-stranded β-sheet structure and looped-out regions of the globular domain. (*c*) HA_2 residues 36–130 comprise an α-helical hairpin in the stem region of the monomer. Three of the long helices, one from each monomer, pack together as a triple-stranded coiled-coil that stabilizes the trimer. (*d*) The membrane end of the molecule contains a five-stranded β-sheet. The central strand is the N terminus of HA_1 and the adjacent strands come from the C-terminal portions of the chain in HA_2. Broken lines suggest the path of the hydrophobic anchoring peptide removed by bromelain cleavage. The site of the oligosaccharide attached at HA_1 8 is shown as a triangle.

roll'' β-structures have been described in other structures (TBSV, SBMV, STNV, CAP). This domain provides the framework for the site which binds to sialic acid, a necessary constituent of the cellular receptor for influenza virus (see below). The remainder of HA_1 "returns" back to the stem region running nearly antiparallel to the initial stretch of HA_1 (Fig. 7*e*).

The N terminus of HA_2 is 22 Å from the C terminus of HA_1 (Fig. 7*e* and *f*) indicating that a conformational change accompanied the cleavage of the two polypeptide chains. The hydrophobic and glycine-rich amino-terminal peptide of HA_2 hereafter called the "fusion peptide," due to its proposed involvement in membrane fusion (see below), forms a helixlike series of bends at the trimer interface before wrapping across the monomer like a belt. The major structural feature of the HA_2 chain is a hairpin loop of two α-helices packed together (Fig. 7*f* and *g*, Fig. 8*c*). The first helix (36–54) is connected by a loop (55–77) to the second helix which in 50 amino acids reaches 76 Å back toward the membrane (Figs. 7*g* and 8*c*). This long helix forms the "backbone" of the stem region. HA_2 terminates in a short α-helix at the membrane end of the molecule from which it can be cleaved by bromelain from the remainder of the hemagglutinin at Gly 175. The intact HA polypeptide would be expected to enter the membrane 10 amino acids later.

7. BIOSYNTHESIS

The overall fold of the monomer resembles a loop in that it begins and ends at the membrane end of the molecule. Those stretches of HA_1 not involved in the globular domain at the top run up and down the molecule, appearing to be braced against the stem structure formed by HA_2. At the membrane end of the molecule below the HA_2 N-terminal "belt" is a five stranded β-sheet, where two hairpin loops of β-chains from HA_2 enclose a central extended chain from HA_1. This is a remarkable fold: the central β-strand occurs 350 amino acids earlier in the sequence than those adjacent to it on either side (Fig. 8*d*). We have speculated that this unlikely fold would be facilitated if the amino-terminal signal sequence anchored HA_1 into the membrane during biosynthesis so that the structure was formed while the molecule was tethered at both ends to the membrane surface (see Fig. 9 for details). What needs to be determined is when the HA signal peptide is detached. In this respect we note that the neura-

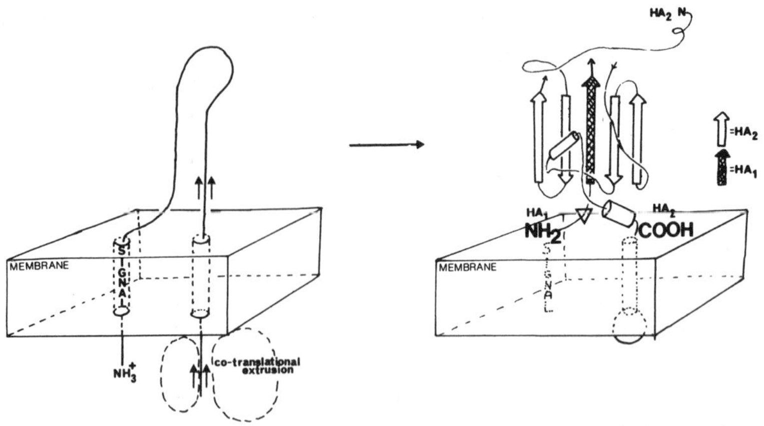

Figure 9. Biosynthesis of the hemagglutinin. The hemagglutinin is translocated across the endoplasmic reticulum membrane co-translationally by a process mediated by the signal peptide. Folding of the unusual β-sheet with its central strand from the N terminus of HA_1 and its C terminus of HA_2 may have been facilitated if the amino terminal of HA_1 was tethered to the membrane surface near the location where the C-terminal portion of the structure emerges from the membrane.

minidase, a second membrane glycoprotein of influenza virus, has no removable signal peptide. The N-terminal hydrophobic (presumably "signal") sequence in that case remains attached to the molecule after biosynthesis and acts as the transmembrane anchor peptide (Fields et al., 1981).

7.1. Trimeric Interactions

Independent contacts stabilize the trimer in the globular and stem regions. The HA_1 globular domains make pairwise contacts near the tip of the molecule, which include protein–carbohydrate interactions (Fig. 4 of Wiley et al., 1981). A triple stranded α-helical coiled-coil provides the major contacts in the HA_2 region. This is formed by threefold contacts along a 40 Å hydrophobic surface (Ile 77, Leu 80, Val 84, Leu 91, Leu 98, 99, 102) on the first half of the long HA_2 helices (Fig. 10*a* and *b*). The helices then diverge, but the hydrophobic core on the threefold axis is extended by threefold contacts of Leu 2 and Phe 3 of the amino-terminal "fusion" peptide of HA_2. Ward and Dopheide (1980a) predicted a coiled-coil in this approximate location from heptad repeats in their HA sequence. The helices then diverge further with hydrophilic and charged residues predominating around the threefold symmetry axis, presumably bridged by solvent molecules.

7.2. Carbohydrate

During biosynthesis the nascent HA polypeptide chain traverses the membrane of the endoplasmic reticulum to emerge in the lumen where oligosaccharide chains are attached to the Asn-X-Ser/Thr sequences on the growing chain (Kiely et al., 1976). Simple oligosaccharides (Fig. 11*a*) containing only *N*-acetylglucosamine and mannose are transferred blockwise from a lipid precursor (Marshall, 1972) to six sites on HA_1 (8, 22, 38, 81,

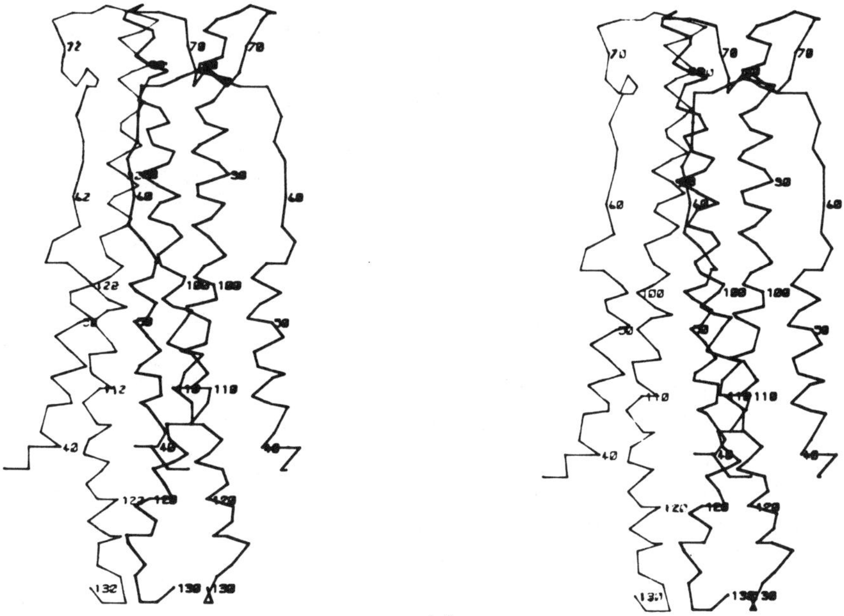

(*a*)

Figure 10. (*a*) Stereo drawing of amino acids 36–130 of the HA_2 chain, illustrating the triple-stranded coiled-coil of α-helices which forms the backbone of the stem of the HA trimer. On page 316: (*b*) A detail of the interaction on the threefold symmetry axis shows the top of three helices, one from each interacting monomer. Ile 77 from each monomer forms a hydrophobic interaction on the threefold axis and salt bridges are formed by the Glu 74 and Arg 76 of adjacent monomers paired with cyclic symmetry.

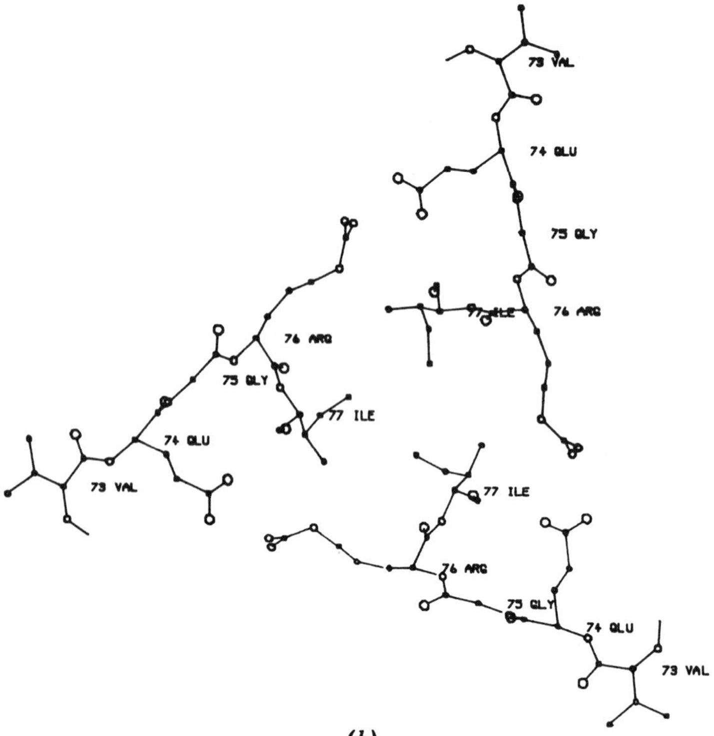

(b)

Figure 10. Continued

165, 285) and one on HA_2 (154). Subsequently, in the smooth ER, all sites except 165 and 285 are further processed by mannosylglycosidases, which remove some mannose residues, and by various glycosyltransferases, which add fucose, galactose, and neuraminic acid to produce "complex" oligosaccharides (Leblond and Bennet, 1977) (Fig. 11*b*). (The terminal neuraminic acid is eventually removed by the virion coded neuraminidase.)

The most striking feature of the carbohydrate structure of the HA is its distribution (Fig. 6). All sites except 165 are on the lateral surfaces of the molecule, none on the distal surface projecting into the solvent as often depicted on drawings of glycoproteins. One site (HA_1, 8), with a host-specific sulfated carbohydrate moiety, is at the extreme membrane end of the molecule. Although electron density can be tentatively identified in the vicinity of each attachment site (Fig. 6) (except HA_1 residue

8 which is on the disordered N terminus), at this stage of our refinement we are only able to build a molecular model for about 25% of the carbohydrate residues at sites 81, 165, and 285 (Wilson et al., 1983).

The oligosaccharide at 165 appears to have a structural role in the HA of the 1968 strain where it stabilizes the oligomeric contact between globular units. Oxygen and nitrogen atoms of *N*-acetylglucosamine residues are within hydrogen bonding distance of polar atoms on the adjacent monomer. A number of potential van der Waals contacts with mannose residues which could stabilize the oligomeric contact are also evident. The conformation of the carbohydrate NAG-NAG-MAN core appears constrained to a plane by hydrogen bonds from the 3 position hydroxyl groups to the ring oxygens of adjacent hexoses and/or the adjacent C_6 hydroxyl.

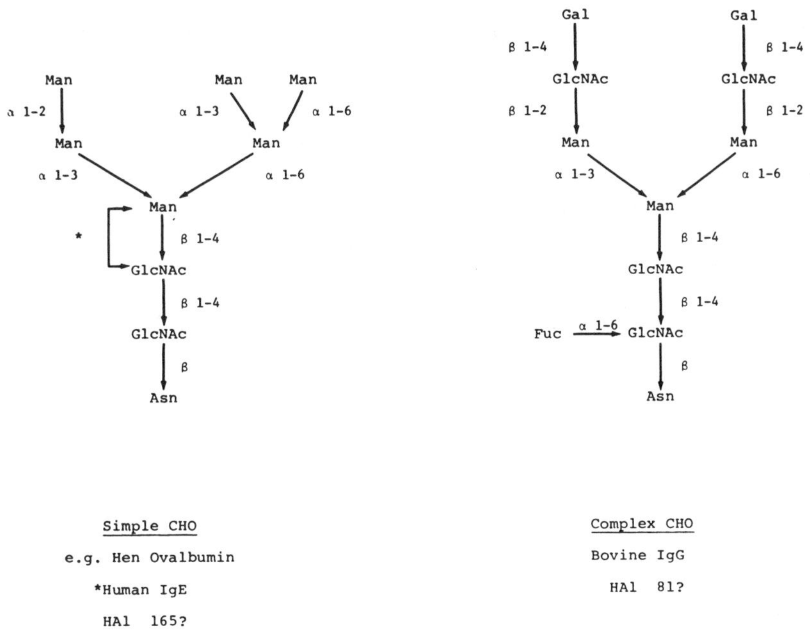

Figure 11. "Simple" and "complex" N-linked oligosaccharide chains. Simple oligosaccharides composed of *N*-acetylglucosamine (GlcNAc) and mannose (Man) are transferred as a unit to Asn-X-Ser/Thr sequences on glycoproteins. Carbohydrate processing enzymes in the smooth endoplasmic reticulum remove mannose residues and add fucose, *N*-acetylglucosamine, galactose, and terminal neuraminic acids. Some viruses contain neuraminidase, which removes the terminal neuraminic acid, resulting in oligosaccharides such as that shown.

In both this respect and in its location at a subunit interface, this oligosaccharide resembles that of the Fc fragment of human IgG (Deisenhofer, 1981), although it differs from that of the Fc fragment of rabbit IgG (Sutton and Phillips, 1983).

The polypeptide chain around the seven carbohydrate attachment sites Asn-X-Ser/Thr on the HA do not show similar main chain stereochemistry. In fact, 38 and 165 are in extended chains; 22, 81, and 285 in bends; and 154 on an α-helix.

The role of carbohydrate in modulating recognition of the protein by the immune system is discussed below.

8. RECEPTOR BINDING SITE

To initiate infection, the virus hemagglutinin binds to a target cell surface receptor, which contains sialic acid. A region at the distal tip of the hemagglutinin was postulated to be the sialic acid receptor binding site on the basis of a concentration of totally conserved residues in a surface pocket: Tyr 98, His 183, Trp 153, Leu 194, Glu 190, and others. The location of this site has recently been confirmed by the selection of mutants by growth

Table 3. Correlation of Receptor Specificity and Amino Acid Substitution at Residue 226

Virus[a]	Receptor Specificity[a] NeuAcα2 → 3Gal	Receptor Specificity[a] NeuAcα2 → 6Gal	Amino Acid at Residue 226
X-31	−	+ + +	Leu
X-31/HS	+ + +	−	Gln
X-31/GPS	+ + +	−	Gln
M1/5	−	+ + +	Leu
M1/H510	−	+ + +	Leu
M1/HS7	+ + +	+ + +	Met
M1/HS8	+ + +	−	Gln

[a] X:31 and M1/5 are wild-type influenza viruses with H3 hemagglutinin A/Aichi/2/68 and A/Memphis/102/72, respectively. X-31/HS and X:31/GPS are variants produced by growth in nonimmune horse and guinea pig sera, respectively. [From Rogers et al. (1983)].

[b] Relative binding preference as deduced from hemagglutinin of derivatized erythrocytes with + + + indicating strong agglutination and − indicating little or no agglutination.

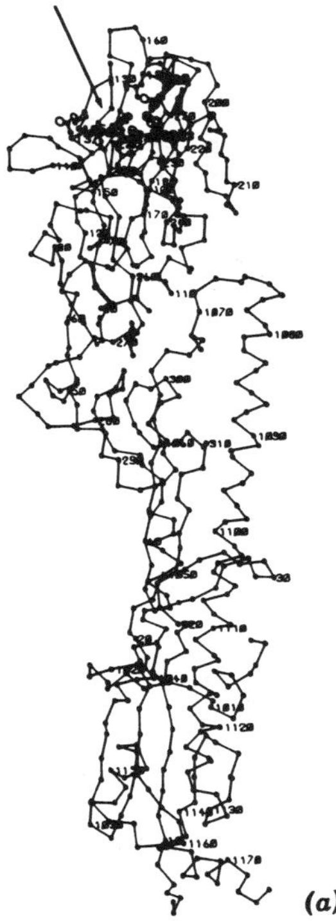

(a)

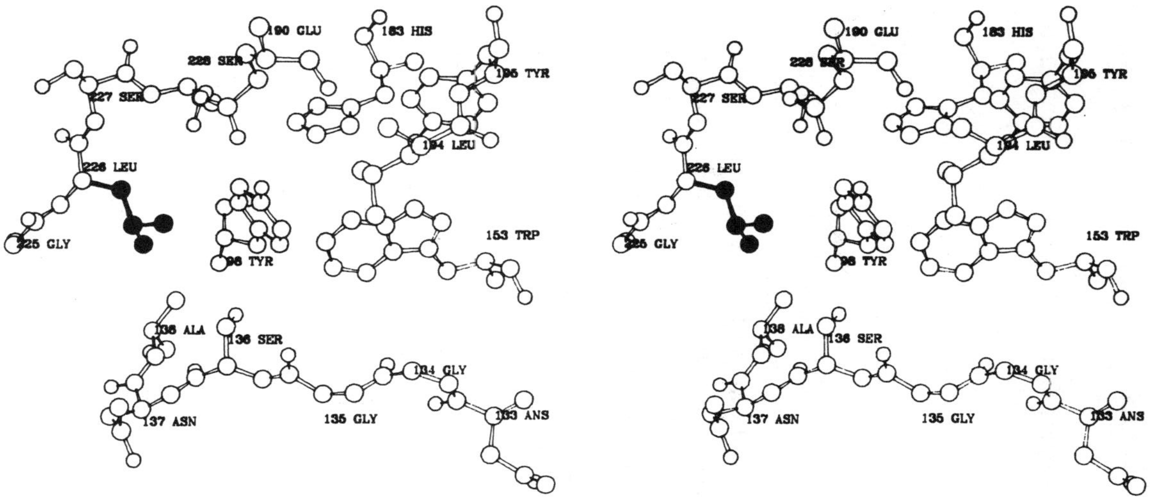

(b)

Figure 12. (*a*) Selected amino acids in and around the sialic acid receptor binding pocket are displayed on a schematic drawing of the α-carbon backbone of the hemagglutinin from the 1968 influenza virus X:30. (*b*) A detail of the same residues shown in Figure 12*a* viewed from the top of the molecule in stereo. Leucine 226 projects into the pocket near the central residue Tyr 98. The site illustrated binds NeuAcα2→6Gal linkages, while glutamine at position 226 alters the site to bind specifically NeuAcα2→3Gal linkages.

in the presence of inhibitors (Rogers et al., 1983). Virus selected in the presence of nonimmune horse or guinea pig sera (a source of NeuAcα2 → 6Gal) binds preferentially to neuraminic acid in 2,3-linkage to galactose, whereas 1968 wild-type virus binds specifically to 2,6-linkages. RNA sequencing of the hemagglutinin gene of these two variants and a third variant with intermediate binding properties showed that alteration in Leu 226, which is in the pocket described above, correlated with the changes in phenotype (Table 3; Fig. 12). HA with leucine at 226 binds the NeuAcα2 → 6Gal preferentially whereas the variant with Gln at 226 binds to NeuAcα2 → 3 Gal. A second variant with methionine 226 had intermediate properties. The structural basis for differences in binding specificity is not proposed at this time in anticipation of the analyses of crystals of hemagglutinins with different binding specificities and of hemagglutinin–sialyllactose complexes, which are in progress.

A similarity of residue types in the binding site of wheat germ agglutinin (Wright, 1980) and the hemagglutinin has been noted (Wilson et al., 1981).

9. MEMBRANE FUSION

Influenza virus, like a number of lipid enveloped animal viruses, enters a cell by binding to a portion of the plasma membrane which is later endocytosed by a clathrin mediated process (Fig. 13). The virus is then found inside of a vesicle termed an endosome, its nucleic acid still being separated from the cytoplasm by two membranes, that of the virus and that of the endosome. A great deal of recent evidence indicates that reduction of the pH in the endosome triggers a virally induced membrane fusion event, which merges the viral and endosomal membranes effecting transfer of the viral genome into the cytoplasm (Fig. 13) (for review, see White et al., 1983). [A candidate for the endosomal proton pump which reduces the pH has recently been isolated (Stone et al., 1983; Forgac et al., 1983)]. Influenza virus fusion is optimal in the pH range pH 5.0 to 5.5, depending on viral strain (White et al., 1981b; Maeda and Ohnishi, 1980). For an individual virus such as the 1968 Hong Kong strain that we have studied, the pH optima is very sharp. If the pH in the external media is reduced to 5.0, the virus will then fuse and infect at the plasma membrane surface (White et al., 1981a). In a confluent monolayer of cells so treated, the membrane fusion activity results in cell–cell fusion. The con-

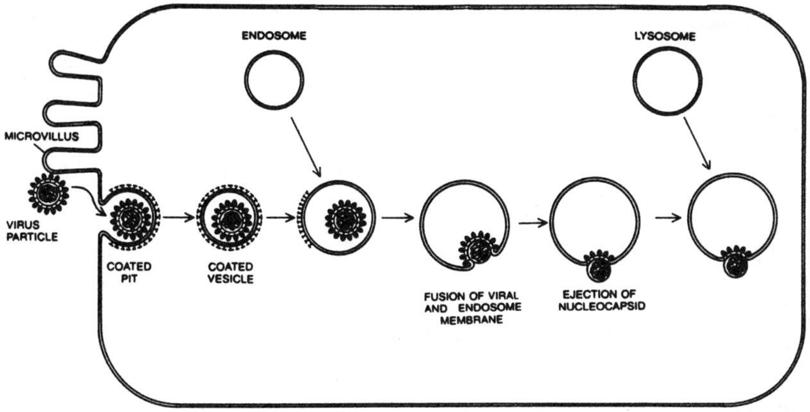

Figure 13. Viral entry. A number of enveloped animal viruses enter cells by being endocytosed into clathrin coated vesicles. Inside the cell the vesicle loses its clathrin coat and fuses with a large vesicle called an endosome. The low pH in endosomal vesicles triggers a membrane fusion reaction, mediated by a viral glycoprotein, which results in the fusion of the viral and endosomal membranes and release of the viral nucleocapsid into the cytoplasm. From Simons et al. (1982).

clusive demonstration that the influenza hemagglutinin is the only protein necessary for membrane fusion was provided by the molecular cloning and expression of the HA gene in monkey kidney cells (White et al., 1982). Such cells with the influenza hemagglutinin on their surface when treated with pH 5.0 buffer fuse into polykaryons.

The hemagglutinin had been previously implicated in membrane fusion through two observations: (1) Virus with precursor HA_0 is not infectious, and cleavage to HA_1 and HA_2 normally carried out by cellular enzymes during viral maturation, is required to activate the HA for membrane fusion (Klenk et al., 1974; Huang et al., 1980; White et al., 1981b). This parallels an observation in Sendai virus where the "fusion" protein must be cleaved before it is active (Homma and Ohuchi, 1973; Scheid and Choppin, 1974). (2) The hydrophobic amino terminal sequence of HA_2 has amino acid sequence homology to the N-terminal sequence of the Sendai virus fusion protein (Gething et al., 1978).

This "fusion peptide" is the longest sequence of highly conserved residues in the HA and after the signal and C-terminal anchor the most hydrophobic (Fig. 14). On the trimer, we have seen that amino acids 2

HA2 AMINO TERMINI

	1	2	3	4	5	6	7	8	9	10	11	12	13	14	15	16	17	18	19	20	21	22	23
A/AICHI/2/68	GLY	LEU	PHE	GLY	ALA	ILE	ALA	GLY	PHE	ILE	GLU	ASN	GLY	TRP	GLU	GLY	MET	ILE	ASP	GLY	TRP	TYR	GLY
A/DUCK/UKR/63 [a]																							
A/DUCK/UKR/63 [b]		ILE																					
A/NT/60/68																							
A/MEM/102/72																							
A/VIC/3/75		ILE																					
A/PR/8/34												GLY			THR								
A/JAP/305/57												GLY			GLN			VAL					
A/WSN/33												GLY			THR								
A/FPV/ROS/34																	LEU	VAL					
A/BELLAMY/42												GLY			THR						—	—	—
A/WEISS/43												GLY			THR						—	—	—
B/LEE/40		PHE								LEU		GLY								ALA		HIS	
B/MD/59		PHE								LEU		GLY								ALA		HIS	
B/HK/8/73		PHE								LEU		GLY								ALA		HIS	
C/JOHAN/1/66		ILE			ILE		(ASP)	LEU	ILE		—	—	—	—	—	—	—	—	—	—	—	—	—
Sendai		PHE	PHE	GLY	ALA	VAL	ILE	GLY	THR	ILE	ALA	LEU	GLY	VAL	ALA	THR	ALA	ALA	GLN	ILE	THR	—	—
SV5			ALA		VAL				LEU	ALA										VAL		—	—
NDV			ILE			ILE			GLY	VAL												—	—

Figure 14. "Fusion peptide" amino acid sequences. The first 12 entries are HA_2 N-terminal sequences from influenza A viruses, the next three are from influenza B viruses, and the last one in the first group is from an influenza C virus. Blank entries indicate exact sequence homology with the 1968 AICHI hemagglutinin (line 1).

The second set of sequences are F_1 N-terminal sequences from paramyxoviruses. These sequences have been aligned to begin at position 2 Phe for maximum overlap with the influenza virus HA sequences. Underlined residues are the same in the 1968 AICHI HA and the Sendai F_1 "fusion peptide." Sources: (influenza A viruses) Verhoeyen et al. (1980), Fang et al. (1981), Both and Sleigh (1980), Sleigh et al. (1980), Min Jou et al. (1980), Winter et al. (1981), Gething et al (1980a), Hiti et al. (1981), Porter et al. (1979), Skehel and Waterfield (1975); (B viruses) Krystal et al. (1982, 1983); (C virus) Herrler et al. (1981); (parainfluenza) Richardson et al. (1980).

and 3 of this sequence are in the stem interface of the trimer 35 Å from the membrane end of the molecule and 100 Å from the distal end which would be bound to the target membrane. Thus, we postulated that the molecule probably undergoes a conformational change triggered by low pH which exposes the N-terminal sequence and possibly alters the 135 Å molecule so that the membranes can approach each other more closely (Wilson et al., 1981).

The first evidence for this conformational change comes from circular dichroism (Skehel et al., 1982) of the bromelain released hemagglutinin (BHA) (lacking the C-terminal anchor peptide) at pH 7.0 and 5.0. A small change in the near UV-CD indicating a change in the environment of some aromatic residues was observed. More illuminating is the observation that at pH 5.0 BHA aggregates, moving from its usual position of 9 S on a sucrose gradient to 30 S (Fig. 15). In the electron microscope, the 30 S aggregates look like protein–protein micelles of the type obtained from intact HA (C-terminal anchor attached) in the absence of detergent. The low pH induced aggregation can be prevented by nonionic detergent (Fig. 15), which direct measurement shows binds to pH 5.0 but not to pH 7.0 treated BHA (Skehel et al., 1982). Furthermore, BHA after brief pH 5.0 treatment binds vesicles reconstituted from pure lipids. These results all indicate that exposure of HA to pH 5.0 causes a conformational change resulting in new hydrophobic properties (Fig. 16).

One indication of the nature of these changes is revealed by trypsin treatment of the pH 5.0 treated BHA. Untreated BHA is unusually resistant to proteolytic digestion; overnight incubation with weight-to-weight trypsin at 37°C has no effect. On the contrary, pH 5.0 treated BHA is readily clipped at amino acid 27 (HA_1) (and secondarily to a lower level at 224) by mild (1 : 20 trypsin/BHA) tryptic treatment for 10 min at 20°C (Skehel et al., 1982). Analysis of the tryptic products of pH 5.0 BHA indicated that the HA comes apart into a soluble protein comprising all

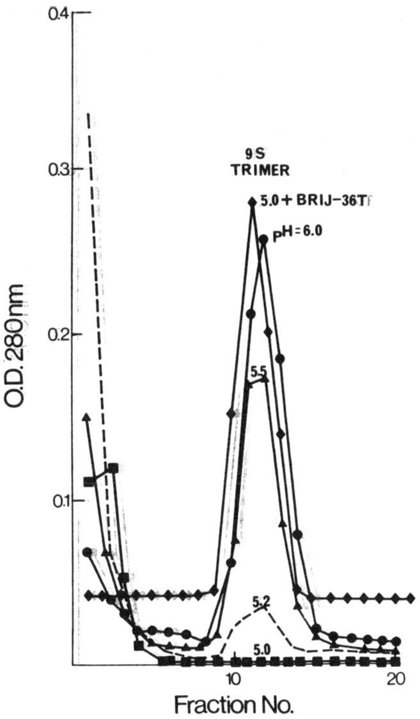

Figure 15. Sucrose density gradient centrifugation of BHA after incubations at various pH values. X:31 BHA solution (1 mg/ml) in phosphate buffered saline were adjusted to various pH values by addition of 0.1 *M* citric acid, layered on 5–20% sucrose gradients, and centrifuged at 35,000 rpm at 20°C for 16 hr. pH: ●, 6; ▲, 5.5; ---, 5.25; ■, 5.0; ◆, BHA solution made 0.5% in BRIJ/3GT (a nonionized detergent) before adjustment to pH 5.0. Fraction 1 is the bottom of the product and fraction 12 contains the pH 7.0 native BHA, 9S trimer spikes. From Skehel et al. (1982).

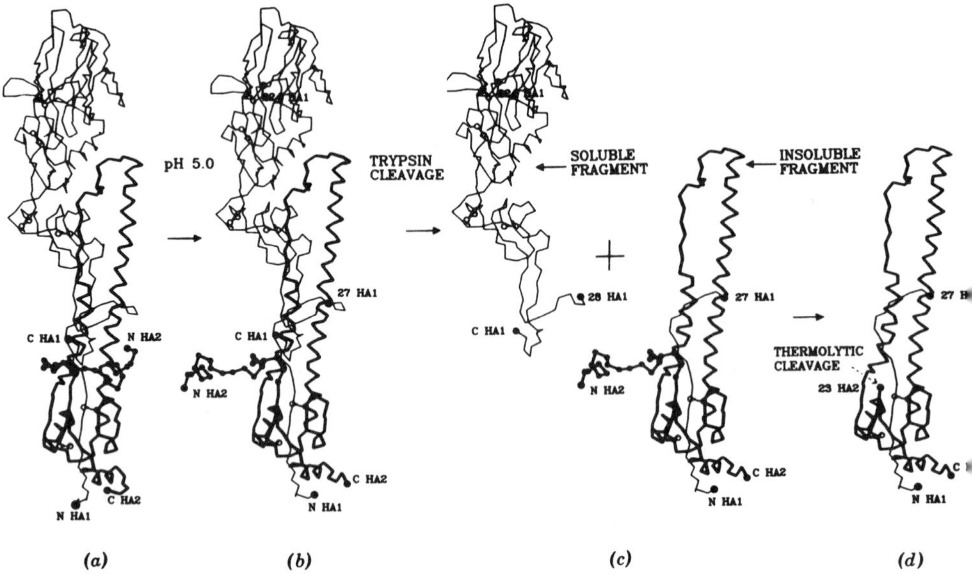

Figure 16. Schematic representation of the effect of incubating BHA at pH 5.0. (*a*) pH 7.0 monomer structure. (*b*) At pH 5.0 the hydrophobic N-terminal of HA_2 is exposed (see text). (c) The low pH conformation of BHA can be cleaved *in vitro* by trypsin at HA_1 27, which leads to release of the "top" globular domain from the stem, indicating a change in tertiary interactions in the stem region of the monomer. (*d*) The hydrophobic properties of the pH 5.0 form are lost after thermolytic removal of residues 1–23 of HA_2.

of HA_1 except HA_1 1–27 which remains disulfide linked to HA_2. This soluble "top" fragment retains antigenic activity and exhibits a strongly β-sheet circular dichroism, likely indicating preservation of the distal globular fold (Fig. 16*c*). Chemical cross-linking experiments and molecular sieving by high performance liquid chromatography indicate that the released top domain is a monomer (Daniels et al., 1983a).

HA_2 and the covalently attached 1–27 of HA_1 remain insoluble indicating that the newly expressed hydrophobic region is in the stem part of the molecule.

Recently, Daniels et al. (1983b) have been able to disaggregate (or release from membranes) this pH 5.0 stem fraction by mild thermolytic digestion. Amino acid sequencing of the product indicates that a single cleavage, removing HA_2 1–23, suffices to eliminate the hydrophobic properties of pH 5.0 treated BHA (Fig. 16*d*). This directly establishes that the HA_2 N-terminal "fusion" peptide is the hydrophobic region exposed by the pH triggered conformational change (Fig. 16*d*).

The mechanistic consequences of the exposure of this hydrophobic peptide *in vitro* are unknown. The fusion peptide could interact with the virus membrane or with other fusion peptides to aggregate proteins or its exposure could coincide with the activation of a cryptic activity (e.g., phospholipase), but to date no evidence indicates any of these possibil-

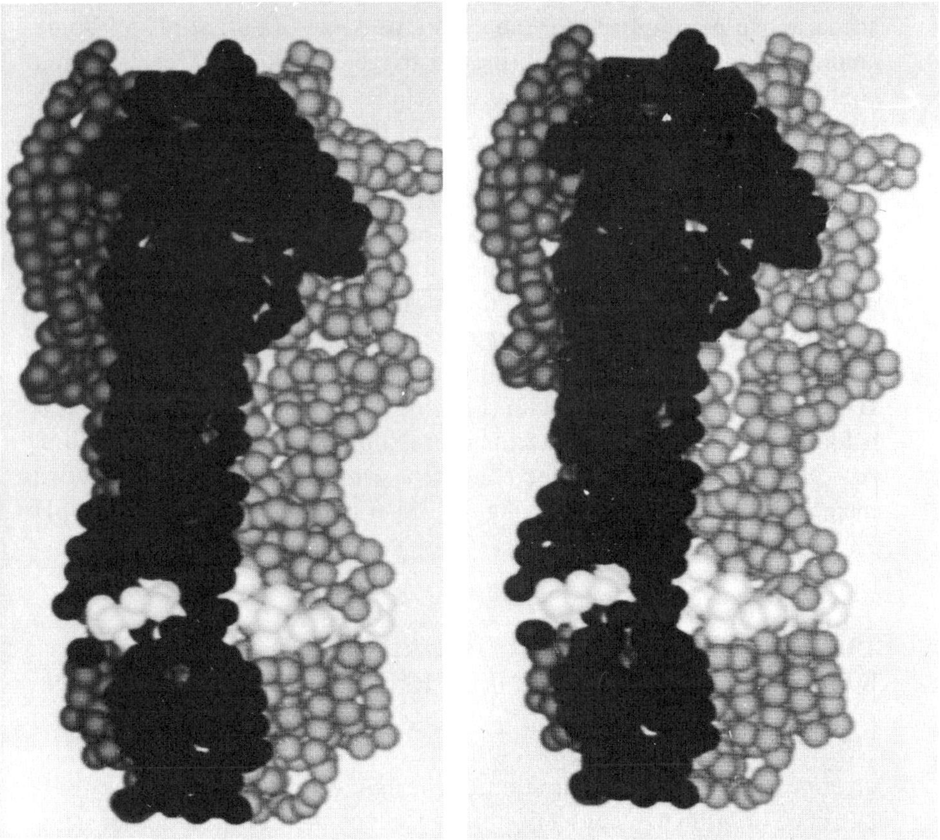

Figure 17. A schematic diagram of the influenza hemagglutinin trimer drawn as spheres at each α-carbon position. The "fusion peptide," which wraps around each monomer like a belt before tucking into the threefold contact, is also shown. Figure 17 was generated by Richard J. Feldman and Thomas K. Porter of the NIH computer graphics facility.

ities. The more likely event, that the fusion peptide interacts with the target cell membrane, is supported by the observation that influenza virus can fuse with artificial liposomes devoid of proteins and sialic acid bearing receptors (Matlin et al., 1981; Maeda et al., 1981; White et al., 1983). Thus, it seems likely that the target membrane can be recognized directly by the "fusion peptide." The HA therefore would become an integral membrane protein of both the virus and cell membranes, held in the former by its anchor peptide and in the latter by the "fusion peptide."

Although the low pH conformation of the HA is unknown, a plausible scenario can be proposed for the events we know about: at pH 5.0 some amino acid(s) bind a proton(s) (the pH trigger is unknown) resulting in a conformational change. This change includes the expulsion of the amino terminal residues of HA_2 from the core of the trimer interface in the stem region of the molecule. This can be accomplished on a model by changing ϕ, ψ torsion angles of residues near HA_2 23. That pulls these residues from the cavity which they occupy and peels the rest of the amino-terminal peptide away from the surface where it is wrapped across the trimer like a belt (Fig. 17). Thus exposed, the fusion peptide is able to insert into the target membrane. The further structural consequences are unknown, but the observation that essentially all of HA_1 can be dissociated from HA_2 by a single proteolytic clip (Fig. 16*b*), and that when dissociated, it is found as a monomer, not a trimer, probably indicates a large structural rearrangement. This would be consistent with the need to bring the two membranes together by removing the 135 Å pillar represented by the pH 7.0 conformation of the HA.

10. IMMUNE RECOGNITION AND THE ANTIGENIC STRUCTURE OF THE INFLUENZA HEMAGGLUTININ

Neutralizing antibodies against influenza virus are all directed at the HA glycoprotein. Antigenic variation of the HA with time causes the reoccurrence of influenza epidemics in man. Since the first isolation of influenza virus in 1933 (H1 strain), two major pandemics have been recorded (in 1957 and 1968), and on both occasions the viruses responsible were characterized by marked differences in the antigenic properties of their HAs. These strains were termed H2 in the case of the Asian strains prevalent between 1957 and 1968, and H3 for the Hong Kong viruses which appeared in 1968 and are still prevalent (WHO, 1971). In each of these pandemic eras, epidemics have been recorded every few years caused by viruses which had undergone "antigenic drift" from the prototype strains (Burnett, 1955).

The results of a number of studies of the amino acid sequences of HAs of natural and laboratory selected antigenic variants of H3 subtype influenza viruses (Webster and Laver, 1980; Min Jou et al., 1980; Sleigh et al., 1980) led to the proposal that certain regions identified on the three-dimensional structure of the HA interact with antibodies and undergo structural modification during antigenic drift (Wiley et al., 1981). Figure 18*a* and *b* shows the amino acid substitutions that occurred between 1968 and the epidemics caused in 1972 by Mem/102/72 and 1975 by Vic/3/75

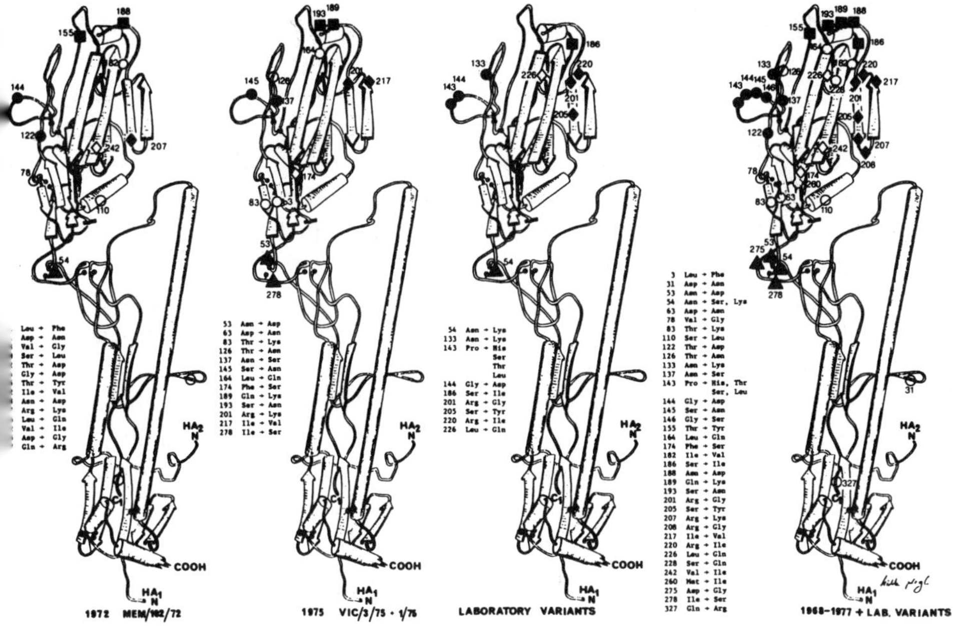

(*a*) (*b*) (*c*) (*d*)

Figure 18. Schematic drawing of a monomer of the AICHI-1968 hemagglutinin showing locations of amino acid substitutions on HA_1. ●, site A; ■, site B; ▲, site C; ◆, site D; ◇, surface; ○, neutral. (*a*) Amino acid changes from AICHI/2/68 to MEM/102/72 (Ward and Dopheide, 1980b, 1981; Sleigh et al., 1980; Gething et al., 1980b; Verhoeyen et al., 1980). (*b*) Amino acid changes from MEM/102/72 to VIC/3/75. (*c*) Laboratory-selected antigenic variants from growth in monoclonal or most-avid-fraction antibody. (*d*) Summary of all known alterations from 1968 to 1977 including laboratory selected variants. A three-dimensional (stereo) drawing of the same view of the molecule is found in Fig. 3 in Wiley et al. (1981).

plotted on the 1968 structure. Fourteen amino acid substitutions are found in the HA_1 chain of A/Mem/102/72 relative to that of A/Aichi/2/68 and a further sixteen in HA_1 of A/Vic/3/75. Figure 18*c* shows single amino acid substitutions of antigenic mutants selected by growth in the presence of monoclonal antibodies (Laver et al., 1980). Each symbol in that case represents a virus with a single amino acid substitution that drastically reduced the affinity of a monoclonal antibody for the HA. Figure 18*d* is a summary of the alterations known from 1968 to 1977 including the laboratory variants in Figure 18*c*. This indicates four antigenic sites on the molecule. Site A is centered on a protruding loop from amino acid 140–146. Site B has a prominent loop above 155 and the external residues 187–198 around an α-helix at the distal end of the molecule. Site C (53, 54, 275, 278) is a bulge in the tertiary structure below the globular domain. Site D (207, 174, etc.) has both surface residues and possibly some amino acids in the region of the interface between monomers. Details of how the charges and steric properties of the observed substitutions could alter antibody binding are tabulated elsewhere (Wiley et al., 1981). On the basis of this analysis it was concluded that both major disease causing viruses of that period had at least one mutation in each of the four antigenic sites described, and may have required these to avoid virus neutralization.

These results were supported by the analysis of the HA of the 1979 epidemic virus A/Bangkok/1/79 which showed further variation in each of the proposed regions (Sleigh et al., 1981). Analysis of more monoclonal antibody selected variants of the 1968 HA (Fig. 19*a*) provided direct proof of site B, confirmed site A, and established that region E could also be recognized by antibodies (Daniels et al., 1983b). Figure 19*b* summarizes the antigenic sites derived from natural and monoclonal antibody selected variation of the H3 strains between 1968 and 1979.

The monoclonal antibody selected change at 63 from Asp to Asn (Fig. 19*a*) is of special significance as it introduces an oligosaccharide site at 63. Experiments with virus grown in the presence of tunicamycin, which inhibits glycosylation of the HA, indicate that the substitution Asp-Asn is not recognized by the antibody, which binds to the variant virus unless the Asn 63 is glycosylated (Skehel, personal communication). This directly demonstrates that carbohydrate can alter the immune recognition of the glycoprotein structure presumably by sterically masking the protein surface. The presence of probable sites of oligosaccharide attachment in other subtypes in regions proposed as antigenic sites in the H3 strains

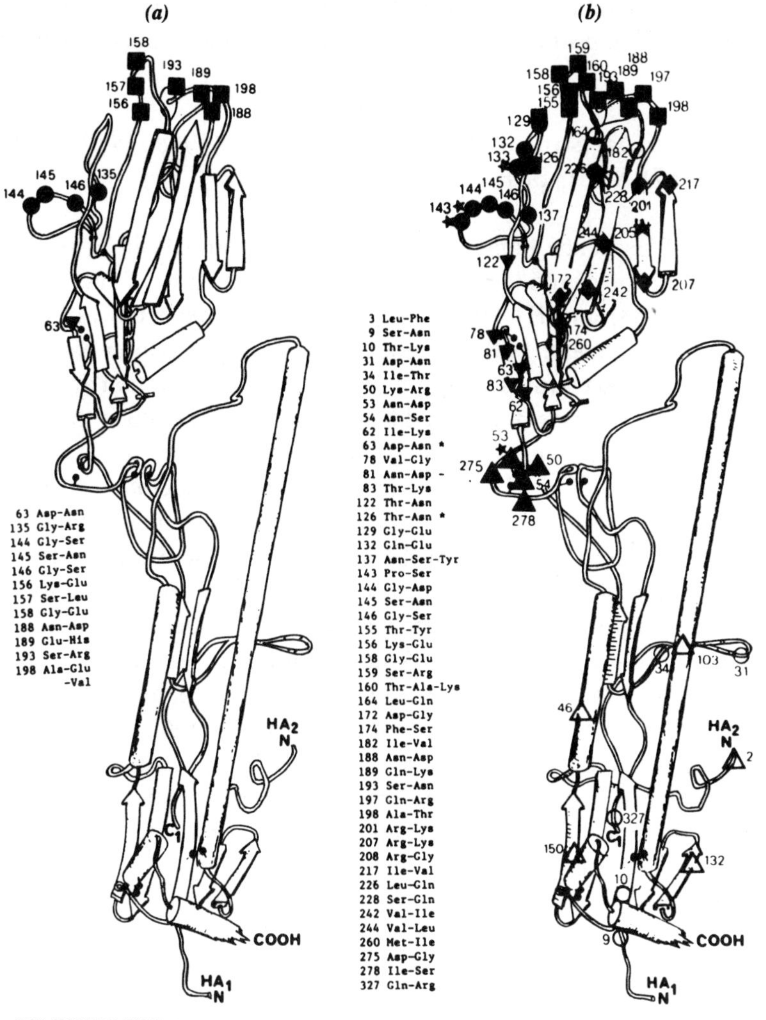

Figure 19. (*a*) Antigenic variants of X:31 (AICHI/68) were selected with monoclonal antibodies. The HA gene sequence determined the amino acid substitution listed. Each symbol represents a single variant virus (see text). (*b*) ●, Site A; ■, Site B; ▲, Site C; ◆, Site D and ▼, Site E. Natural sequence variation of the 1968 hemagglutinin observed between 1968 and 1979 (monoclonal variants of MEM/72 are shown as stars). From Daniels et al. (1983b).

indicates the likelihood that carbohydrate modulates recognition of the HA by covering portions of antigenic sites in some strains (Wiley et al., 1981).

10.1. Comparison of the HAs of the H3 (1968–1979) and H1 (1933–1957) Subtypes

Because of sequence homology (35% in HA_1 and 53% in HA_2) and the conservation of disulfide bonds, the overall structure of the HAs from H3 (1968–1983) and H1 (1933–1956) viruses are expected to be similar (Ward and Dopheide, 1981; Verhoeyen et al., 1980; Winter et al., 1981). The architectural features, loops, and helices, and so on, that are antigenic sites in the H3 strains are embedded in conserved residues that are also found in the H1 strains, indicating that even those structural details are preserved. Thirty-five HA genes from distinct monoclonal antibody selected variants of the H1 strain A/PR/8/34 were partially sequenced to define the antigenic structure of the H1 HA molecule (Caton et al., 1982). A comparison of the regions recognized by antibodies on the 1968 structure with the positions recognized by mouse monoclonal antibodies on the 1934 structure indicates a striking similarity (cf. Fig. 20 with Fig. 19*b*) discussed in detail elsewhere (Caton et al., 1982; Daniels et al., 1983b). One interesting difference is that the region 165, 166, 167, which is an extension of site B in the 1934 HA, is covered by an oligosaccharide chain in the 1968 HA and therefore not recognized in that strain.

A similar but more limited comparison has been made with B strains of influenza where again the same regions of the molecule are implicated in antigenic binding and again oligosaccharide chains are sometimes found masking regions of antigenic significance (Krystal et al., 1982, 1983). Large (greater than four amino acids) insertions are found in the HAs of B viruses relative to A viruses at positions 48, 135, 160, and 196, which are in or near surface features to which antibodies are thought to bind in influenza A viruses (Fig. 21). These insertions indicate the plasticity of these regions of the molecule which can accept extensive alteration without otherwise impairing the molecule's functions during infections.

Whether all the antigenic regions illustrated are recognized by a sufficient proportion of the antibodies in immune human sera to play a role in the cycles of suppression of the disease and its reoccurrence by antigenic drift is not known. The suggestion has been made that one or a

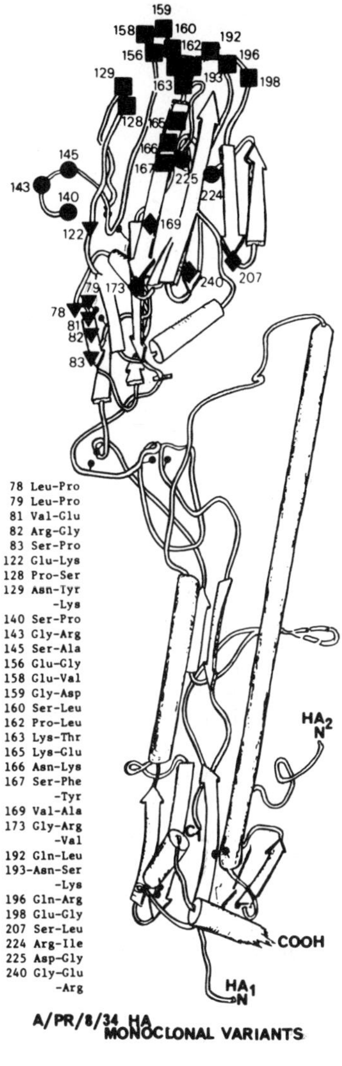

Figure 20. Monoclonal antibody selected variants of the A/PR/8/34 H_1 hemagglutinin (Caton et al., 1982) shown on the 1968 H3 structure for comparison with the antigenic structure of the H3 strains in Figure 19*b*. From Daniels et al. (1983b).

small number of antigenic sites may be dominant (Laver et al., 1980) and this can be rationalized in structural terms by the prominent accessibility of certain surface loops that may favor the formation of high affinity antibodies (Wiley et al., 1981) leading to immunodominant sites by "antigenic selection" (Siskind and Banacerraf, 1969). One study of postinfection human sera, however, demonstrates that approximately 50% of

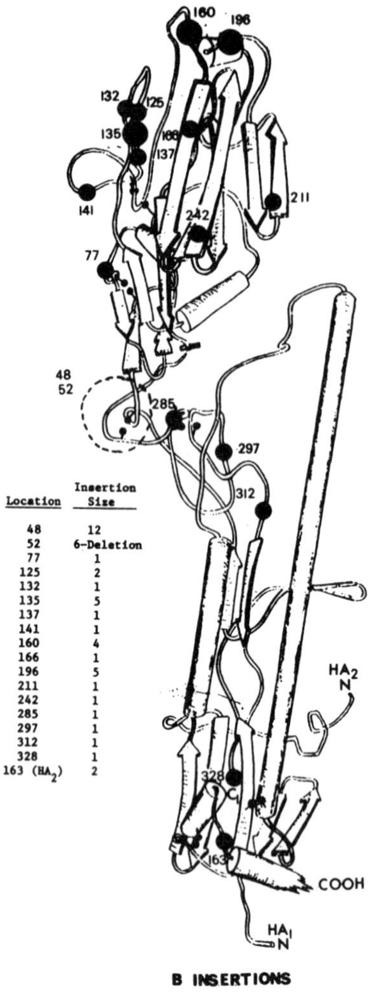

Location	Insertion Size
48	12
52	6-Deletion
77	1
125	2
132	1
135	5
137	1
141	1
160	4
166	1
196	5
211	1
242	1
285	1
297	1
312	1
328	1
163 (HA_2)	2

Figure 21. The location of amino acid insertions in the HAs of B strain of influenza virus relative to the sequence and structure of the A strain A/AICHI/2/68 (illustrated on a schematic drawing of the 1968 HA structure. •, insertions of one or two amino acids; ●, insertions of four or more amino acids. Most of the insertions occur in regions of the strain B HAs corresponding to antigenically important regions of the A HA structure.

human sera containing antibodies against A/Victoria/3/75 had the capacity to neutralize a monoclonal antibody selected variant of that virus, which seems to indicate that human sera considered collectively recognize more than just a single dominant site (Natali et al., 1981). What is required to answer the question are the abundances and affinities of antibodies in human antisera which are directed at individual sites on the HA, a problem that can now be addressed using the available collections of monoclonal antibody selected variants with defined differences in the sequence of their HAs.

10.2. Vaccines

Various strategies to induce antibodies against viral membrane glycoproteins have been employed and are discussed in detail elsewhere (e.g., Tyrrell and Smith, 1979). Recently, antibodies have been produced in laboratory animals by the injection of small synthetic peptides with the same amino acid sequence as a region of a protein antigen (Walter et al., 1980; Sutcliffe, 1980; Muller et al., 1982). For example, a number of peptides ranging from 7 to 40 amino acids in length with sequences from the H3 influenza hemagglutinins have been shown to induce antibodies which cross-react with the influenza HA and virus (Muller et al., 1982; Green et al., 1982). This implies that these small synthetic peptides are able to adopt a structure similar to their conformation in the HA protein while interacting with the surface immunoglobulins of antibody secreting cells. The loss in conformational entropy that this implies for a flexible peptide is presumably compensated for by the favorable free energy of binding to the antibody. The possibility that the antipeptide antibodies locally distort a protein antigen during recognition has not been ruled out. The proportion of antibodies that recognize the parent protein in antipeptide antisera and their affinity for the parent protein are not yet established, although in one case low affinity antibodies prompted the suggestion that careful selection of appropriate peptides may be necessary (Neurath et al., 1982). The use of peptides to induce protection against infection in some laboratory animals has been reported (Bittle et al., 1982).

Acknowledgments. We thank William Willard and Andrew Cherenson for artwork. D.C.W. acknowledges support from the National Institutes of Health, Grant no. AI-13654, and from the National Science Foundation (Grant no. PCM-77-11378, for computing hardware).

REFERENCES

Amzel, L. M., and Poljak, R. J. (1979). *Annu. Rev. Biochem.* **48**, 961–997.

Bittle, J. L., Houghten, R. A., Alexander, H., Shinnick, T. M., Sutcliffe, J. G., Lerner, R. A., Rowlands, D. J., and Brown, F. (1982). *Nature (London)* **298**, 30–33.

Both, G. W., and Sleigh, M. J. (1980). *Nucl. Acid. Res.* **8,** 2561–2575.

Brand, C. M., and Skehel, J. J. (1972). *Nature (London)* **238**, 145–147.

Bricogne, G. (1976). *Acta Crystallogr. A* **32**, 832–847.

Burnet, F. M. (1955). *Principles of Animal Virology*, Academic Press, New York, p. 282.

Caton, A. J., Brownlee, G. G., Yewdell, J. W., and Gerhard, W. (1982). *Cell* **31**, 417–427.

Daniels, R. S., Douglas, A. R., Skehel, J. J., Waterfield, M. D., Wilson, I. A., and Wiley, D. C. (1983a). In *The Origin of Pandemic Influenza Viruses*, C. M. Chu and W. G. Laver, Eds. in press.

Daniels, R. S., Douglas, A. R., Gonsalves-Scarano, F., Palu, G., Skehel, J. J., Brown, E., Knossow, M., Wilson, I. A., and Wiley, D. C. (1983b). In *The Origin of Pandemic Influenza Viruses*, C. M. Chu and W. G. Laver, Eds., in press.

Diesenhofer, J. (1981). *Biochemistry* **20**, 2361–2370.

Fang, R., Min Jou, W., Huylebroeck, D., Devos, R., and Fiers, W. (1981). *Cell* **25**, 315–323.

Fields, S., Winter, G., and Brownlee, G. G. (1981). *Nature (London)* **290**, 213–217.

Forgac, M., Cantley, L., Wiedenmann, B., Altstiel, L., and Branton, D. (1983). *Proc. Natl. Acad. Sci. U.S.A.* **80**, 1300–1303.

Garavito, R. M., and Rosenbusch, J. P. (1980). *J. Cell. Biol.* **86**, 327–329.

Gething, M. J., White, J. M., and Waterfield, M. D. (1978). *Proc. Natl. Acad. Sci. U.S.A.* **75**, 2737–2740.

Gething, M. J., Bye, J., Skehel, J. J., and Waterfield, M. D. (1980a). *Nature (London)* **287**, 301–306.

Gething, M. J., Bye, J., Skehel, J., and Waterfield, M. (1980b). In *Structure and Variation in Influenza Virus,* W. G. Laver, and G. M. Air, Eds. Elsevier, Amsterdam, pp. 63–68.

Green, N., Alexander, H., Olson, A., Alexander, S., Shinnick, T. M., Sutcliffe, J. G., and Lerner, R. A. (1982). *Cell* **28**, 477–487.

Harrison, S. C. (1968). *J. Appl. Crystallogr.* **1**, 84–89.

Helenius, A., McCastin D., Fries, E., and Tanford, C. (1978). *Meth. Enzymol.* **56**, 734–749.

Henderson, R., and Unwin, P. N. T. (1975). *Nature (London)* **257**, 28–32.

Herrler, G., Nagele, A., Meir-Ewert, H., Bhown, A., and Compans, R. W. (1981). *Virology* **113**, 439–451.

Hiti, A. L., Davis, A. R., and Nayak, D. P. (1981). *Virology* **111**, 113–124.

Homma, M., and Ohuchi, M. (1973). *J. Virol* **12**, 1457–1465.

Huang, R. T. C., Wahn, K., Klenk, H.-D., and Rott, R. (1980). *Virology* **104**, 294–302.

Kiely, M. L., McKnight, G. S., and Schimke, R. T. (1976). *J. Biol. Chem.* **251**, 5490–5495.

Klenk, H.-D., Wollert, W., Rott, R., and Scholtissek, C. (1974). *Virology* **57**, 28–41.

Krystal, M., Elliott, R. M., Benz, E. W., Young, J. F., and Palese, P. (1982). *Proc. Natl. Acad. Sci. U.S.A.* **79**, 4800–4804.

Krystal, M., Young, J. F., Palese, P., Wilson, I. A., Skehel, J. J., and Wiley, D. C. (1983). *Proc. Natl. Acad. Sci. U.S.A.* **80**, 4527–4531.

Laver, W. G., Air, G. M., Dopheide, T. A., and Ward, C. W. (1980). *Nature (London)* **283**, 454–457.

Leblond, C. P., and Bennett, G. (1977). In *International Cell Biology*, B. R. Brinkely and K. R. Porter, Eds. Rockefeller Univ. Press, New York, pp. 326–340.

Maeda, T., and Ohnishi, S. (1980). *FEBS Lett.* **122**, 283–287.

Maeda, T., Kawasaki, K., and Ohnishi, S. (1981). *Proc. Natl. Acad. Sci. U.S.A.* **78**, 4133–4137.

Marshall, R. D. (1972). *Annu. Rev. Biochem.* **41**, 673–702.

Mathews, F. S., Argos, P., and Levine, M. (1971). *Cold Spring Harbor Symp. Quant. Biol.* **36**, 387–395.

Matlin, K. S., Reggio, H., Helenius, A., and Simons, K. (1981). *J. Cell Biol.* **91**, 601–613.

Michel, H. (1982). *J. Mol. Biol.* **158**, 567–572.

Min Jou, W., Verhoeyen, M., Devos, R., Saman, E., Fang, R., Huylebroeck, D., Fiers, W., Threfall, G., Barber, C., Carey, N., and Emtage, S. (1980). *Cell* **19**, 683–696.

Monaco, H. C. (1978). Ph.D. thesis, Harvard Univ., Cambridge, Massachusetts.

Muller, G. M., Shapira, M., and Arnon, R. (1982). *Proc. Natl. Acad. Sci. U.S.A.* **79**, 569–573.

Natali, A., Oxford, J. S., and Schild, G. C. (1981). *J. Hyg. Camb.* **87**, 185.

Neurath, A. R., Kent, S. B. H., and Strick, N. (1982). *Proc. Natl. Acad. Sci. U.S.A.* **79**, 7871–7875.

Porter, A. G., Barber, C., Casey, N. H., Hallewell, R. A., Threlfall, G., and Emtage, J. S. (1979). *Nature (London)* **282**, 471–477.

Richardson, C. D., Scheid, A., and Choppin, P. W. (1980). *Virology* **105**, 205–222.

Rogers, C. N., Paulson, J., Daniels, R. S., Skehel, J. J., Wilson, I. A. and Wiley D. C., (1983). *Nature (London)* **304**, 76–78.

Scheid, A., and Choppin, P. W. (1974). *Virology* **57**, 475–490.

Simons, K., Garoff, H., and Helenius, A. (1982). *Sci. Am.* **246**(2), 58–66.

Siskind, G. W., and Banacerraf, B. (1969). *Adv. Immunol.* **10**, 1–50.

Skehel, J. J., and Waterfield, M. D. (1975). *Proc. Natl. Acad. Sci. U.S.A.* **72**, 93–97.

Skehel, J. J., Bayley, P. M., Brown, E. B., Martin, S. R., Waterfield, M. D., White, J. M., Wilson, I. A., and Wiley, D. C. (1982). *Proc. Natl. Acad. Sci. U.S.A.* **79**, 968–972.

Sleigh, M. J., Both, G. W., Brownlee, G. G., Bender, V. J., and Moss, B. A. (1980). In *Structure and Variation in Influenza Virus*, W. G. Laver and G. M. Air, Eds. Elsevier, Amsterdam, pp. 69–78.

Sleigh, M. J., Both, G. W., Underwood, P. A., and Bender, V. J. (1981). *J. Virol.* **37**, 845–853.

Stone, D. K., Xie, X. S., and Racker, E. (1983). *J. Biol. Chem.* **258,** 4059–4062.

Sutcliffe, J. G., Shinnick, T. M., Green, N., Liu, F. T., Niman, H. L., and Lerner, R. A. (1980). *Nature* (*London*) **287**, 801–805.

Sutton B. J. and Phillips D. C. (1983) *Biochem, Soc. Trans.* **11,** 130–132.

Takahashi, K. (1977). *J. Biochem.* **81**, 395–402.

Tyrrell, D. A. J., and Smith, J. W. G. (1979). *Br. Med. Bull.* **35**, 77–85.

Verhoeyen, M., Fang, R., Min Jou, W., Devos, R., Huylebroeck, D., Saman, F., and Fiers, W. (1980). *Nature* (*London*) **286**, 771–776.

Walter, G., Scheidtmann, K.-H., Carbone, A., Laudano, A., and Doolittle, R. F. (1980). *Proc. Natl. Acad. Sci. U.S.A.* **77**, 5197–5200.

Ward, C. W., and Dopheide, T. A. (1981). *Biochem. J.* **195**, 337–340.

Ward, C. W., and Dopheide, T. A. (1980a). *Aust. J. Biol. Sci.* **33**, 449–455.

Ward, C. W. and Dopheide, T. A. (1980b). In *Structures and Variation in Influenza Virus*, W. G. Laver and G. M. Air, Eds. Elsevier, Amsterdam, pp. 27–38.

Webster, R. G., and Laver, W. G. (1980). *Virology* **104**, 139–148.

White, J., Kartenbeck, J., and Helenius, A. (1981a). *J. Cell. Biol.* **87**, 264–272.

White, J., Matlin, K., and Helenius, A. (1981b). *J. Cell. Biol.* **89**, 674–679.

White, J., Helenius, A., and Gething, M. J. (1982). *Nature* (*London*) **300**, 658–659.

White, J., Killian, M., and Helenius, A. (1983). *Qt. Rev. Biophys.*, **16,** 151–195.

Wiley, D. C., and Skehel, J. J. (1977). *J. Mol. Biol.* **112**, 343–347.

Wiley, D. C., and Flanagan, M. T., and Skehel, J. J. (1977a). *Topi. Infect. Dis.* **3**, 131–134.

Wiley, D. C., Skehel, J. J., and Waterfield, M. D. (1977b). *Virology* **79**, 446–448.

Wiley, D. C., Wilson, I. A., and Skehel, J. J. (1981). *Nature* (*London*) **289**, 373–378.

Wilson, I. A., Skehel, J. J., and Wiley, D. C. (1980). In *The Structure and Variation of Influenza Virus*, W. G. Laver and G. M. Air, Eds. Elsevier, Amsterdam, pp. 339–349.

Wilson, I. A., Skehel, J. J., and Wiley, D. C. (1981). *Nature* (*London*) **289**, 366–373.

Wilson, I. A., Ladner, R. C., Skehel, J. J., and Wiley, D. C. (1983). *Biochem. Soc. Trans.* **11**, 145–147.

Winkler, F. K., Schutt, C. S., and Harrison, S. C. (1979). *Acta Crystallogr. A* **35**, 901–911.

Winter, G., Fields, S., and Brownlee, G. G. (1981). *Nature* (*London*) **292**, 72–75.

World Health Organization (1971). *Bull. W.H.O.* **45**, 119–124.

Wright, C. S. (1980). *J. Mol. Biol.* **139**, 53–60.

9

Structural Investigations of Hexon, the Major Coat Protein of Adenovirus

ROGER M. BURNETT
College of Physicians and Surgeons
Columbia University
New York, New York

CONTENTS

1. INTRODUCTION

1.1. Adenovirus

Adenovirus is the group name for related viruses, first isolated by Rowe (1953), that are found in mammalian respiratory tracts. The immunological properties of the different capsid proteins vary within the group and serve to define more than 80 different serotypes, which have been isolated from nine species. The 31 human adenoviruses have been classified into subgroups according to their varying capacity for agglutinating erythrocytes (Rosen, 1960), or for inducing tumors in newborn hamsters (Huebner, 1967).

Adenoviruses of different types have different epidemiological behavior (Monto, 1980): accounting for 5% of common respiratory disease occurring in children under 5 (ad1, 2, 3, 5, 6); epidemics of acute respiratory disease in military recruits and other institutionalized populations (ad3,4,7, 14, 21); epidemic keratoconjunctivitis or "shipyard eye" (ad8); pharoconjunctival fever occurring in outbreaks with a common source such as a swimming pool (ad3, 7); and, rarely,pneumonia (ad7, 12). Adenoviruses are unusually resistant to acidic and other adverse conditions, and this is demonstrated by the unusual persistence of infectious ad8 in ophthalmic solution, or on instruments, leading to iatrogenic dissemination through eye examinations (Jawetz, 1959). Variation in virion stability with serotype has been demonstrated by the differing sensitivity of capsids to disruption with formamide (Stasny et al., 1968). Ad7 was more resistant than ad4, and ad5 was the most stable which must reflect a corresponding variation in the bonding of the capsid proteins. Comparative structural studies will indicate how modifications have occurred to counter particular environmental conditions.

Adenoviruses have been studied extensively for reasons other than their ability to cause disease. They infect mammalian cells, and the integration of their DNA into the host cell genome provides a model system for eukaryotic gene expression. They can also transform *in vitro* cultured cells, and can induce tumors in nonpermissive hosts. The field has been reviewed by Philipson et al. (1975). A further property, which has been extensively exploited, is that infected cells contain viral structural proteins in 10 to 100-fold excess over those assembled into complete virions (White et al., 1969). The excess viral proteins have been shown to be immunologically and morphologically similar to the structural components of intact virions, and may be separated and used in studies of the constituent parts of the virus (reviewed by Ginsberg, 1979). Because these proteins are soluble under approximately physiological conditions, which is unusual, their individual interactions could be studied *in vitro*, although little work has yet been carried out in this area. Most structural studies have been performed on the nononcogenic serotypes 2 and 5 from the subgroup consisting of ad1, 2, 4, 5, 6.

The adenovirus capsid has the shape of an icosahedron, with 12 vertices and 20 triangular faces (Horne et al., 1959), and is shown schematically in Figure 1. Ad2 has an estimated particle weight of 175×10^6, of which 28×10^6 is DNA and the remainder is due to 11–14 different proteins. The structural proteins have been labeled polypeptides II to XII on the basis of their decreasing molecular weight as revealed by SDS-polyacrylamide gel electrophoresis (Maizel et al., 1968a). Polypeptide II, the hexon subunit, is the largest of the structural polypeptides, because poly-

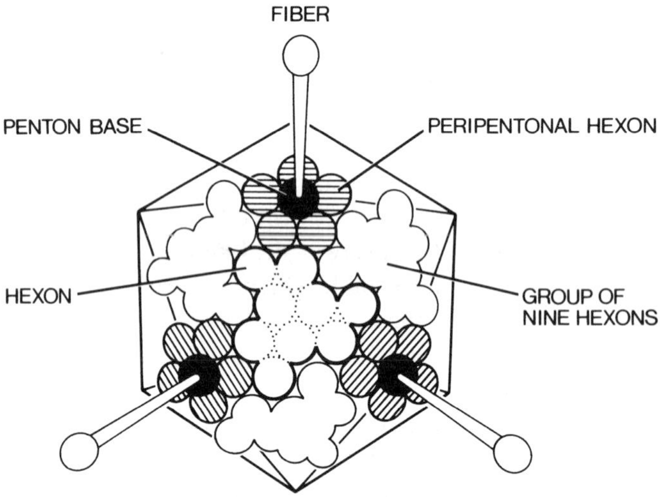

Figure 1. A symbolic representation of the major external components of the adenovirus capsid. The penton is a complex of the fiber with the penton base and is surrounded by five peripentonal hexons. Each group-of-nine forms part of a triangular facet, interlocking with its neighbor to form a capsid edge. The group-of-nine is left-handed when viewed from outside the capsid. Illustration by John Mack.

peptide I was subsequently identified as penton that was incompletely dissociated into its constituent base (III) and fiber (IV). The 240 capsomeres forming the triangular faces and edges each have six neighbors, and the 12 capsomeres at the vertices have five neighbors. These morphological units have therefore been named "hexons" and "pentons," respectively (Ginsberg et al., 1966). The penton consists of a penton base with a projecting fiber, and microcrystals of the latter have been obtained (Mautner and Pereira, 1971). The virion may be progressively dissociated, first losing the pentons, then the "peripentonal" hexons adjacent to the penton base at each vertex, and finally the "groups-of-nine" to leave the core (Smith et al., 1965; Laver et al., 1968; Prage et al., 1970). The core also has an icosahedral shape, and contains a DNA–protein complex enveloped in a protein shell (Brown et al., 1975). The group-of-nine consists of the nine hexons that remain in a triangular facet of the virus after the peripentonal hexons have been lost. The dissociation pattern suggests that the contacts made by the peripentonal hexons and those at the faces and edges are different, and has even led to speculation that different classes of hexon may exist (Shortridge and Biddle, 1970).

At a late stage after infection with adenovirus, cell nuclei contain crystalline inclusions that have been shown to consist of virions (Kjellen et al., 1955; Morgan et al., 1960; Henry et al., 1971). The crystals have been analyzed in the electron microscope by negatively staining thin sections through the crystals in different planes (Boulanger et al., 1974). Virions were found to pack in the cubic space group F23, with their 23 symmetry coincident with that of the crystal to give four particles per unit cell of edge 1210 ± 60 Å.

The accumulation of excess viral protein in infected cells is also so great that these too give rise to crystals, which are often found in close proximity to both individual and crystalline virions (Morgan et al., 1957, 1960). The protein crystals were first postulated to arise from capsid material (Boulanger et al., 1970), but immunological evidence suggesting the presence of basic adenovirus core proteins (Henry and Atchison, 1971) led to a revised model (Lifchitz et al., 1975). A helical polyhedral tube with facets of the same size as those of the capsid is folded about a tubular fiber of the basic protein. A separate image analysis study agreed with these conclusions (Carstens and Marusyk, 1975). Tubules are also seen on infection by other animal viruses, and those from polyoma have been extensively analyzed (Kiselev and Klug, 1969). Tubule formation has been explained as a fault to be expected in the self-assembly of a spherical virus (Caspar and Klug, 1962).

Adenovirus structural components have a tendency to aggregate in ordered arrays. Although most reports of crystallization derive from studies on purified material, various crystalline sheets can be formed from the crude extract from infected cells (Carstens and Marusyk, 1977). These include helical penton aggregates similar to those reported by Pereira et al. (1975), fiber crystals similar to those of Mautner and Pereira (1971), and "holey" sheets resembling those of Pereira and Wrigley (1974).

1.2. Hexon

Hexon was the first animal viral protein to be crystallized (Pereira et al., 1968). The first weak diffraction patterns indicated a space group of I23 or $I2_13$ with a cell edge of 213 Å (Macintyre et al., 1969). Subsequent work by Franklin et al. (1971a) on rhombic dodecahedral crystals of ad2 hexon and Cornick et al. (1971) on tetrahedral crystals of ad5 hexon determined the space group of both forms to be $P2_13$. The cell edge was reported as 149.9 Å, with each asymmetric unit containing one-third of the hexon molecule. The diffraction patterns were very similar for both types, but crystal density measurements gave differing molecular weights for hexon. The original incorrect assignment of space group was ascribed by Cornick et al. (1971) to the choice of the $P2_13$ face diagonal as principal axis. Dohner and Hudemann (1972) later showed for ad1, 2, 5, and 6 that the crystal habit was dependent on the crystallization conditions, but was independent of serotype. These serotypes all belong to the same adenovirus subgroup, and the results indicate that this group contains hexons that are structurally very similar even though they are immunologically distinct.

It should be noted that in space group $P2_13$ the hexon molecules are not arranged in the same mutual orientation as those within the viral capsid. Ad1 hexons have been reported to form two-dimensional sheets in the same orientation as within the capsid, but this was not observed for other types (Adam and Nasz, 1979).

The subunit composition of hexon was controversial, as was the molecular weight, until it was established that ad2 hexon is a trimer of approximately 360,000 daltons (Grutter and Franklin, 1974; Jornvall et al.,

1974a). The sequence of the ad2 hexon polypeptide and the corresponding gene have both been under investigation (Jornvall et al., 1974a, 1978). The combined approach, necessitated by the exceptionally long chain, has shown the ad2 subunit to consist of 966 residues with a molecular weight of 108,113 (Jornvall et al., 1981; Jornvall and von Bahr-Lindstrom, 1981). Part of the sequence of ad5 hexon has also been determined and indicates that the two sequences are very similar (von Bahr-Lindstrom et al., 1982). Only 23 residues out of the 529 sequenced are different, and of these 18 are clustered in three regions with the remaining five all due to single base substitutions. Two deletions in the ad5 sequence, of approximately 60 residues, reduce its molecular weight by 7000, confirming the suggestion of Ginsberg (1979) that a difference in length would explain some of the earlier confusion. The most recent sequence data gives a chain length of 967 residues for ad2 hexon (H. Jornvall, personal communication, and Dayhoff *et al.*, 1982).

The controversy over the exact molecular weight of hexon is just one example of an unfortunate tendency in the literature to assume that differences between serotypes are negligible, when considering possible explanations for slightly different properties. In this case, the confusion was compounded by the observation that the diffraction patterns from ad2 hexon (Franklin et al., 1971a) and ad5 hexon (Cornick et al., 1971) were very similar. Electron density maps of ad2 hexon (Burnett et al., 1978, unpublished results) reveal that hexons pack in the unit cell so that the top of the molecule, which presumably contains the type-specific regions, is not involved in hexon–hexon contacts. Thus, different top regions can be accommodated within the same crystal lattice. Additionally, the polypeptide chain is very faint at the top, suggesting that it is rather flexible. The combination of the two factors leads to similar diffraction patterns, which mimic isomorphism.

Several unique features are apparent in the ad2 amino acid sequence. Prolines are common in the amino- and carboxyl-terminal thirds of the molecule, where they often occur in a pronounced pairwise association separated by one to five other residues, but they are underrepresented in the middle third. Repetitive sequences of particular residues occur, but consecutive hydrophobic residues are unusual. Charged residues are common, and there are two acidic regions, one of which may be removed after limited tryptic digestion (Jornvall and Philipson, 1980). There are 84 three- and four-residue repeats in the sequence, but no long repeated segments are found. A further conclusion from the combined approach is that the hexon gene does not contain intervening sequences.

Other chemical investigations of hexon have included the identification of one reactive SH group per polypeptide in the native state, compared with seven after denaturation with detergent. The availability of a single reactive group played a vital role in the solution of the first heavy atom derivative (Burnett et al., 1978). A reactive serine has also been identified (Devaux and Boulanger, 1980) that is accessible only in the unfolded polypeptide, and which can be blocked to prevent folding of the individual polypeptide chains into the native hexon trimer. The accessibility of the polypeptide to modification is consistent with data showing that the hexon polypeptide requires 3–4 min for synthesis, is then released from the ribosome, and requires a further 3–4 min for assembly (Horwitz et al., 1969).

Temperature-sensitive mutants have been isolated that are unable to assemble hexon from its component polypeptides or cannot transport the assembled hexon into the cell nucleus where virion assembly occurs (Ginsberg et al., 1974; Williams et al., 1974; Leibowitz and Horwitz, 1975). The mutations occur either in the hexon gene (Oosterom-Dragon and Ginsberg, 1981) or in the gene for the adenovirus nonstructural 100K protein (Kauffman and Ginsberg, 1976) indicating that both proteins are required for correct hexon assembly. Subsequent work, using monoclonal antibodies against hexon that are highly conformation specific, establishes that the 100K protein is required for trimerization of hexon polypeptides into their protease-resistant final form (Cepko and Sharp, 1982).

Hexon is subject to degradation during storage, which may be simulated by digestion with trypsin. The resultant fragments do not separate, but remain linked by noncovalent bonds, and display approximately normal morphological, antigenic, and crystallographic characteristics (Pereira and Skehel, 1971). A small highly acidic fragment of approximate MW 3000 can be removed after trypsin treatment of ad2 hexon (Jornvall and Philipson, 1980) and the resultant material can only be crystallized by seeding the solution with a small native crystal (J. L. White, J. van Oostrum, and R. M. Burnett, unpublished), indicating that the changes are limited to a region that is not involved in intermolecular contacts in the crystal. The result is consistent with an alteration at the top of the molecule, in the type-specific region, since ad5 hexon is resistant to trypsin digestion as it lacks 60 residues corresponding to the trypsin sensitive region of ad2 hexon (von Bahr-Lindstrom et al., 1982).

A minor class of ad2 hexons, "special" hexons, have been isolated from infected cells (Pettersson, 1971). These have a lower electrophoretic mobility, but a similar size and shape compared to the major class of normal hexons. Their presence may be explained by the subsequent identification of a class with similar properties, called "slow" hexons, which lack 50–60 residues at the N terminus (Boulanger et al., 1978a). Slow hexons were not found in the virion, and were presumed to result from proteolysis of N termini that had not been protected by the acetylation that is found in normal hexons (Jornvall et al., 1974b).

Antibodies against hexon have been used as structural probes, identifying both group- and type-specific determinants α and ϵ. These are antibody recognition sites that are either common to the whole adenovirus group, except for the avian and some bovine viruses, or are specific to one serological type. α is located on the inside of the virion, whereas ϵ is on the outside (Norrby, 1969a). It should be noted that "specificity" in this context denotes a range of actual sites, and that only monoclonal antibodies are truly specific for a particular molecular location and conformation.

It is remarkable that, within only 3 years after the discovery of adenovirus, vaccines already had been developed and then abandoned upon discovery of a connection with cancer (Hilleman, 1966). The vaccines were found to be contaminated with oncogenic SV40 virus; adenovirus itself was revealed to be oncogenic; and the two viruses were discovered to form hydrids. Crystalline preparations of fiber and hexon confer immunity to human adenovirus (Couch et al., 1973) and thus these capsid proteins are a suitable alternative vaccine to the complete virus, which had originally been used.

Hexon from ad2, 3, 5 has been degraded with chymotrypsin, subtilisin, and papain to leave a "core" (Pettersson, 1971). The group-specific determinants were removed but the core retained its type-specific properties. Trypsin does not affect the antigenic properties of hexon, but maleic anhydride inactivates the type-specific determinants. Slow hexons possess an anomalously high type-specific antigenicity. Ad2 hexon cores of MW 85,000, resulting from papain digestion, aggregate into two classes of trimers differing slightly in electric charge (Boulanger, 1975).

A comprehensive investigation of the physical properties of hexon has determined values for the following parameters: extinction coefficient $E_{1cm}^{1\%}$ = 14.3 ± 0.3 at 279 nm; partial specific volume $\bar{\nu}$ = 0.738 ± 0.004 $cm^3\ g^{-1}$; sedimentation coefficient $s_{20,w}^{0}$ = 13.0S; Stokes' radius = 64 ± 0.6 Å (Grutter and Franklin, 1974). Spectroscopic investigations have employed circular dichroism and optical rotatory dispersion as probes of the secondary structure within hexon. The conformation has been estimated as 27% α-helical and 26% β-sheet (Day et al., 1972). Boulanger and Loucheux (1972) obtained results in qualitative agreement with these figures and also studied the effect of sodium dodecyl sulfate on hexon. The spectrum was then found to possess a strong helical component, in agreement with results from nonhelical proteins, which show a similar transition to helicity on exposure to anionic detergents (Jirgensons, 1973).

The spectroscopic results are in accord with our tracing of the chain within a 2.9 Å electron density map which shows three short α-helices, a modest amount of β-sheet, and extensive regions where chains are roughly parallel. A comparative analysis of the circular dichroism spectra of free hexon and groups-of-nine has indicated that tryptophan residues are involved in the bonding between hexons (Boulanger et al., 1978b).

An initial analysis of the small-angle X-ray scattering from hexon showed it to be consistent with that from a cylinder of height 125 Å and diameter 80–84 Å (Tejg-Jensen et al., 1972). A later investigation by Berger et al. (1978) compared the theoretical scattering curves for various models of hexon to the experimental curves. The results showed that a hollow cylindrical model was only representative of hexon to a first approximation, the best fit to the data being given by a model of height 130 Å and inner and outer diameters of 15 and 90 Å. A more complex model of hexon, which had been proposed by Nermut (1975) from high resolution electron microscopy, was also tested. Hexon was described as a "conical triangular prism" of height 110 Å, with an equilateral triangular base of side 98 Å, and an approximately circular top of diameter 75 Å containing a large hole. This model also gave poor agreement with the observed scattering curves.

Preliminary 6 Å electron density maps, calculated using phases determined from the native data and three heavy atom derivatives (Burnett et al., 1978), were used to derive an idealized representation of hexon. This model had an hexagonal base of side 89 Å and an equilateral triangular top of side 96 Å. Two vertical edges ran from adjacent corners of the base to each apex of the triangular top. A hole at the top, of diameter 10 Å and depth 12 Å, and a conical cavity at the base, of diameter 35 Å narrowing to a diameter of 10 Å at a height of 50 Å, defined the hollow interior of the molecule (Fig. 2). The theoretical curve from the crystallographic model agreed very well with the experimental observations.

The study by Berger et al. (1978) indicated that, although small-angle X-ray scattering could not provide a detailed model of a molecule, it could distinguish between various models proposed from electron microscopy. It was quite powerful in excluding incorrect models as illustrated by the rejection of the model proposed by Nermut (1975), which placed a large internal cavity incorrectly at the triangular end.

Hexon has been the subject of several electron microscopic investigations, both on individual particles and on aggregates such as the group-of-nine or larger arrays. The first reports indicated that hexon was a hollow cylinder (Dales,1962) or a short hollow polygonal rod (Wilcox et al., 1963), although some workers preferred the more conservative description of a sphere (Valentine and Pereira, 1965). Later work showed more confidence in assigning a polygonal shape and a central hole (Pettersson et al., 1967) and it was reported that the polygons were occasionally six-sided (Smith, 1965; Norrby and Ankerst, 1969). It is curious that more recent descriptions have tended toward a circular, rather than a polygonal,

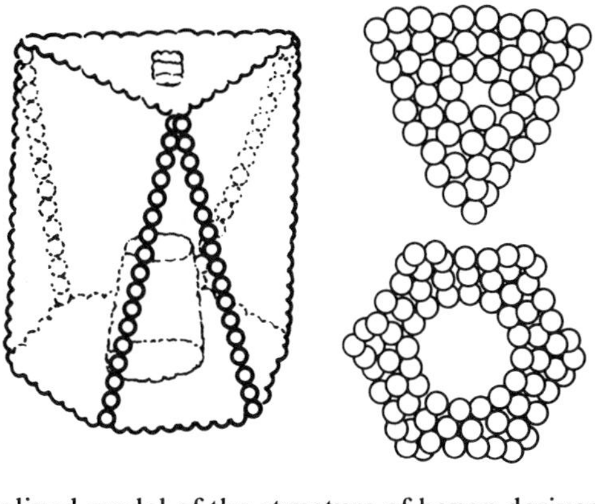

Figure 2. An idealized model of the structure of hexon derived from X-ray crystallography. The model consists of 1200 spheres arranged in 18 layers, the top and bottom of which are shown on the right. This was used in a small-angle X-ray scattering study that tested the ability of the method to discriminate between various models proposed for hexon. Redrawn from Berger et al. (1978).

description concurrent with the shift of the chemical evidence from a hexameric toward a trimeric molecule. We can now see that the original description was correct, because the micrographs are showing the pseudo-hexagonal symmetry of the hexon base.

A more detailed picture of hexon resulted from studies on groups-of-nine (Franklin et al., 1971b) in which the axial hole was not only clearly visible, but its relative position moved in accord with the tilt angle of the goniometer stage of the electron microscope. It was suggested that the individual polypeptides were arranged along the cylinder as six cigar-shaped objects, in accord with the current ideas on the number of subunits.

Franklin et al. (1971b) also calculated a Patterson function using 10 Å X-ray diffraction data from ad2 hexon crystals. The rotation function (Rossmann and Blow, 1962) and a vector superposition method (Simpson et al., 1965) were used to search for molecular symmetry axes, and strong indications were found for a diad perpendicular to the threefold molecular axis. This finding was apparently at variance with the preferential deposition of one side of the groups-of-nine onto electron microscope support films, which revealed a polarity (E. C. Follet, reported in Norrby, 1969b). No indication of molecular sixfold symmetry was found and it was concluded that the apparent 32 symmetry arose from the pattern of subunit clustering.

Rotation and translation functions were also used to analyze 5 Å diffraction data from ad2 hexon (M. Leijonmarck, O. Ronnquist, P.-E. Werner, and L. Philipson, unpublished) and the resultant description of molecular symmetry produced a model of hexon which is quite recognizable in the final structure (Fig. 3).

It can now be seen that all these features in the vector maps arise from the symmetry and relative orientation of the triangular top and the hexagonal base, which give the molecule a rectangular appearance when it is viewed normal to a basal face below an apex of the triangular top. Franklin et al. (1971b) also postulated, on the basis of a cluster of strong reflections, that hexon may contain β-sheet parallel to the molecular axis. This too has been confirmed by the chain tracing in our 2.9 Å resolution electron density map.

The groups-of-nine have proved to be very useful objects in extending the model of hexon because they make a small array of molecules available for study from the two axial directions. Crowther and Franklin (1972)

Figure 3. A model of hexon deduced from interpretations of rotation and translation functions in vector space at 5 Å resolution. Three dimers of α elements (white) are arranged in a tier below three dimers of β elements (black). The α_2 diad axes pass through the threefold molecular axis, but the β_2 axes do not intersect it. The α and β elements represent the location and orientation of the repetitive elements, but not their size. The model was a gift from Dr. P.-E. Werner on the inception of our studies on hexon.

were thus able to use the powerful technique of rotational filtering (Crowther and Amos, 1971) on electron micrographs of groups-of-nine and to confirm that hexons have threefold and not sixfold symmetry. This result was in accord with a trimer of polypeptides, each of MW 120,000 (Horwitz et al., 1970), although the molecular weight remained controversial until the work of Grutter and Franklin (1974). The individual hexons were found to be arranged on a p3 net, rather than the p6 net expected for a hexamer clustered $T = 25$ surface lattice. It followed that hexons at the edges of the capsid were related by local twofold axes and thus that hexons were capable of different types of bonding at the edge and in the face. It was observed that adenovirus was therefore not a simple virus in the sense of Caspar and Klug (1962). This was an additional complication in the description of the capsid, because the vertex protein penton was already known to be distinct from hexon.

Groups-of-nine have a hand, given by the relationship of the outer dimers to the central trimer, which imparts its characteristic propeller shape (Figs. 1 and 7). A left-handed appearance corresponds to an anticlockwise arrangement of the dimers, but the absolute hand can only be determined if it is known from which side the groups-of-nine are being viewed. Pereira and Wrigley (1974) were able to make this assignment when they reconstituted empty viral capsids, lacking vertices, from purified groups-of-nine. They defined the view from outside the capsid as the standard direction. Rings of five nonamers are pyramidal and have a different distribution of stain depending on whether they lie with their convex or concave side on the support film. The latter accumulate more stain at the center, between protein and film, and are composed of apparently left-handed groups-of-nine. The view in this case is effectively from outside the capsid, and therefore the left-handed arrangement corresponds to that with the absolute hand. Note that the hand provides an easy way of determining which side of the group-of-nine is in contact with the microscope grid, even if the resolution is insufficient to see details of the individual hexons, because a right-handed appearance indicates that the group is inverted.

The majority of groups-of-nine are seen as left-handed using hydrophobic carbon support film and it was suggested by Pereira and Wrigley (1974) that the inside surface of the groups-of-nine was also hydrophobic. Nermut and Perkins (1979) later determined that over 90% of groups-of-nine at pH 7.0 were left-handed on carbon films, but that the same proportion were right-handed after the films were rendered positively charged by glow discharging in the presence of amylamine (Dubochet et al., 1971). However at pH 4.4, close to the isoelectric point, an equal distribution of the two hands was seen, suggesting that there is little difference in relative charge between the top and base at this pH.

Pereira and Wrigley (1974) also noted a nonrandom dissociation pattern of groups-of-nine which they explain by proposing a bonding pattern consisting of a central trimer, surrounded by three differently bonded dimers. This explained their observation of two-dimensional crystals of hexon containing a regular pattern of holes, but was at variance with a p3 net.

The most comprehensive electron microscopic study of hexon has been undertaken by Nermut who has published a fine series of electron micrographs showing the adenovirus capsid, groups-of-nine, and individual hexons (Nermut, 1975). A model of hexon was proposed from this investigation, but unfortunately it was assumed that negative stain accumulated around the upper surface of the groups-of-nine rather than their lower surface. In addition, a triangular rather than a hexagonal shape was assigned to the hexon molecule, probably due to its trimeric nature. The resulting triangular prism model has been described earlier. The micrographs of the groups-of-nine show a clear distinction between the appearance of left- and right-handed images, corresponding to the lower and upper surfaces, respectively. The images were invaluable in placing the crystallographic structure (Burnett et al., 1978) into its biological context within the viral capsid (Burnett et al., 1979). Computer modeling was later used to simulate the effect of negative stain on both individual hexons and groups-of-nine (Nermut and Perkins, 1979). This approach corrected the previous assumptions about the position of negative stain and led to a model in basic agreement with the crystallographic model. These studies on the capsid, and others on the core, have been reviewed by Nermut (1980).

2. STRUCTURE DETERMINATION OF HEXON

2.1. Overall Strategy

Particular care has been necessary in several areas of the crystallographic analysis of hexon because it is a large molecule in a space group of high symmetry. The large unit cell of hexon mandated the use of photographic methods for measurement of the diffracted intensities. At the outset of this investigation, oscillation photography was a new method that was still undergoing development. For this reason, advantage was taken of the high symmetry of the cubic space group which theoretically permits all three positional parameters to be obtained from a single precession photograph. This follows from the direction of the three-fold axes, one of which lies along a body diagonal of the cubic cell. Three different projections of the hexon molecule can therefore be obtained from a zero level precession photograph.

We therefore planned to screen for heavy atom derivatives in the usual way using precession photographs, but to use these to calculate difference Patterson projection maps from which the binding sites could be identified. This approach posed two critical demands. The first was that the data should be as accurate as possible. The second was that, for this approach to work in a high symmetry space group, single site binding would be essential in order to minimize the number of vectors.

Although some attempts were made to use direct methods (Steitz, 1968; Neidle, 1973) to locate the binding sites, the solution was eventually obtained through identification of peaks on the Harker lines in the vector map. Finally, oscillation photography was used to collect three-dimensional data from native and derivative crystals. The identification and solution of the initial heavy atom derivatives formed part of the thesis work of Dr. M. G. Grutter (1976). A detailed description of all the crystallographic methods employed will appear elsewhere.

It is interesting to note that although the demands of data collection are not as severe for hexon as those for a complete virus, a revolutionary change has occurred since the beginning of this investigation. It was originally regarded as fortunate that hexon crystallized with one subunit per asymmetric unit. It is now apparent that the power of the molecular replacement method, to produce accurate phases from averaged structures, far outweighs any initial advantage to be gained from a smaller asymmetric unit.

2.2. Crystallization

Adenovirus type 2 was propagated and purified according to the procedures of Pettersson et al. (1967) with the modifications described by Grutter and Franklin (1974). Hexon was concentrated to approximately 10 mg ml^{-1} in 0.01 *M* sodium phosphate buffer containing 0.02% sodium azide at pH 7.0. Crystallization was performed under conditions determined to favor the rhombic dodecahedral over the tetrahedral form (Franklin et al., 1971a). A 5 mg ml^{-1} solution of hexon, in 0.5 *M* sodium citrate at pH 3.2, was placed in sealed glass capillary tubes and stored at room temperature. Rhombic dodecahedral crystals, 0.1–0.2 mm in size, usually appeared and these grew within 1–2 weeks to a size of 0.4–0.8 mm. Small variations of the above conditions resulted in either small crystals, needles, thin plates, or no crystal growth. The sensitivity of the exact crystal form to the crystallization conditions was in agreement with the results of Franklin et al. (1971a) for ad2 and Dohner and Hudemann (1972) for ad1, 2, 5, and 6. The space group was found to be $P2_13$, with a unit cell edge of 150.6 Å, for both the rhombic dodecahedral and tetrahedral forms.

2.3. Data Collection

The relatively large size of the hexon unit cell is correlated with a correspondingly weak diffraction pattern, which requires an intense X-ray source and a long recording time for accurate measurement. An Elliott GX6 rotating anode generator was used as an X-ray source. Although focusing mirrors (Harrison, 1968) are not mandatory for hexon, a highly uniform X-ray beam with low background scatter was essential in obtaining accurate diffraction data from the long exposures required.

The oscillation method (Arndt et al., 1973; Arndt and Wonacott, 1977) is the most efficient method of data collection for large unit cells, and was used for the collection of three-dimensional data. In summary, the use of a graphite monochromator, special beam collimation, and application of absorption corrections to oscillation films have allowed us to achieve conventional R factors on amplitude $R_f = \sum[F - \bar{F}]/\sum F$ of 2% for symmetry related reflections on precession films and 4–5% for complete data sets on oscillation films.

Each oscillation film contains approximately 5000 reflections, of which 50% are fully recorded. An R_f of 4% was obtained for the 120,000 fully recorded reflections from native crystals, which reduced to a unique set of 27,000 reflections representing 95% of the available data to a spacing of 2.9 Å. The partially recorded reflections were also used, as the increase in error level was slight and attributable almost entirely to the lower individual photon counts. This gave a significant increase in the number of measurements of a particular reflection and increased the fraction measured to 97%. The derivative data gave a somewhat poorer R_f of around 6%, using all reflections. Anomalous differences were found to be significant if they were measured between Friedel related reflections on the same film. The magnitude of the anomalous effect is of the same order as the error level of the data, but anomalous difference Fourier maps revealed the positions of the major heavy atom sites in the best derivatives. Phase refinement with the correct, and the incorrect, hand was used to determine the value of these measurements for further work.

2.4. Heavy Atom Derivatives

Because there is one accessible SH group per hexon subunit (Grutter and Franklin, 1974), organomercurial compounds were first explored as possible heavy atom derivatives. It was hoped that this approach would yield a single-site derivative amenable to solution from difference Patterson projection maps. The derivatives were prepared in the normal manner by adding the reagent to the mother liquor. Several mercury compounds, such as *p*-chloromercuribenzenesulfonic acid (PCMBS), were found to give useful derivatives at the crystallization pH of 3.2 and all gave roughly similar diffraction patterns with changes of intensity of 20–30%. As had been expected, all the mercury compounds tested gave a similar distribution of binding sites.

Some difficulty was experienced in finding other suitable derivatives at pH 3.2 as platinum compounds usually disrupted the crystals and other compounds failed to bind. The search for derivatives was more successful at pH 5.0, and platinum ethylenediamine dichloride [$Pt(en)Cl_2$], K_2PtCl_4, and $KAu(CN)_2$ were identified as useful. The K_2PtCl_4 derivative deviates from isomorphism and was only useful initially. Some difficulty was experienced in finding other derivatives, but 3-chloromercuri-2-methoxypropyl urea (CMMPU) and a double derivative (PCAU), prepared from PCMBS and $KAu(CN)_2$, were eventually found to be suitable. These completed the set of five derivatives used for phase determination at 2.9 Å.

The modified salt concentration and pH have a slight effect on the native diffraction pattern, and also decrease the PCMBS binding, and so all data collection was carried out at pH 5.0. For PCMBS, the pH was adjusted after modification of the crystals because otherwise a derivative was obtained with only half the normal occupancy.

2.5. Heavy Atom Location and Phase Refinement

The heavy atom positions in the PCMBS derivative were originally located from difference Patterson projection maps using data derived from precession films of the (h0l) zone. This is a centric zone in $P2_13$ and allows the determination of all three positional parameters because the threefold axes lie at an angle to the projection axis and are thus noncrystallographic symmetry axes in projection. The major binding site was identified from peaks on the two Harker lines in this projection.

The solution of this two-dimensional difference Patterson, in which the highest nonorigin peak was only 2.5 times the noise level, gave us great confidence in our system of data collection and reduction. It may be noted that this procedure failed at 5 Å but succeeded at 3.6 Å resolution. Sites in the other derivatives were then identified from difference Fourier maps.

Refinement of the heavy atom binding sites and phasing of the native structure factors was performed using a program due to Dr. G. C. Ford (Blow and Crick, 1959; Cullis et al., 1961; North, 1965; Matthews, 1966). A newer method, placing more reliance on the experimentally determined errors for each reflection, also proved useful (Sygusch, 1977). Fourier inversion was performed using the Fast Fourier Transform program of Ten Eyck (1973) modified by Dr. G. C. Ford for generality, and later by us to permit calculation of 256 × 256 sections in the limited core of the Univac computer. A comprehensive program set for the use of the molecular replacement method (Bricogne, 1976) was adapted for the Univac and used to compute, by interpolation, the electron density for sections in arbitrary orientations. A convenient skew-planes procedure is necessary for hexon because the molecule lies along a threefold body diagonal in the unit cell. Maps are usually calculated normal to this axis to ensure that the threefold relationship of the subunits is readily apparent.

Cycles of heavy atom refinement, difference map inspection, and identification of new binding sites were performed in projection at 3.6 Å. When the data became available, three-dimensional refinement was carried out first at 3.6 Å and then at 2.9 Å, using the PCMBS, $Pt(en)Cl_2$, and $KAu(CN)_2$ derivatives. Anomalous scattering data were used in the refinement and phasing for the three-derivative map.

Data for two further derivatives, CMMPU and PCAU, were later added to the phasing set. Also, anomalous data were found to have a disproportionate effect on the native phases. The effect did not occur when the Sygusch phasing method was used and lessened with more derivatives. Anomalous data were not used in phasing the five-derivative map.

2.6. Envelope Model of Hexon at 6 Å Resolution

A 6 Å electron density map was calculated using phases derived from the 2.9 Å refinement with three derivatives. Map sections were calculated

(a)

(b) (c)

normal to the threefold molecular axis, using the skew procedure described above.

The subunits of hexon are very tightly interwoven and the intermolecular contacts are not extensive, and so there was no difficulty in assigning the correct trimer of subunits to the molecule. In space group $P2_13$ there is a potential ambiguity, because the various threefold axes relate subunits differently.

The main interest of the low resolution map lay in the information it could give concerning the overall molecular shape. Because hexon comprises almost 60% of the viral protein, it was thought likely that its shape alone would be the determining factor in its packing in the adenovirus capsid, and that other structural proteins could be ignored. An envelope model is sufficient to investigate packing, and so this was delineated on each section between the solvent and the protein density corresponding to that of one hexon molecule. Envelopes on two adjacent sections were averaged and the shapes cut from plywood which were glued together to give a model of the hexon molecule (Fig. 4). Two additional models were constructed so that the interactions between hexon molecules in the viral capsid could be studied.

Hexon was revealed to consist of two major components with differing morphological symmetry. The most obvious feature of the "top" of the molecule is the set of three towers that rise 40 Å above a triangular section of thickness 24 Å. The top rests on a more massive "base" of height 52 Å with a pseudo-hexagonal shape. The upper 42 Å of the base has a pseudo-hexagonal shape, which rounds off to give a roughly circular bottom 7 Å deep. Three small protrusions extend 3–4 Å down from the circular section. The base thus has six extensive vertical faces of height 40 Å and width 52 Å that are related by the pseudo-sixfold symmetry. Exact threefold symmetry relates the three identical faces within each of the two pseudo-related groups. The two different faces alternate about the base and provide extensive contact surfaces for hexons arranged in a flat array (Burnett et al., 1978, 1979).

Figure 4. (*Opposite*) An envelope model of hexon derived from a 6 Å resolution electron density map. Each layer in the model represents a 4 Å section through the molecule. (a) The complete hexon has its smaller top resting on a more bulky base. (b and c) The model is split apart at the boundary between the top and base. (b) A view of the base from underneath revealing its pseudo-hexagonal shape and internal cavity. (c) The three towers, rising from the triangular top, bestow a spiky appearance on the adenovirus capsid.

A small aperture of 2–4 Å diameter between the towers widens to form a cavity of diameter 35 Å at the junction of the top and the base. It then constricts to a diameter of 15 Å at the midpoint of the base before emerging as a larger opening. The hole is most probably an example of structural economy, analogous to the use of tubes instead of bars in framework. The reduction in mass is shown most dramatically in the base, which despite its bulky appearance contributes only 55% to the total for hexon. The overall reduction in mass is approximately 15%.

Hexon can conveniently be symbolized by a triangular top resting on a hexagonal pediment (see Figs. 7 and 8). This method of representation is a slight distortion as it implies a discontinuity between the top and base, whereas the molecule merely has a slight waist at the junction. The model used in the analysis of the low-angle X-ray scattering (Berger et al., 1978) shown in Figure 2 is better in this regard. The apices of the triangular top are rotated by about 10° from positions directly above the midpoints of three of the six basal edges. The rotation is anticlockwise when the molecule is viewed from above.

The pseudo-hexagonal shape of the base permits hexons to form the flat continuous surface seen on each facet of the viral capsid (Figs. 1 and 8). The appearance of hexon in a left-handed group-of-nine corresponds to a polygonal base containing a large hole, and the appearance in a right-handed group to the top with three towers arranged as a triangle (see Nermut, 1975). Because Pereira and Wrigley (1974) had shown that the groups-of-nine were left-handed when viewed from outside the virus, hexons are oriented in the capsid with their bases inward. This argument depends on the original reasonable assumption that the groups-of-nine are stained by an accumulation of electron dense material between the molecules and the microscope grid. The orientation also agrees with the spiky appearance of the capsid in electron micrographs which would be given by an array of hexon towers. The spiky appearance is independent of assumptions concerning the placement of stain and so the two pieces of evidence taken together determine unambiguously that the hexon base is inward. The under surface of a group-of-nine is formed by the bases, and is comparatively flat and featureless. Some electron micrographs, showing the groups-of-nine from the side, suggest this difference in the texture of the two surfaces (Pereira and Wrigley, 1974).

A paradoxical aspect of the images of groups-of-nine published by Crowther and Franklin (1972) lies in the shape of the hexons. The images

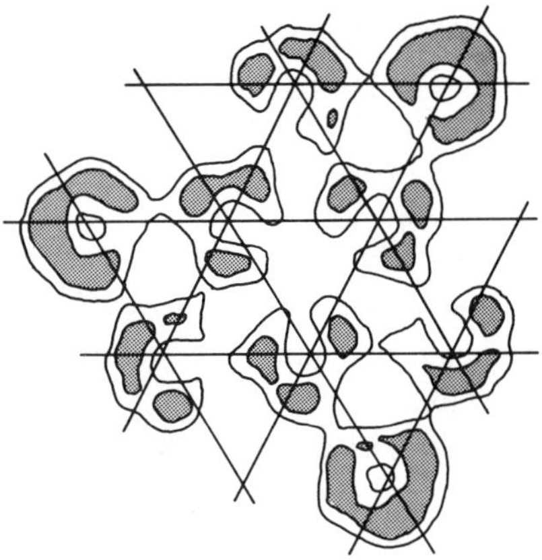

Figure 5. A rotationally filtered electron micrograph of a group-of-nine. The characteristic propeller shape is seen with the three outer dimers arranged to point in a clockwise fashion about the central trimer. This gives a right-handed group but reveals hexon bases, particularly clearly in the outer hexons, as explained in the text. The filtering operation yields an averaged view of the hexons in the central trimer and of each of the hexons in the dimers. Hexons were revealed to have threefold symmetry and to be arranged on a p3 net. Redrawn from Crowther and Franklin (1972).

are right-handed and thus should represent inverted groups-of-nine with the triangular hexon tops downward. The hexons however appear to have slightly polygonal shapes and to contain holes. The rotationally filtered images show triplets of density for the central hexons (Fig. 5). Smeared density occurs at the outer hexons and, because this was thought to correspond to a lack of preservation, specimens with minimum smearing were preferred.

A reevaluation of the filtered images, using our hexon model, reveals first that the right-handed images have been inverted in reproduction and thus are correctly left-handed, consistent with the authors' note that they are of the preferred type. Second, the orientation of the triplets of density relative to the p3 net is inconsistent with that which would be given by the triangular tops. Rather, it corresponds to that of the three small protrusions that extend 3–4 Å below the hexon base. A ring of density would be revealed by stain rising to a height of 10 Å about an upright hexon, and an increasingly hexagonal appearance would be obtained as the stain rises above this point.

The images therefore do not reveal outer hexons that are "squashed flat," but rather a greater height of stain around their bases. The stain penetrates to the middle of the groups-of-nine to a lesser extent, and only covers the base of the inner hexons up to a height where the small protrusions are revealed. The original selection for "preserved specimens" with minimum smearing can now be seen as a selection for minimum staining, so that the exact threefold symmetry of the molecule and not the pseudo-hexagonal symmetry of the base was emphasized. An argument can therefore be made for minimum staining when examining viral structural proteins, because these may have morphological symmetry, required for packing, which is greater than their true molecular symmetry.

2.7. Electron Density Map of Hexon at 2.9 Å Resolution

Hexon has been investigated at higher resolution by using the full sets of data to calculate 2.9 Å resolution electron density maps. Initially the three original derivatives were used, and later those from CMMPU and PCAU were added. The primary working map contained 120 skew sections of electron density calculated normal to the molecular axis, but the density was also contoured at a smaller scale on sections parallel to the molecular axis. The "parallel" map contains almost the whole molecule, and also shows the chains of β-sheet more clearly because these run parallel to the molecular axis, but the threefold operation relating the subunits is hard to visualize in this view.

The initial 2.9 Å map calculated with three derivatives was tantalizing as it appeared to contain many stretches of polypeptide chain complete with features representing side chains at the appropriate intervals. It was soon apparent that the map was uninterpretable, and two additional derivatives were added to the phasing set before a second map was calculated (Fig. 6). The second map showed a great improvement and we were able to trace long stretches of the polypeptide chain and locate over 850 residues. These are arranged in some ten segments of chain in which we have a high degree of confidence, and may be connected to give four segments with a very much lower degree of certainty.

The most obvious features of the tracing are the paucity of α-helix, with only one of the three helices present having an appreciable length. There is a modest amount of β-sheet, and some of this defines the vertical basal faces involved in hexon–hexon interactions. A significant proportion of the chain appears to run roughly parallel to neighboring chains.

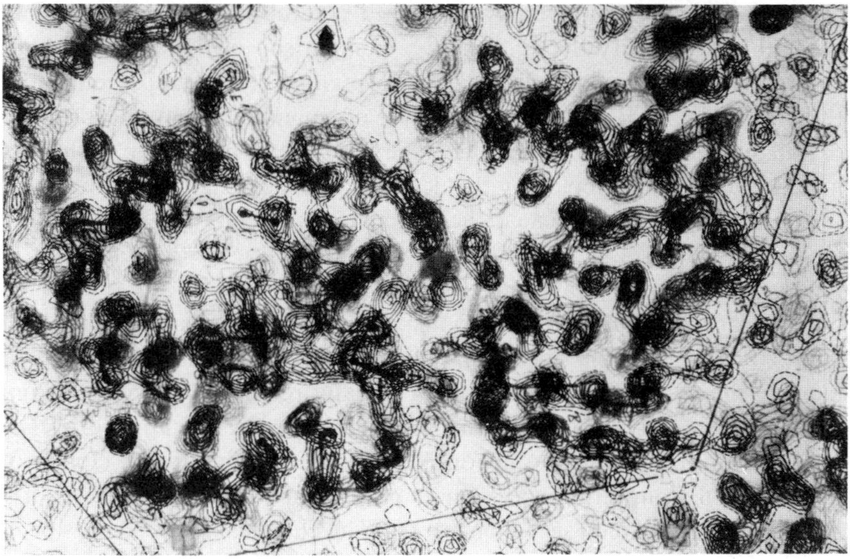

Figure 6. An 8 Å section through an electron density map of hexon at 2.9 Å resolution. The section is through the base of the molecule, and points of reference are the hexagonal outline at the bottom and the molecular threefold axis at the top. Two symmetry related α-helices are shown at top left and top right, and the Cα backbone can be seen marked in the density. At bottom left, four strands of β-sheet rise vertically to form the outer layer of an eight-strand structure that also can be seen.

These results are in accord with the predictions of 17% α-helix and 26% β-sheet from circular dichroism measurements (Day et al., 1972).

However, the tracing is still insufficient to delineate each subunit without ambiguity. There are three main difficulties: the first is that the density is very faint near the top of the molecule in the "towers"; the second is that in other parts of the map there are extremely dense uninterpretable regions; the third difficulty is that several chains often run in parallel and when these enter a difficult patch it is often impossible to determine the relationships between the ingoing and outcoming strands. Further definition of the chain awaits the results of model building in order to improve the phases.

The chemical sequence (Jornvall et al., 1981) will be of help in achieving unambiguous connectivity, but a common origin is required to correlate the sequence with the chain tracing. This problem is not trivial for a chain of 967 amino acids and we have explored several possible avenues. One method is to identify the sulfhydryls. Comparison of the sequence containing the reactive sulfhydryl with the regions in the map around the PCMBS and CMMPU binding sites gave no unambiguous fit. The sulfhydryl in question may be unreactive at the low soaking pH, or the sulfhydryl may not have been labeled by the derivatives used. The latter alternative seems quite plausible, because the mercury compounds dissociated to the ionic form at pH 3.2, and is lent further credibility by the reduction in binding when the crystals were soaked at pH 5.0.

The traced stretches were also inspected for large residues that could be correlated with the chemical sequence. We have also attempted to identify the short internal region, specifically removed by trypsin, from difference maps between the partially digested protein (supplied by Dr. H. Jornvall) and the native hexon. The location of this region could not be identified from difference Fourier projection maps, and analysis of a three-dimensional data set is in progress. The chemical sequence has been checked for regions that may be identifiable in the map, but no useful stretches such as a row of large residues have been identified. The β-sheet region, forming part of the hexon–hexon interface, was also a possibility. The outer of the two layers forms part of the sheet, interacts with neighboring hexons, and is presumably hydrophobic on one side and hydrophylic on the other, which should be reflected in an alternating sequence. This approach was defeated by the small number of residues involved in each strand and their ambiguous connectivity.

The only course remains model building, starting from the Cα positions that have already been located. This method was successful in the case of yeast hexokinase B, where the sequence was unknown (Anderson *et al.*, 1978a,b). It should work for hexon, despite its size, because progress should be rapid as soon as a few stretches of sequence have been identified.

Although the plastic minimaps were adequate for preliminary work, they are not suitable for detailed fitting of the peptide sequence to the electron density. The traditional path using Kendrew–Watson model parts and a Richards' optical comparator is ruled out for hexon by the size of the molecule. At the normal scale of 2 cm $Å^{-1}$, the molecule would be over 2 m in height! A recently installed Computer Graphics System, comprising a VAX 11/750 computer and an Evans and Sutherland MPS color display, will be used for model building. The program FRODO (Jones,

1978) will be used for the display and manipulation of the model in the electron density map.

3. STRUCTURE OF ADENOVIRUS CAPSID

3.1. Nomenclature

An immediate difficulty in describing subassemblies of the adenovirus capsid is that of nomenclature. The term "hexon trimer" has been used to refer to either the three polypeptides within the hexon molecule or to an aggregate of three hexons. The system adopted here will be to reserve the existing terms dimer, trimer, and so forth for the quaternary structure of a molecule. Groups of molecules will be referred to as duos, trios, quartets, quintets, sextets, septets, octets, and nonets. An exception will be made when a name is already well established, as with the group-of-nine for a particular nonet of hexons.

3.2. Hexon Packing in the Group-of-Nine

Because the form of hexon must result from its structural function, its pseudo-hexagonal shape implies that the molecules are close packed in a hexagonal array on the virus surface. Indeed the center-to-center distance for close-packed hexons is 89 Å which would give an edge length for the icosahedral capsid of approximately 445 Å, assuming that the packing portion of the penton base is approximately the same size as that of a hexon. This is in excellent agreement with the value of 440 Å found from micrographs of freeze-etched adenovirus, but is somewhat greater than the length of 425 Å measured on images of whole freeze-dried particles (Nermut, 1975). There is a large discrepancy between these values and the center-to-center distance of 100 Å reported from small-angle neutron and X-ray scattering (Devaux et al., 1983).

The constraints imposed by the flat nature of the adenovirus capsid facets and the assumption of threefold symmetry are powerful, but are still not sufficient to determine the precise orientation of hexons within the capsid. Fortunately, micrographs of the groups-of-nine provide more detail than those of the complete virion and have allowed us to position 180 of the 240 hexons in the capsid. The positioning of the remaining 60 peripentonal hexons will be discussed in Section 3.3.

In the discussion that follows it is necessary to distinguish between the two basal faces of hexon that are engaged in hexon–hexon interactions in a close-packed array. These faces are "morphologically" related by the pseudo-hexagonal shape, but are chemically distinct. There are three "arms" projecting from the towers at a height of 90 Å which define the apices of a triangle of edge 90 Å. The arms give a useful reference point, and we have defined the extensive contact faces below them as the "A" hexon basal faces. The other three are defined as the "B" faces (Fig. 7). Due to their shape, there are only three ways in which hexons can interact as a pair if they are constrained to lie in a close-packed array. These are the *A–B*, *A–A*, and *B–B* interfaces, and there is no significant difference between the center-to-center distances for pairs of hexon models arranged to show these interactions. This demonstrates the comparative precision of the morphological pseudo-symmetry and confirms that packing considerations alone cannot define the capsid interactions.

Excellent micrographs of the groups-of-nine are available (Nermut, 1975), and these show sufficient detail to recognize the hexon bases in the left-handed views and the tops in the right-handed views. The latter suggest strongly that hexons interact asymmetrically via the *A–B* interface

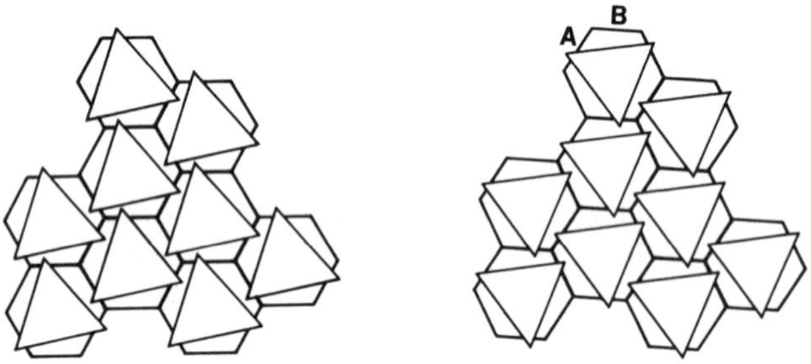

(a) *(b)*

Figure 7. A symbolic representation of the groups-of-nine seen from above, or outside the capsid. Hexons are represented as a triangular top resting on an hexagonal base. The anticlockwise offset of the apices by 10° from the exact midpoints leads to an alternating pattern of large and small cavities between hexon tops. Two arrangements are possible for a left-handed group-of-nine. (a) This is the arrangement found in electron micrographs, and has the larger cavity in the middle. (b) This is the other possible arrangement, with the smaller cavity at the center. The apices of the triangular top are approximately above the midpoints of three of the vertical faces engaged in hexon–hexon contacts. These contact regions are defined as the *A* faces, and those pseudo-related as the *B* faces.

and form a p3 net within the group-of-nine. This conclusion is in agreement with the results of Crowther and Franklin (1972), who first demonstrated the presence of a p3 net by rotationally filtering micrographs of the groups-of-nine.

There is a remaining ambiguity because two possible arrangements of hexon are consistent with the p3 net. Because the molecule does not have exact sixfold symmetry, there are two different threefold axes present that relate hexon molecules, besides the molecular threefold axis that relates subunits. If the $A \rightarrow B$ interface is defined as clockwise, then a trio of hexons can be arranged clockwise about a central threefold axis. There will be a corresponding anticlockwise arrangement $B \leftarrow A$. These clockwise and anticlockwise trios and threefold axes alternate on the p3 net (Fig. 7).

Because the arms have an offset of 10° in the anticlockwise sense, the clockwise trio has a larger cavity at its center than the anticlockwise trio. These cavities are formed between three towers from different hexons and may be recognized on micrographs of right-handed groups-of-nine. The ambiguity of the exact arrangement of hexons within the groups-of-nine can thus be resolved by noting that the larger of the two cavities is seen at its center (Burnett *et al.*, 1981). A group-of-nine thus consists of three clockwise trios arranged on the p3 net to give the correct shape, and this will give a clockwise trio at the center (Fig. 7*a*).

3.3. Orientation of the Peripentonal Hexons

An icosahedral shell of hexons may be built up by rotating a group-of-nine by 60° about its normal, and then tilting it by the appropriate dihedral angle before interlocking it with its neighbor to give a ring of five around each vertex (Figs. 1 and 8). The rotation of 60° brings similar hexon faces into contact at the edges to give a row of alternating A–A and B–B interfaces. Thus, planar hexon interactions are associated in the capsid with the asymmetric A–B interface, and nonplanar interactions with the symmetric A–A and B–B interfaces.

Hexon cannot be an exact hexamer, because it must be able to engage in interactions at the edge different to those formed within a face. The pseudo-hexagonal symmetry facilitates close-packing, but the lack of exact symmetry provides alternative bonding possibilities.

Unlike the asymmetric case, hexon models do not reveal immediately

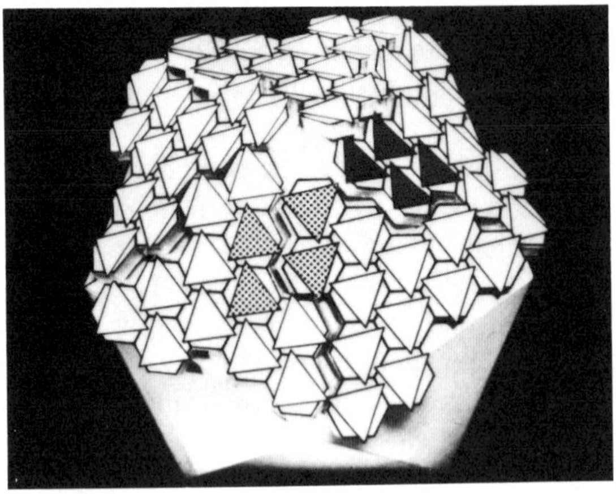

Figure 8. A model of the adenovirus capsid showing five facets arranged about the vertex of the underlying icosahedron. Each facet consists of 12 hexons arranged on a p3 net. Asymmetric *A–B* interfaces occur within the facet, but the mutual orientation of the facets is such that symmetric *A–A* and *B–B* interfaces occur at the edges. In the capsid, the height of hexon (116 Å) permits more extensive hexon–hexon contacts across the edges than those suggested by this model. The symmetry of hexon leads to a $T = 4$ surface lattice for which alternative asymmetric units can be defined. The asymmetric unit marked on the left is nonplanar and spans an edge. The quartet has a maximum dimension equivalent to three rows of close-packed hexons, and this enables it to contribute to both the ring of peripentonal hexons about a vertex, and to the trio of hexons filling in the center of a facet. An alternative planar asymmetric unit is marked on the right, and this forms one-third of a capsid facet. Here all hexons are linked by asymmetric planar interactions, whereas the former quartet embraces both types of interaction.

how the symmetric interactions result in the angle formed between two facets. However, the faces do provide a local twofold axis of bonding symmetry, as required for an isologous association (Monod *et al.*, 1965), instead of only the complementarity required for the asymmetric interaction. This suggests their direct involvement in bonding at the edge, rather than an indirect association mediated by another structural protein.

The geometry of the arrangement with a large cavity at the center of the group-of-nine gives a set of interfaces at the capsid edge between two groups-of-nine: *A–A*; *B–B*; *A–A*; *B–B*; *A–A* (Fig. 8). The other arrangement, with a small cavity at the center, would have a correspondingly altered set: *B–B*; *A–A*; *B–B*; *A–A*; *B–B*.

There is one remaining question at this point in the analysis of hexon packing. The peripentonal hexons have not been positioned, images of the virion are not sufficiently detailed to visualize their orientation directly, and no convenient artifacts such as the groups-of-nine are available. The picture is further clouded by the assumption that the peripentonal hexons must engage in different forms of interaction than those seen either in the facet or at the edge, because they dissociate from the virion before the groups-of-nine.

The analysis that follows assumes that all the hexons are identical and ignores any special nature of the peripentonal hexons. The final test of a satisfactory model of the capsid will remain the explanation of all relevant biochemical data. It will be shown later how polypeptide IX can stabilize the groups-of-nine and causes a dissociation that is different than that to be expected from a p3 net. This need not be considered here except to note that the dissociation of a virus may not necessarily reflect its basic symmetry if other structural proteins are involved.

The first assumption is that the peripentonal hexons are arranged with exact fivefold symmetry about the vertex. This is reasonable not only from structural considerations, but is also consistent with the observation that starlike aggregates of 12 pentons can form (Norrby, 1966), which have been called dodecons (Gelderblom *et al.*, 1967). The symmetry of the pentons in this arrangement is the same as that within the icosahedral capsid, but without the hexons. Because the peripentonal hexons should have surfaces complementary to those of the penton base, they must themselves be arranged with fivefold symmetry. There are then two possible ways of arranging the peripentonal hexons in a ring of five, if they are to be close packed. The two arrangements are both characterized by symmetric interactions within the ring, but one has *A–A* interfaces and the second has *B–B* interfaces. If these rings are placed in position at a vertex of a shell formed from groups-of-nine, it can be seen that the *A–A* ring leads to each peripentonal hexon making two *A–B* interfaces with the other hexons in the corresponding group-of-nine. The *B–B* ring leads to a pair of symmetric interfaces *A–A* and *B–B*.

Because the facets of adenovirus appear to be flat, and not truncated at the vertices, symmetric interfaces, which are associated with an edge, seem unlikely at this location. Thus, each peripentonal hexon would appear to lie on an extension of the p3 net. The question of why the peripentonal hexons dissociate before the others immediately arises, but this is deferred until the placement of polypeptide IX is considered.

A recent determination of the molecular weight of the penton base and fiber suggests that the penton base is a trimer (Devaux *et al.*, 1982). If this is correct, then another protein must be associated with the penton base so as to conserve the symmetry shown by dodecon formation. The references to the surface of the penton base in the preceding argument would then refer to that of the complex.

With the location of the peripentonal hexons, all 240 hexons have been placed within the capsid, which is seen to consist of 20 facets each comprising a small p3 net of 12 hexons.

3.4. Overall Symmetry of Hexons in the Capsid

The overall symmetry of the 240 hexons in the capsid can now be considered. Inspection of a model showing the hexons located and oriented by the preceding analysis shows a nonplanar icosahedral asymmetric unit consisting of four hexons which repeats with exact 532 symmetry to complete a shell with holes at the vertices (Fig. 8). Two of these quartets form an edge, related by the icosahedral twofold operation. The surprising part of this arrangement is that a hexon from each of the three edges of a capsid facet lies adjacent to the central threefold axis and this trio fills in the central portion of the facet. The emphasis on groups-of-nine had focused attention on their planarity, and the capsid appeared too large for hexons to participate in both planar and nonplanar interactions simultaneously. It is now clear that the interlock at the edge permits this to occur.

A second asymmetric unit is also seen which is planar, and three of these form one facet. The nonplanar asymmetric unit is the preferred description because it emphasizes that formation of the edges leads inevitably to formation of a shell.

3.5. Placement of Polypeptide IX within the Capsid

Polypeptide IX is a small protein of MW 14,300 (Alestrom *et al.*, 1980) which has been postulated to stabilize the capsid because it was found in association with aggregates of hexons (Maizel *et al.*, 1968b). Everitt *et al.* (1973) later showed that virions could be dissociated with pyridine to give groups-of-nine that contain only hexons and polypeptide IX. The location of polypeptide IX within the capsid has been explored by various techniques including enzymatic iodination (Everitt *et al.*, 1975) and immunological methods (Everitt *et al.*, 1975; Boulanger *et al.*, 1979). The two groups of workers agreed that polypeptide IX was buried within the capsid, but differed on whether it was located on the inner or outer surface of the group-of-nine.

Boulanger *et al.* (1979) determined that there were approximately 14 copies of polypeptide IX per group-of-nine and argued that it therefore must lie at the 15 equivalent hexon contact points within the group. This model placed polypeptide IX on the inner surface of the group-of-nine because it appeared inaccessible to antibody when part of the virion. The location was in conflict with the results of Everitt *et al.* (1975), who found that the virion could be precipitated with anti-IX serum. The model also failed to explain how a molecule distributed in the same manner as the hexons could give rise to the observed dissociation pattern of the group-of-nine.

A deletion mutant of Ad5 has since been found by Colby and Shenk (1981) that produces virions lacking polypeptide IX. The mutant capsids are more sensitive to heat than wild-type, and do not yield groups-of-nine upon disruption with pyridine. The existence of viable mutants lacking polypeptide IX confirms its role as a capsid cement, and disproves the idea that groups-of-nine are stages in the assembly of adenovirus. This model for assembly has been attractive because groups-of-nine can be induced to form empty shells (Pereira and Wrigley, 1974). However, the model has never provided an explanation for the mechanism by which the characteristic propeller shape of the group-of-nine arises from hexons with an apparent propensity to form a p3 net.

Once the exact arrangement of hexons within the group-of-nine had been determined, it could be inspected for specific features that might explain its nonrandom dissociation pattern. Pereira and Wrigley (1974) had found that preparations of groups-of-nine contained varying quantities of smaller fragments, but that these showed a decidedly nonrandom pattern of dissociation. Apart from individual hexons, septets, pentets, trios, and duos were present, but octets, sextets, and quartets were almost absent.

The large cavity, present at the center of the group-of-nine (Fig. 7), which had served to resolve the ambiguity in the hexon arrangement, also suggested a possible answer to the enigma. Polypeptide IX placed at this location, and the three related locations within the group-of-nine, would stabilize the central trio and strengthen the attachment of the three outer duos. In order to conserve symmetry, a trimer of polypeptide IX should be placed within each large cavity. Inspection of models of hexon, arranged as a group-of-nine, shows that the large cavity comprises three niches formed between adjacent hexons. If a copy of polypeptide IX is placed within each niche, then a trimer would occupy the cavity. The size of the cavity is consistent with the volume of the monomer, which can be calculated from its molecular weight if an average partial specific volume is assumed. However, the monomer must have an elongated shape, which is consistent with the observed axial ratio of 3.5 (Lemay and Boulanger, 1980).

A model for polypeptide IX binding that has maximum consistency with the biochemical data can be derived from this observation. Polypeptide IX binds at the niche formed between a duo of hexons at the right side of the $A \rightarrow B$ interface, and slightly stabilizes this interaction. The geometry of the group-of-nine allows a trimer of polypeptide IX to form in the large cavities, and these have an additional binding component, which serves to stabilize trios of hexon (Fig. 9). The monotony of the p3

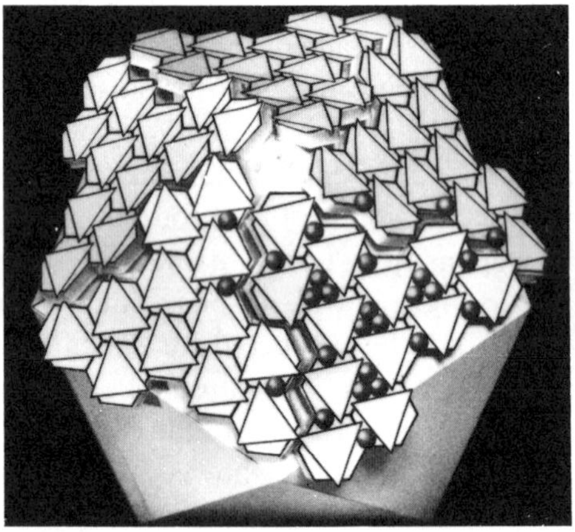

Figure 9. A model of the adenovirus capsid showing the proposed binding sites for polypeptide IX. A monomer binds in the niche formed between the triangular tops of neighboring hexons engaged in asymmetric *A–B* interactions. Trimers are formed in the four large cavities within each facet. Polypeptide IX does not stabilize the edges because binding sites are not formed by symmetric hexon interactions. The peripentonal hexons are slightly stabilized by two monomers, but are not as tightly bound as the hexons within a group-of-nine which are stabilized by trimers.

net is then broken, as illustrated by the dissociation pattern. The model is consistent with the tendency of polypeptide IX to aggregate (Everitt and Philipson, 1974), and with a redetermination of the stoichiometry of polypeptide IX within the group-of-nine by radioactive labeling of methionine (C. L. Cepko, personal communication). A value of 15.4 was obtained using molecular weights calculated from the sequences of Ad2 hexon (Jornvall *et al.*, 1981) and polypeptide IX (Alestrom *et al.*, 1980).

Monoclonal antibodies against Ad2 polypeptide IX show no reactivity with the intact virion, and a low response to groups-of-nine (Cepko *et al.*, 1981). The model explains these results because trimers of polypeptide IX would be buried within the group-of-nine but the outer monomers, buried within the capsid edge formed by interlocking groups-of-nine, would be exposed to solution upon capsid dissociation.

As soon as the orientation of the peripentonal hexons had been deduced, the locations of additional binding sites for polypeptide IX could be identified. Each peripentonal hexon gives rise to an additional two monomer binding sites at the capsid edge, but trimer formation cannot occur. Thus, polypeptide IX cannot confer additional stability at the peripentonal sites beyond the contribution that the monomer makes to the binding at all the $A \rightarrow B$ interfaces.

The model for polypeptide IX placement, therefore, also explains why the peripentonal hexons are easier to dissociate than the groups-of-nine. This result gives confidence that the arguments leading to the orientation of the peripentonal hexons and to the placement of polypeptide IX within the capsid are correct.

4. CONCLUSIONS

4.1. Triangulation Numbers and Complex Structure Units

Crick and Watson (1956) proposed that the regular form of viruses was due to their construction from identical subunits, and this also explained the small size of the viral genome (Crick and Watson, 1957). These ideas were later formalized in the principles of Caspar and Klug (1962) who demonstrated how identical structure units could be arranged in sets of 60, each with strict icosahedral symmetry, to form shells larger than those attainable by one set alone. The members of each set would lie in an identical environment, but members of a different set would lie in a different but related environment that would require a corresponding structural or bonding modification. This principle has become known as quasi-equivalence as there would be T related positions associated with the T sets of structure units.

Clustering of individual structure units about particular local symmetry elements would result in the morphological units seen in the electron microscope. The features of known viruses were successfully explained by this approach, and pentamer–hexamer clustering was found to be common, giving 12 pentamers at the vertices and $10(T - 1)$ hexamers elsewhere.

Adenovirus was originally considered to have a $T = 25$ surface lattice which would associate the 252 morphological units seen in the electron microscope (Horne *et al.*, 1959) with 240 hexameric and 12 pentameric groups of structure units. This description has since been shown to be incorrect for two reasons. The pentons and hexons are chemically distinct, and hexons are trimeric rather than hexameric.

Thus, Caspar–Klug principles have been thought not to apply to complex viruses such as adenovirus, but there remains the apparent paradox of a successful prediction of morphology from a description now known to be formally incorrect. This difficulty is usually ignored, and most textbooks continue to use the incorrect $T = 25$ description.

The preceding analysis of the symmetry of the adenoviral capsid allows the resolution of the paradox. If the hexon is taken to be the structure unit, then the surface lattice can be described as $T = 4$, because hexons are placed in only four different environments. The hexons however, will be located at the local sixfold axes of a $T = 25$ surface lattice. The hexons thus form an empty shell, the structure of which obeys Caspar–Klug principles of quasi-equivalence when the entire hexon molecule is taken to be the structure unit.

The two different descriptions of adenovirus yield similar morphological features on the capsid: 1500 structure units arranged on a $T = 25$ lattice and clustered into pentamers and hexamers would have a similar appearance to hexons and pentons closepacked on a p3 net.

The original Caspar–Klug principles allowed for the possibility of more than one polypeptide chain forming the structure unit. However, due to the early successful explanation of even complex viruses, there has been a tendency to equate the structure unit with a single polypeptide chain.

The first detailed structures to be determined crystallographically have been simple plant viruses, for which this approach has been adequate, but the problems raised by the complex viruses have not yet been faced. Complex arrangements have arisen in studies using electron microscopy, and symmetry in virus architecture has been discussed by Mattern (1977). The particular problem of capsids formed from nonidentical chains has been treated by Dunker (1974). However, the identification of structure units with individual chains has been maintained, with clustering used to explain morphological features as before.

A satisfactory resolution to this problem is found if the structure unit is defined to be either "primitive" or "complex". A primitive structure unit comprises one polypeptide chain and this suffices to describe the simple plant viruses such as tomato bushy stunt virus (Harrison *et al.*, 1978), southern bean mosaic virus (Abad-Zapatero *et al.*, 1980, see also Chapter 2) and satellite tobacco necrosis virus (Liljas *et al.*, 1982, see also Chapter 3). A complex structure unit comprises more than one polypeptide chain and thus may be a homopolymer or a heteropolymer depending on its subunit composition. The original definition of the structure unit as the smallest functionally equivalent building block of the capsid (Caspar *et al.*, 1962) can then be maintained.

In the case of adenovirus, hexon is clearly the appropriate entity for which Caspar–Klug principles should apply. Hexon is a highly stable molecule that remains essentially intact even after trypsin has cut the chain in several places (Pereira and Skehel, 1971; Pettersson, 1971). The tight intersubunit binding is revealed by the extensive interweaving of the chains seen in the 2.9 Å electron density map. Correct hexon folding can only occur with the aid of the nonstructural adenovirus 100K protein, which acts as a scaffold (Cepko and Sharp, 1982). This also points to the problem of assembling a highly stable structure from the very long chain of 967 residues. The high stability of hexon is exactly the quality sought in a structure unit, which should be more stable than the assembly of which it forms a part.

There is one further advantage to be gained from the use of a complex structure unit. There appears to be a limit to the length of polypeptide chain employable, as only three protein subunits are known that are somewhat longer than hexon. The largest, excluding some polyprotein sequences of up to 2209 residues, is the β-chain of DNA-dependent RNA polymerase from *E. coli* with approximately 1400 residues (Dayhoff *et al.*, 1982).

The effect of this limitation on subunit size is to require high triangulation numbers for the construction of large shells from primitive structure units. As an example, *Sericesthis* and *Tipula* iridescent insect viruses (Wrigley, 1969, 1970) would require no less than 9360 primitive units arranged on a $T = 156$ surface lattice to form their icosahedral shells of 1562 morphological units.

Crane (1950) has pointed out the difficulty of forming a perfect structure of any appreciable size if each unit is incorporated sequentially into the assembly with an equal probability of error. The approach of the production line is far more efficient, where subassemblies are first manufactured, and only perfect examples of these are selected for the next stage of assembly. Using Crane's value of a 99% probability for a successful incorporation of a structure unit, there is a one-in-a-thousand chance for the construction of a perfect T = 25 capsid from 1500 polypeptides. However, if a trimer is first formed from a polypeptide of twice the length, then there is an 100-fold increase in the number of perfect capsids.

These figures may seem hypothetical, but cells infected with adenovirus contain several forms of incorrectly processed hexons. These include hexons that have not been acetylated at the N terminus (Boulanger *et al.*, 1978a), incorrectly folded hexons (Cepko and Sharp, 1982), hexons that are not transported from the cytoplasm to the nucleus (Kauffman and Ginsberg, 1976), in addition to hexons that are not incorporated into virions. Assembled capsids do not appear to contain imperfect hexons.

Transport deficient mutants are especially interesting because they are caused by mutations in either hexon or in the 100K protein (Cepko and Sharp, 1982). Thus, there is a positive mechanism for the exclusion of incorrectly folded hexons from the cell nucleus, where viral assembly occurs, in order to prevent their incorporation into capsids. This may mean that hexons mimic a host cell protein, and have evolved a transportation sensitivity to compensate for inaccuracies arising from processing by host cell mechanisms. Alternatively, this may indicate a relationship between viral proteins and those of the host cell nucleus, in particular those concerned with nucleic acid. Hexon has been reported to bind to DNA and to inhibit DNA polymerase *in vitro* (Levine and Ginsberg, 1968).

4.2. Hexon as a Model for Mammalian Viral Coat Proteins

One of the major goals of the structural investigation of hexon has been to determine if it is possible to investigate large assemblies, such as adenovirus, by studying their component parts. It is now clear that the form of the isolated hexon accurately depicts its function in the capsid. The resultant picture of the packing of hexons in the capsid shows how a large structural unit can utilize icosahedral symmetry to limit the number of positions at which different bonding is required.

The analysis of hexon packing in adenovirus has corrected earlier ideas that hexons at the edges, in the faces and at the peripentonal positions were those that are topologically distinct. All four hexons in the asymmetric unit are seen to form part of the face as well as the edge, and the only distinguishing feature of the peripentonal hexon is that it is at the apex of a facet. The model for the placement of polypeptide IX has explained the dissociation pattern of hexons from the capsid and from the groups-of-nine, and shows how ideas concerning the special nature of the peripentonal hexons arose. Most importantly, the analysis has indicated no special requirement for modified hexons in the capsid and provides convincing evidence for the identity of hexons in solution with those in the capsid.

A concern from the beginning of this investigation had been that the soluble hexons from the infected cell, which we crystallized, might in some way be different from those in the capsid. This is reminiscent of the long-standing debate concerning the nature of proteins in solution and in a crystal. The analogy of the capsid as a small crystal is now seen to hold even at the level of maintaining the form of hexon when it is incorporated into the capsid. However, one may expect protein–protein interactions within the capsid to be far more extensive than those in a crystal. The assumption of a rigid body will probably not be true at the level of the individual side chains involved in the hexon–hexon interfaces in the capsid.

The conclusion from our studies on hexon is that the piecemeal approach to large assemblies can be a powerful method for increasing our understanding of biological systems that are too complex for direct analysis with current techniques.

Quasi-equivalent binding in hexon occurs through chemically nonidentical faces. This is in contrast to that found in the small plant viruses where different binding occurs through modification of the same structure. This raises the question of the origin of specificity. Early ideas suggested that high specificity must be due to many weak forces, rather than a few strong ones (Crane, 1950). Thus, complementarity of surfaces would generate a strong bond between two objects due to the combined effect of many weak van der Waals forces. Binding would then be tight when the objects were in the correct position, but feeble at other times, as required for high specificity.

If these ideas are correct, then complementarity could arise in a multitude of ways. The work of Breslow (1982) on artificial enzymes supports this hypothesis. Cyclodextrins were used as mimics of enzymes, and catalytic rates were improved by adjusting the geometry of the binding cavity in these ringlike molecules to maximize the surface contacts made by the substrate. A certain bonding potential can thus be realized by adjusting the contact surfaces, and does not require the same constellation of residues.

The chemical identity of the quasi-equivalently linked structure units in the plant viruses therefore arises because the units are primitive. In the general case, where more complex packing is required, the quasi-equivalence must arise from chemically nonequivalent surfaces. This is demonstrated by the ability of pentons to form dodecons by self-association, or to form the vertices of the capsid by interactions with hexon.

Although complex viruses have structural requirements that dictate the presence of several components, the arrangement of the basic structure must obey the same general principle of simplicity that yields the stable icosahedral form for viruses composed of only one type of protein. Once the framework has been established, and this is less difficult to envisage for a $T = 4$ adenovirus than for one with $T = 25$, then other components can be added. No particular conceptual difficulty is presented by the addition of pentons to their preformed locations in an empty shell, or in the addition of polypeptide IX to stabilize and complete the capsid.

The presence of polypeptide IX illustrates a feature of construction familiar on a human scale, which is that the structural units cannot be too strongly attracted to each other. A certain looseness is required for the units to assume their correct positions, but this must be eliminated in the final structure if it is to be stable. The analogy of a capsid cement for polypeptide IX is therefore apt, and further examples are likely to be encountered in other systems.

Evidence has been presented to support Crane's production line model for the assembly of the adenovirus capsid. Adenovirus is undoubtedly not unique in producing incorrectly processed structural components, but it is probably the most highly studied of the animal viruses. A pool of excess structural proteins is a common feature of mammalian cells infected with virus (Fenner *et al.*, 1974) and probably reflects a similar mixture of components to that seen in adenovirus. A positive selection mechanism for assembly would permit a virus to survive in a cell whose mechanisms are optimized for the cell rather than for the virus.

As an example of the extension of these ideas, it is interesting to consider the papovavirus family. The members of this family have 72 morphological units and thus have been presumed to consist of 420 subunits arranged on a $T = 7d$ surface lattice. Polyoma and SV40 both have three structural proteins (VP1, VP2, and VP3), of which VP1 comprises approximately 75% of the total virion protein.

A recent X-ray analysis of polyoma capsids (Rayment *et al.*, 1982, see also Chapter 7) has shown that all the morphological units are pentamers, which agrees with the stoichiometry of 360 subunits but is in disagreement with the 60 hexamers and 12 pentamers predicted by Caspar–Klug principles. These results have caused some consternation as they appear to cast doubt on the applicability of Caspar–Klug principles to systems other than the simple plant viruses.

The difficulty with polyoma can however be resolved if it is assumed that the structure unit is complex and consists of a pentamer of VP1. This is not an arbitrary assignment to avoid a technical problem, because doughnut-shaped particles resembling the morphological units have been seen when polyoma virions are disrupted (Friedmann, 1971). The particles could be reconstituted into empty shells and thus there is good evidence for their functional role as well as for their independent existence.

By analogy with adenovirus, the 72 morphological units may be divided into two classes, with 12 in one class at the vertices and 60 elsewhere. A trio of pentamers would form a group-of-three, centered about the exact threefold axis in the facet, and this would interdigitate with its neighbors to form the capsid edges. Thus, the 60 pentamers would be arranged on a $T = 1$ shell, with their locations defined by a $T = 7d$ surface lattice. The pentamers would require a second set of bonds so that they could fit into the 12 vertex locations. The presence of two alternative sets of bonds has been already demonstrated by the formation of two classes of cylindrical tubes with the units arranged in a close-packed array (Kiselev and Klug, 1969). The identity of these bonds with those in the capsid has not been shown but they must reflect the different possibilities for bonding. A very recent paper (Baker *et al.*, 1983) has reexamined the cylindrical tubes and found that they both consist of pentamers paired similarly to the vertex pentamer with its neighbors in the capsid. This is an interesting result because dimers of VP1 have been found on partial dissociation of polyoma with detergent (Etchison and Walter, 1977). It would be of great interest to subject the material used in the various structural analyses to exacting chemical analysis.

Polyoma pentamers are analogous to the pentons of adenovirus in their ability to act as a vertex, as illustrated by the formation of small particles of polyoma (Mattern *et al.*, 1967) and SV40 (Anderer *et al.*, 1967) having the size of 12 units. These presumably correspond to the dodecons formed from adenovirus pentons. Bacteriophages that have been studied in detail have a separate protein at the vertex, and this may also be generally true for the mammalian viruses. Examples from the bacteriophages are T4 (Muller-Salamin *et al.*, 1977) and ϕCbK from *Caulobacter crescentus* (Leonard *et al.*, 1972). It is also of interest that isometric phage heads appear to form a framework from a large protein following Caspar–Klug principles, which then matures on the insertion of a small protein (Eiserling, 1979). Examples of this process can be found in λ, ϕCbK, and T4.

VP2 and VP3 had originally been suggested to be the vertex proteins because 60 copies of each are present in the related SV42 capsid (Estes *et al.*, 1971). Their role may rather be to stabilize the capsid, or even one or both of the two classes of structure units. The stoichiometry is unfortunately consistent with several possibilities for either or both VP2 and VP3. It seems unlikely that a monomer can stabilize a pentameric structure unit, and so binding at the 60 nonvertex sites probably does not occur. However, a pentamer could be attached at each of the 12 vertices. Alternatively, 20 trimers could lie at the threefold axes where they would stabilize trios of pentamers in the same manner as polypeptide IX in adenovirus.

The available chemical information is somewhat contradictory because dissociation gives particles containing one copy of VP2 and VP3 with five or six copies of VP1 (Friedmann and David, 1972), which could be a stabilized pentamer. However, other studies indicate that empty capsids, containing only VP1 cross-linked by disulfide bonds, can be found (Etchison and Walter, 1977). An exact dissection of the capsid, and an accurate knowledge of the stoichiometry, would be very rewarding because they would aid greatly in determining the architecture of the capsid.

The primary conclusion of our work on adenovirus is that the complexity of animal viruses is not achieved by abandoning the same general principles of construction as those used by the simple plant viruses. Rather, the appropriate structure unit is less obvious, but when identified will reveal the symmetry that must underlie all spherical capsids.

Acknowledgments. I would like to take this opportunity to acknowledge the contributions of many people to our studies on hexon. First, I thank Dr. Richard M. Franklin for encouraging us to undertake the project, and the Swiss National Science Foundation for its support. Drs. Per-Erik Werner and Paul B. Sigler generously communicated their own crystallographic results, Dr. Hans Jornvall kept us informed of the sequencing, and Dr. Alan J. Wonacott was unstinting in his advice on oscillation photography. Dr. Lennart Philipson has been a source of constant encouragement throughout the investigation.

Finally, and most important, I wish to acknowledge the assistance of my collaborators. Dr. Markus Grutter made an enormous contribution to establishing the project, which has been continued with the able efforts of Drs. Zora Markovic, Janice L. White, and Jan van Oostrum. Dr. J. N. Jansonius, his associates, and the staff of the Biozentrum all contributed in countless ways to our progress.

My colleagues in the Biochemistry Department at Columbia University have done much to provide a stimulating environment in which the later ideas concerning adenovirus could develop. Our current work is supported by NIH grant AI-17270, Biomedical Research Support Grants, and an Irma T. Hirschl Career Scientist Award.

REFERENCES

Abad-Zapatero, C., Abdel-Meguid, S. S., Johnson, J. E., Leslie, A. G. W., Rayment, I., Rossmann, M. G., Suck, D., and Tsukihara, T. (1980) *Nature (London)* **286,** 33–39.

Adam, E., and Nasz, I. (1979). *Acta Microbiol. Acad. Sci. Hung.* **26,** 15–21.

Alestrom, P., Akusjarvi, G., Perricaudet, M., Mathews, M. B., Klessig, D. F., and Pettersson, U. (1980). *Cell* **19,** 671–681.

Anderer, F. A., Schlumberger, H. D., Koch, M. A., Frank, H., and Eggers, H. J. (1967). *Virology* **32,** 511–523.

Anderson, C. M., McDonald, R. C., and Steitz, T. A. (1978a). *J. Mol. Biol.* **123,** 1–13.

Anderson, C. M., Stenkamp, R. E., and Steitz, T. A. (1978b). *J. Mol. Biol.* **123,** 15–33.

Arndt, U. W., and Wonacott, A. J., Eds. (1977). *The Rotation Method in Crystallography*. North-Holland, Amsterdam.

Arndt, U. W., Champness, J. N., Phizackerly, R. P., and Wonacott, A. J. (1973). *J. Appl. Crystallogr.* **6,** 457–463.

Baker, T. S., Caspar, D. L. D., and Murakami, W. T. (1983). *Nature* (*London*) **303,** 446–448.

von Bahr-Lindstrom, H., Jornvall, H., Althin, S., and Philipson, L. (1982). *Virology* **118,** 353–362.

Berger, J., Burnett, R. M., Franklin, R. M., and Grutter, M. (1978). *Biochim. Biophys. Acta* **535,** 233–240.

Blow, D. M., and Crick, F. H. C. (1959). *Acta Crystallogr.* **12,** 794–802.

Boulanger, P. (1975). *J. Virol.* **16,** 1678–1682.

Boulanger, P., and Loucheux (1972). *Biochem. Biophys. Res. Commun.* **47,** 194–201.

Boulanger, P. A., Torpier, G., and Biserte, G. (1970). *J. Gen. Virol.* **6,** 329–332.

Boulanger, P., Torpier, G., and Rimsky, A. (1974). *Intervirology* **2,** 56–62.

Boulanger, P., Devaux, C., and Lemay, P. (1978a). *Virology* **84,** 456–468.

Boulanger, P. A., Devaux, C., and Loucheux-Lefebvre, M.-H. (1978b). *FEBS Lett.* **85,** 52–56.

Boulanger, P., Lemay, P., Blair, G. E., and Russell, W. C. (1979). *J. Gen. Virol.* **44,** 783–800.

Breslow, R. (1982). *Science* **218,** 532–537.

Bricogne, G. (1976). *Acta Crystallogr. A* **32,** 832–847.

Brown, D. T., Westphal, M., Burlingham, B. T., Winterhoff, U., and Doerfler, W. (1975). *J. Virol.* **16,** 366–387.

Burnett, R. M., Grutter, M. G., Markovic, Z., and White, J. L. (1978). *Acta Crystallogr. A* **34,** S51.

Burnett, R. M., Grutter, M. G., Markovic, Z., and White, J. L. (1979). *J. Supramol. Struct.* **S3,** 92.

Burnett, R. M., Grutter, M. G., and White, J. L. (1981). In *Structural Aspects of Recognition and Assembly in Biological Macromolecules*, M. Balaban, Ed., Vol. 2. International Science Services, Rehovot, Israel, pp. 865–868.

Carstens, E. B., and Marusyk, R. G. (1975). *J. Gen. Virol.* **29,** 249–256.

Carstens, E. B., and Marusyk, R. G. (1977). *J. Gen. Virol.* **34,** 541–545.

Caspar, D. L. D., and Klug, A. (1962). *Cold Spring Harbor Symp. Quant. Biol.* **27,** 1–24.

Caspar, D. L. D., Dulbecco, R., Klug, A., Lwoff, A., Stoker, M. G. P., Tournier, P., and Wildy, P. (1962). *Cold Spring Harbor Symp. Quant. Biol.* **27,** 49–50.

Cepko, C. L., and Sharp, P. A. (1982). *Cell* **31,** 407–415.

Cepko, C. L., Changelian, P. S., and Sharp, P. A. (1981). *Virology* **110,** 385–401.

Colby, W. W., and Shenk, T. (1981). *J. Virol.* **39,** 977–980.

Cornick, G., Sigler, P. B., and Ginsberg, H. S. (1971). *J. Mol. Biol.* **57,** 397–401.

Couch, R. B., Kasel, J. A., Pereira, H. G., Haase, A. T., and Knight, V. (1973). *Proc. Soc. Exp. Biol. (N.Y.)* **143,** 905–910.

Crane, H. R. (1950). *Sci. Monthly* **70,** 376–389.

Crick, F. H. C., and Watson, J. D. (1956). *Nature (London)* **177,** 473–475.

Crick, F. H. C., and Watson, J. D. (1957). In *The Nature of Viruses*, G. E. W. Wolstenholme and E. C. P. Millar, Eds. (Ciba Foundation Symposium). Churchill, London, pp. 5–13.

Crowther, R. A., and Amos, L. A. (1971). *J. Mol. Biol.* **60,** 123–130.

Crowther, R. A., and Franklin, R. M. (1972). *J. Mol. Biol.* **68,** 181–184.

Cullis, A. F., Muirhead, H., Perutz, M. F., Rossmann, M. G., and North, A. C. T. (1961). *Proc. R. Soc. London. Ser. A* **265,** 15–38.

Dales, S. (1962). *J. Cell. Biol.* **13,** 303–322.

Day, L. A., Franklin, R. M., Pettersson, U., and Philipson, L. (1972). *Eur. J. Biochem.* **29,** 537–541.

Dayhoff, M. O., Hunt, L. T., Barker, W. C., Orcutt, B. C., Yeh, L. S., Chen, H. R., George, D. G., Blomquist, M. C., Fredrickson, J., and Johnson, G. C. (1982). Protein Sequence Database from the *Atlas of Protein Sequence and Structure*. National Biomedical Research Foundation, Washington, D.C.

Devaux, C., and Boulanger, P. A. (1980). *Virology* **102,** 94–106.

Devaux, C., Zulauf, M., Boulanger, P., and Jacrot, B. (1982). *J. Mol. Biol.* **156,** 927–939.

Devaux, C., Timmins, P. A., and Berthet-Colominas, C. (1983). *J. Mol. Biol.*, **167,** 119–132.

Dohner, L., and Hudemann, H. (1972). *Arch. Ges. Virusforsch.* **38,** 279–289.

Dubochet, J., Ducommun, M., Zollinger, M., and Kellenberger, E. (1971). *J. Ultrastruct. Res.* **35,** 147–167.

Dunker, A. K. (1974). *J. Virol.* **14,** 878–885.

Eiserling, F. A. (1979). In *Comprehensive Virology*, H. Fraenkel-Conrat and R. R. Wagner, Eds., Vol. 13. Plenum, New York, pp. 543–580.

Estes, M. K., Huang, E. S., and Pagano, J. S. (1971). *J. Virol.* **7,** 635–641.

Etchison, D., and Walter, G. (1977). *Virology* **77,** 783–796.

Everitt, E., and Philipson, L. (1974). *Virology* **62,** 253–269.

Everitt, E., Sundquist, B., Pettersson, U., and Philipson, L. (1973). *Virology* **52,** 130–147.

Everitt, E., Lutter, L., and Philipson, L. (1975). *Virology* **67,** 197–208.

Fenner, F., McAuslan, B. R., Mims, C. A., Sambrook, J., and White, D. O. (1974). In *The Biology of Animal Viruses*, 2nd ed. Academic Press, New York, p. 339.

Franklin, R. M. Pettersson, U., Akervall, K., Strandberg, B., and Philipson, L. (1971a). *J. Mol. Biol.* **57,** 383–395.

Franklin, R. M., Harrison, S. C., Pettersson, U., Philipson, L., Branden, C. I., and Werner, P.-E. (1971b). *Cold Spring Harbor Symp. Quant. Biol.* **36,** 503–510.

Friedmann, T. (1971). *Proc. Natl. Acad. Sci. U.S.A.* **68,** 2574–2578.

Friedmann, T., and David, D. (1972). *J. Virol.* **10,** 776–782.

Gelderblom, H., Bauer, H., Frank, H., and Wigand, R. (1967). *J. Gen. Virol.* **1,** 553–560.

Ginsberg, H. S. (1979). In *Comprehensive Virology*, H. Fraenkel-Conrat and R. R. Wagner, Eds., Vol. 13. Plenum, New York, pp. 409–457.

Ginsberg, H. S., Pereira, H. G., Valentine, R. C., and Wilcox, W. C. (1966). *Virology* **28,** 782–783.

Ginsberg, H. S., Ensinger, M. J., Kauffman, R. S., Mayer, A. J., and Lundholm (1974). *Cold Spring Harbor Symp. Quant. Biol.* **39,** 419–426.

Grutter, M. G. (1976). Ph.D. thesis, University of Basel, Switzerland.

Grutter, M. G., and Franklin, R. M. (1974). *J. Mol. Biol.* **89,** 163–178.

Harrison, S. C. (1968). *J. Appl. Crystallogr.* **1,** 84–90.

Harrison, S. C., Olson, A. J., Schutt, C. E., Winkler, F. K., and Bricogne, G. (1978). *Nature (London)* **276,** 368–373.

Henry, C. J., and Atchison, R. W. (1971). *J. Virol.* **8,** 842–849.

Henry, C. J., Slifkin, M., Merkow, L. P., and Pardo, M. (1971). *Virology* **44,** 215–218.

Hilleman, M. R. (1966). In *Viruses Inducing Cancer*, W. J. Burdette, Ed. Univ. Utah Press, Salt Lake City, pp. 372–402.

Horne, R. W., Brenner, S., Waterson, A. P., and Wildy, P. (1959). *J. Mol. Biol.* **1,** 84–86.

Horwitz, M. S., Scharff, M. D., and Maizel, J. V., Jr. (1969). *Virology* **39,** 682–694.

Horwitz, M. S., Maizel, J. V., Jr., and Scharff, M. D. (1970). *J. Virol.* **6,** 569–571.

Huebner, R. J. (1967). In *Perspectives in Virology*, M. Pollard, Ed., Vol. 5. Academic Press, New York, pp. 147–166.

Jawetz, E. (1959). *Br. Med. J.* **1,** 873–876.

Jirgensons, B. (1973). *Optical Activity of Proteins and Other Macromolecules*. Springer-Verlag, New York, pp. 121–123.

Jones, T. A. (1978). *J. Appl. Crystallogr.* **11,** 268–272.

Jornvall, H., and von Bahr-Lindstrom, H. (1981). *J. Biol. Chem.* **256,** 6187–6198.

Jornvall, H., and Philipson, L. (1980). *Eur. J. Biochem.* **104,** 237–247.

Jornvall, H., Pettersson, U., and Philipson, L. (1974a). *Eur. J. Biochem.* **48,** 179–192.

Jornvall, H., Ohlsson, H., and Philipson, L. (1974b) *Biochem. Biophys. Res. Commun.* **56,** 304–310.

Jornvall, H., von Bahr-Lindstrom, H., Akusjarvi, G., Philipson, L., Fowler, A., Apella, E., Alestrom, P., and Pettersson, U. (1978). Collected Abstracts, EMBO Workshop: *The Molecular Biology of Adenovirus*, Orenas, Sweden.

Jornvall, H., Akusjarvi, G., Alestrom, P., von Bahr-Lindstrom, H., Pettersson, U., Appella, E., Fowler, A. V., and Philipson, L. (1981) *J. Biol. Chem.* **256,** 6181–6186.

Kauffman, R. S., and Ginsberg, H. S. (1976). *J. Virol.* **19,** 643–658.

Kiselev, N. A., and Klug, A. (1969). *J. Mol. Biol.* **40,** 155–171.

Kjellen, L., Lagermalm, G., Svedmyr, A., and Thorsson, K.-G. (1955). *Nature (London)* **175,** 505–506.

Laver, W. G., Pereira, H. G., Russell, W. C., and Valentine, R. C. (1968). *J. Mol. Biol.* **37,** 379–386.

Leibowitz, J., and Horwitz, M. S. (1975). *Virology* **66,** 10–24.

Lemay, P., and Boulanger, P. (1980). *Ann. Virol. Inst. Pasteur* **131,** 259–275.

Leonard, K. R., Kleinschmidt, A. K., Agabian-Keshishian, N., Shapiro, L., and Maizel, J. V., Jr. (1972). *J. Mol. Biol.* **71,** 201–216.

Levine, A. J., and Ginsberg, H. S. (1968). *J. Virol.* **2,** 430–439.

Lifchitz, A., Rimsky, A., Torpier, G., and Boulanger, P. A. (1975). *Acta Crystallogr. B* **31,** 441–445.

Liljas, L., Unge, T., Jones, T. A., Fridborg, K., Lovgren, S., Skoglund, U., and Strandberg, B. (1982). *J. Mol. Biol.* **159,** 93–108.

Macintyre, W. M., Pereira, H. G., and Russell, W. C. (1969). *Nature (London)* **222,** 1165–1166.

Maizel, J. V., Jr., White, D. O., and Scharff, M. D. (1968a). *Virology* **36,** 115–125.

Maizel, J. V., Jr., White, D. O., and Scharff, M. D. (1968b). *Virology* **36,** 126–136.

Mattern, C. F. T. (1977). In *The Molecular Biology of Animal Viruses*, D. P. Nayak, Ed., Vol. 1. Dekker, New York, pp. 1–39.

Mattern, C. F. T., Takemoto, K. K., and DeLeva, A. M. (1967). *Virology* **32,** 378–392.

Matthews, B. W. (1966). *Acta Crystallogr.* **20,** 82–86.

Mautner, V., and Pereira, H. G. (1971). *Nature (London)* **230,** 456–457.

Monod, J., Wyman, J., and Changeux, J.-P. (1965). *J. Mol. Biol.* **12,** 88–118.

Monto, A. S. (1980). In *Maxcy-Rosenau Public Health and Preventive Medicine*, J. M. Last, Ed., 11th ed. Appleton-Century-Crofts, New York, pp. 121–131.

Morgan, C., Godman, G. C., Rose, H. M., Howe, C., and Huang, J. S. (1957). *J. Biophys. Biochem. Cytol.* **3,** 505–508.

Morgan, C., Godman, G. C., Breitenfeld, P. M., and Rose, H. M. (1960). *J. Exp. Med.* **112,** 373–382.

Muller-Salamin, L., Onorato, L., and Showe, M. K. (1977). *J. Virol.* **24,** 121–134.

Neidle, S. (1973). *Acta Crystallogr. B* **29,** 2645–2647.

Nermut, M. V. (1975). *Virology* **65,** 480–495.

Nermut, M. V. (1980). *Arch. Virol.* **64,** 175–196.

Nermut, M. V., and Perkins, W. J. (1979). *Micron* **10,** 247–266.

Norrby, E. (1966). *Virology* **28,** 236–248.

Norrby, E. (1969a). *Virology* **37,** 565–576.

Norrby, E. (1969b). *J. Gen. Virol.* **5,** 221–236.

Norrby, E., and Ankerst, J. (1969). *J. Gen. Virol.* **5,** 183–194.

North, A. C. T. (1965). *Acta Crystallogr.* **18,** 212–216.

Oosterom-Dragon, E. A., and Ginsberg, H. S. (1981). *J. Virol.* **40,** 491–500.

Pereira, H. G., and Skehel, J. J. (1971). *J. Gen. Virol.* **12,** 13–24.

Pereira, H. G., and Wrigley, N. G. (1974). *J. Mol. Biol.* **85,** 617–631.

Pereira, H. G., Valentine, R. C., and Russell, W. C. (1968). *Nature* (*London*) **219,** 946–947.

Pereira, H. G., Wrigley, N. G., and Mautner, V. (1975). In *Proceedings of the International Symposium on Macromolecules*, E. Mano, Ed. Elsevier, Amsterdam, pp. 451–463.

Pettersson, U. (1971). *Virology* **43,** 123–136.

Pettersson, U., Philipson, L., and Hoglund, S. (1967). *Virology* **33,** 575–590.

Philipson, L., Pettersson, U., and Lindberg, U. (1975). *Molecular Biology of Adenoviruses*, Virology Monographs 14, S. Gard and C. Hallauer, Eds. Springer-Verlag, New York.

Prage, L., Pettersson, U., Hoglund, S., Lonberg-Holm, K., and Philipson, L. (1970). *Virology* **42,** 341–358.

Rayment, I., Baker, T. S., Caspar, D. L. D., and Murakami, W. T. (1982). *Nature* (*London*) **295,** 110–115.

Rossmann, M. G., and Blow, D. M. (1962). *Acta Crystallogr.* **15,** 24–31.

Rosen, L. (1960). *Am. J. Hyg.* **71,** 120–128.

Rowe, W. P., Huebner, R. J., Gillmore, L. K., Parrott, R. H., and Ward, T. G. (1953). *Proc. Soc. Exp. Biol.* (*N.Y.*) **84,** 570–573.

Shortridge, K. F., and Biddle, F. (1970) *Arch. Ges. Virusforsch.* **29,** 1–24.

Simpson, P. G., Dobrott, R. D., and Lipscomb, W. N. (1965). *Acta Crystallogr.* **18,** 169–179.

Smith, K. O., (1965). *J. Immunol.* **94,** 976–989.

Smith, K. O., Gehle, W. D., and Trousdale, M. D. (1965). *J. Bacteriol.* **90,** 254–261.

Stasny, J. T., Neurath, A. R., and Rubin, B. A. (1968). *J. Virol.* **2,** 1429–1442.

Steitz, T. A. (1968). *Acta Crystallogr. B* **24,** 504–507.

Sygusch, J. (1977). *Acta Crystallogr. A* **33,** 512–518.

Tejg-Jensen, B., Furugren, B., Lindqvist, I., and Philipson, L. (1972). *Monatsh. Chem.* **103,** 1730–1736.

Ten Eyck, L. F. (1973). *Acta Crystallogr. A* **29,** 183–191.

Valentine, R. C., and Pereira, H. G. (1965). *J. Mol. Biol.* **13,** 13–20.

White, D. O., Scharff, M. D., and Maizel, J. V., Jr. (1969). *Virology* **38,** 395–406.

Wilcox, W. C., Ginsberg, H. S., and Anderson, T. F. (1963). *J. Exp. Med.* **118,** 307–314.

Williams, J. F., Young, C. S. H., and Austin, P. E. (1974). *Cold Spring Harbor Symp. Quant. Biol.* **39,** 427–437.

Wrigley, N. G. (1969). *J. Gen. Virol.* **5,** 123–134.

Wrigley, N. G. (1970). *J. Gen. Virol.* **6,** 169–173.

Index

Shepherd College Library
QP 801.P64B554 1984
rusc
Biological macromolecules and assemblies
3 4550 200098991 7

QP
801
.P64
B554
1984
v.1

162815

DEC 19 '86

QP
801
.P64
B554
1984
v.1

162815

Biological macromolecules
and assemblies

DATE	ISSUED TO
DEC 19 '86	Dawn Grove Gardiner 314